8th International Conference on Future Environment and Energy (ICFEE 2018)

IOP Conference Series: Earth and Environmental Science
Volume 150

Phuket, Thailand
10 – 12 January 2018

ISBN: 978-1-5108-6336-1
ISSN: 1755-1307

Printed from e-media with permission by:

Curran Associates, Inc.
57 Morehouse Lane
Red Hook, NY 12571

Some format issues inherent in the e-media version may also appear in this print version.

Printed by Curran Associates, Inc. (2018)

For permission requests, please contact the Institute of Physics
at the address below.

Institute of Physics
Dirac House, Temple Back
Bristol BS1 6BE UK

Phone: 44 1 17 929 7481
Fax: 44 1 17 920 0979

techtracking@iop.org

Additional copies of this publication are available from:

Curran Associates, Inc.
57 Morehouse Lane
Red Hook, NY 12571 USA
Phone: 845-758-0400
Fax: 845-758-2633
Email: curran@proceedings.com
Web: www.proceedings.com

Table of contents

Volume 150

8th International Conference on Future Environment and Energy (ICFEE 2018)

10–12 January 2018, Phuket, Thailand

Accepted papers received: 1 May 2018
Published online: 22 May 2018

Preface

Papers

Resources and Environmental Engineering

Chemical Engineering

Energy Engineering

IOP Conf. Series: Earth and Environmental Science **150** (2018) 011001 doi:10.1088/1755-1315/150/1/011001

PREFACE

It is our great pleasure to welcome you to 2018 8th International Conference on Future Environment and Energy (ICFEE 2018) which will be held in Phuket, Thailand, 10-12 January, 2018. ICFEE 2018 is dedicated to issues related to Future Environment and Energy.

The major goal and feature of the conference is to bring academic scientists, engineers, and industry researchers together to exchange and share their experiences and research results, and discuss the practical challenges encountered and the solutions adopted. Professors from Thailand, Malaysia, Germany, Australia and USA are invited to deliver keynote speeches regarding latest information in their respective expertise areas. It will be a golden opportunity for students, researchers and engineers to interact with the experts and specialists to get their advice or consultation on technical matters, sales and marketing strategies.

This proceeding present a selection from papers submitted to the conference from universities, research institutes and industries. All papers were subjected to peer-review by conference committee members and international reviewers. The papers selected depended on their quality and their relevancy to the conference. The volume tends to present to the readers the recent advances in the field of Future Environment and Energy and various related areas, such as Resources and Environmental Engineering, Chemical Engineering, Energy Engineering, Emission Reduction and Sustainable Development, etc.

We would like to thank all the authors who have contributed to this volume, and also the organizing committee, reviewers, speakers, chairpersons, sponsors and all the conference participants for their support to ICFEE 2018.

Prof. Orawan Siriratpiriya, D.Sc.
Environmental Expert
The Aquatic Resources Research Institute (ARRI)
Chulalongkorn University, Thailand
January 17, 2018

IOP Conf. Series: Earth and Environmental Science **150** (2018) 011002 doi:10.1088/1755-1315/150/1/011002

Organizing Committee

Conference Chair

Prof. Orawan Siriratpiriya, Environmental Research Institute of Chulaongkorn University, Thailand

Prof. Ngai Weng Chan, Universiti Sains Malaysia, Malaysia

Prof. Vladan Babovic, National University of Singapore

Program Chair

Prof. Je-Lueng Shie, National I-Lan University, Taiwan

Assoc. Prof. Anja Pfennig, HTW University of Applied Sciences Berlin, Germany

Publication Chair

Qiang Li, Nanyang Technological University, Singapore

International Technical Committee

Prof. Ken Nagasaka, Tokyo University of Agriculture and Technology, Japan

Prof. Enrico Sciubba, University of Roma Sapienza, Italy

Prof. Beng Soo Ong, California State University, USA

Prof. Roberto San Jose, Technical University of Madrid, Spain

Prof. Subhransu Sekhar Dash, Srm University, India

Prof. Enrique Romero Cadaval, University of Extremadura, Spain

Prof. Jian Wang, Kingston University London, UK

Prof. Hideaki OHGAKI, Kyoto University, Japan

Prof. Hui Tong Chua, University of Western Australia

Prof. Seung-Hoon Yoo, Seoul National University of Science & Technology, South Korea

Prof. Ali Cheknane, Amar Telidji University of Laghouat, Algeria

Prof. Jeong-Hun Park, Chonnam National University, South Korea

Prof. Lucjan Pawłowski, Lublin University of Technology, Poland

Assoc. Prof. Henryk Wasag, Lublin University of Technology, Poland

Assoc. Prof. Pedro Roncero, University Castilla La Mancha, Spain

Assoc. Prof. Lim Teik Thye, Nanyang Technological University, Singapore

Dr. Xiaolin Wang, University of Tasmania, Australia

Dr Ahmad El-kharouf, University of Birmingham, UK

Dr. Gerardo Reyes, Lakehead University, Canada

Dr. Viktor Kochkodan, Hamad bin Khalifa University, Qatar

Dr. Hen Friman, Holon Institute of Technology, Holon, Israel

Dr. Arpapong Changjan, Pathumwan Institute of Technology, Thailand

Dr. A Bergmann, University of Dundee, UK

Peer review statement

All papers published in this volume of *IOP Conference Series: Earth and Environmental Science* have been peer reviewed through processes administered by the proceedings Editors. Reviews were conducted by expert referees to the professional and scientific standards expected of a proceedings journal published by IOP Publishing.

IOP Conf. Series: Earth and Environmental Science **150** (2018) 012001 doi:10.1088/1755-1315/150/1/012001

Water Footprint Analysis of Paddy Rice and the Nexus of Water-Land-Rice in Taiwan: *2005-2014*

T C Wu[1]

1 Associate Professor, Department of Land Management and Development, Chang Jung Christian University, Taiwan (R.O.C.)
E-mail: tcwu-1@mail.cjcu.edu.tw

Abstract. This paper explores the water footprint (WF) of paddy rice and the nexus of water-land–food (rice) in Taiwan. The research results indicate that the average annual rice WF for the years 2005-2014 was about 7,580 m^3/ton, of which 80% was blue, 17% was green, and 3% was grey. This average annual footprint was about 5.7 times larger than the 2000-2004 average annual WF of rice for countries around the globe of 1325 m^3/ton, of which 48% was green, 44% was blue, and 8% was grey. The blue WF is the most important source of water for rice production in Taiwan. The water consumption of the second crop is higher than that of the first crop. The water use efficiency in the southern region of Taiwan is the best, while the northern part of Taiwan exhibits relatively high inefficiency. The rates of change in cultivated land and rice production in Taiwan are decreasing in a stable manner. However, the annual rate of change in the rice WF is unstable. The nexus of land, water, and food should be taken into consideration to protect water availability, maintain agricultural production, and avoid land degradation. The results could offer useful information for agriculture policy and water resource management.

1. Introduction

Rice is one of the major crops in Taiwan. In order to increase food security and the self-sufficiency ratio of rice, large irrigation projects are often constructed to meet the water demands for rice production [1], thereby allowing large populations to achieve food security. As a result, rice is among the crops that consume the largest amounts of water. Recently, concerns regarding the water-food nexus have significantly increased due to population growth, which has itself increased the demands for water and food. As water is becoming scarcer and scarcer, mitigation strategies and the conscious use of water resources are key concerns [2].

The concept of a water footprint (WF) was previously proposed as an indicator to evaluate the utilization of water resources relative to human consumption [3]. From the perspective of water consumption and pollution, WF evaluation has become one of the priorities for water sustainability and provides decision-making support for water resources management [4].

The WF of a product is defined as the volume of freshwater consumed and polluted in order to produce the product. The WF accounts not only for the direct water use of a consumer or producer but also for indirect water use [5]. A given total WF usually consists of a green WF (rainwater that does not run-off or recharge the groundwater), blue WF (irrigation water withdrawn from the ground or surface water), and grey WF (the volume of freshwater that is required to assimilate the load of pollutants) [1], [6].

The first study to analyse crop cultivation in terms of WFs was conducted by [7], who calculated WFs of cotton consumption. A similar study concerning rice was subsequently conducted by [8] when they investigated freshwater consumption for the 13 most important rice producing countries. According to the results of [8], the average annual rice WF for the investigated countries was 1325 m^3/ton (48% green, 44% blue, and 8% grey). In 2013, Yoo et al. calculated an annual rice WF for Korea equal to 844.5 m^3/ton [9], which was a slightly different result from 829 m^3/ton reported by [8]. Given the similar assumptions of the two studies, the differing results were probably influenced by climate in the different temporal contexts. Besides, [10] calculated the annual rice WF of Argentina and found a similar result of 845 m^3/ton (43.5% green and 56.3% blue) in the northern regions of the country and 987 m^3/ton (36.5% green and 63.5% blue) in the southern regions. Given the climate differences, the biggest variability was shown to be in blue water use. [2]

Recently, the development of WF research has proceeded rapidly. In previous studies, the WF methodology has been applied to many different fields related to the uses of water. Applications of the methodology to explore regarding agricultural products have been popular, with various studies having considered different products and countries. Such as, Chapagain and Hoekstra (2007) assessed the WF of coffee and tea consumption in the Netherlands [11]. The WF methodology has also been applied to other products consumed by people, such as the cotton consumed for clothes production [4], and tomato [12]. Likewise, the WF methodology has also been applied to account for the WFs of different diets [13], [14]. Similarly, the WFs of different regions and countries have also been evaluated [15], [16]. WFs have also been used to assess the production of hydropower energy [17] and biofuels [18], [19], amongst other applications [20]-[23].

For the most part, previous studies of this sort conducted global WF analyses of crop production, but a few recent studies focused, instead, on regional WFs, with only a few of studies having provided complete calculations of total WFs. A 2016 study by Lovarelli et al. reviewed the literature published from 2000-2015 regarding the WFs associated with the production of various crops [2]. According to that review, in early studies the main goal was to assess the water trade of products on a global scale, whereas in subsequent years, the goal was the rigorous quantification of the three components for specific crops in specific geographical areas. In the most recent assessments, similarities in the methodology and the tools employed emerged. Out of 96 scientific articles using WFs as indicators of agricultural production, 78% of the studies aimed to quantify WFs, while the remaining 22% analysed methodology, uncertainty, future trends, and comparisons using other footprints. It emerged that the largest percentage of studies that quantified WFs concerned cereals (33%), among which maize and wheat were the most investigated crops. In 46% of studies, all three components of the total WFs were assessed, while in 18%, no indication of any subdivisions was given; in the remaining 37%, only the blue component or the green and blue components were quantified [2].

Furthermore, land and water resources are central to agriculture and rural development, and are intrinsically linked to the global challenges of food insecurity, poverty, and climate change adaptation and mitigation, as well as to the degradation and depletion of natural resources that affect the livelihoods of millions of rural people around the world [24]. Current projections indicate that the world population will increase from 6.9 billion people today to 9.1 billion in 2050. In addition, economic progress, notably in the emerging countries, consistently translates into increased demand for food and diversified diets. The worldwide demand for food will thus surge as a result, and it is projected that food production will increase by 70% around the world and by 100% in developing countries. At the same time, both land and water resources, the basis of our food production, are finite and already under heavy stress, and future agricultural production will thus need to be more productive and more sustainable at the same time [24].

This paper estimated the blue, green, and grey WFs of rice production in Taiwan from 2005-2014 in order to explore the differences in the use of cultivated water resources in different regions of Taiwan, as well as to analyse its historical changes and composition characteristics. This paper is divided into four parts. In Section 2, the data and methodology, including the WF approach and the definition of its components, are introduced. In Section 3, the results are presented, analysed, and discussed. Finally, in Section 4, the conclusions are drawn.

2. Data and methodology

2.1. Data

The research data acquired for parameter construction were collected from 2005 to 2014. Survey items included the cultivation region and cultivation period of rice, fertilizer applications, and harvest yields. Rice can be planted twice a year in Taiwan. The growing seasons of rice are divided into two periods, the first season crop period (2/1~6/30) and second season crop period (7/1~11/30).

Crop evapotranspiration (ET_0) and effective rainfall (P_{eff}) values were calculated using local temperature, humidity, sunlight, and rainfall data from 2005 to 2014 from the Central Weather Bureau, while actual irrigation was analysed using period average data. The crop coefficient (K_c) is an indicator of water consumption during the entire growth period. In this study, the crop coefficient was set based on previous local studies [25]-[28].

2.2. Methodology

The WF of a product is the result of the quantification of three water volume components: (i) the green water component (WF_{green}, m^3), (ii) blue water component (WF_{blue}, m^3), and (iii) grey water component (WF_{grey}, m^3). Each of them represents an essential element of water use. [2]

Based on the "Water Footprint Manual" [29] and its update, "The Water Footprint Assessment Manual" [5], WF was calculated as follows:

$$WF = WF_{green} + WF_{blue} + WF_{grey} \tag{1}$$

First, the green WF is defined as the consumption of water from precipitation that is stored in the soil and does not run off or recharge the groundwater and, thus, is available for the evapotranspiration of plants.

CROPWAT 8.0 software was applied to calculate crop evapotranspiration and effective rainfall values, which were then used to establish the green and blue WFs of rice. Given that the water contained in the crop accounts for only 0.1%-1% of the total evapotranspiration, it was disregarded, and WF_{green} was set equal to the sum of green water evaporation alone [29] and was calculated as follows:

$$WF_{green} = \frac{10*min(ET_{c,\ tot},\ P_{eff,\ tot})\times A}{Y} \tag{2}$$

where WF_{green} is defined as the green WF (m^3/ton), $ET_{c,tot}$ is total evaporation (mm/period), $P_{eff,tot}$ is the effective total precipitation (mm/period), A is the planting area (ha), and Y is the crop yield (tons).

Secondly, the blue WF is an indicator of the surface water or groundwater consumption, which includes the evaporated water, water incorporated into the product, and lost return flow. WF_{blue} calculates the agricultural irrigation. Equation (3) was used to directly calculate WF_{blue}:

$$WF_{blue}(m^3/ton) = \frac{IR}{Y} \tag{3}$$

where IR is the amount of irrigation water per crop period (m^3).

Thirdly, the grey WF of a process step indicates the degree of freshwater pollution that can be associated with the process step. The grey WF is defined as the volume of freshwater that is required to assimilate the load of pollutants based on natural background concentrations and existing ambient water quality standards [5]. WF_{grey} calculates the use of water to dilute pollution, which mainly results from the use of nitrogen fertilizer [28]. WF_{grey} (m^3/yr) was calculated as follows:

$$WF_{grey} = \frac{1000\times\alpha\times AR\times A}{Y(C_{max}-C_{nat})} \tag{4}$$

where AR is the area of nitrogen fertilizer usage (kg/ha); α is the leaching factor (%), which is set at 10% [1]; C_{max} is the maximum concentration of nitrogen for a given water body (mg/L), which is set at 10 based on Taiwan EPA groundwater pollution control standards [30]; and C_{nat} is the natural background concentration of nitrogen (mg/L), which is set at 0.

3. Results and discussion

3.1. Comparison of WFs for rice in different growth periods

The research results indicate that the average annual rice WF from 2004-2015 was about 7,580 m^3/ton (Figure 1), of which 80% was blue,17% was green, and 3% was grey. This average annual WF was about 5.7 times larger than the 2000-2004 average annual WF of rice for countries around the globe of 1325 m^3/ton, of which 48% was green, 44% was blue, and 8% was grey [1]. As indicated by the above results, the blue WF is the most important source of water for rice production in Taiwan. The WF from 2005-2008 increased by about 5%. The fallow area was increased by 5%, the cultivated acreage was reduced by 6%, and the production was reduced by 0.7%. However, the amount of irrigation water consumed was not reduced.

Figures 2-4 show the differences in the composition of WFs between the first crop and second crop for each year from 2004-2015 in Taiwan. The average WF of rice production for the first crop was 5,212 m^3/ton, while the average WF of rice production for second crop was 10,588 m^3/ton. In other words,

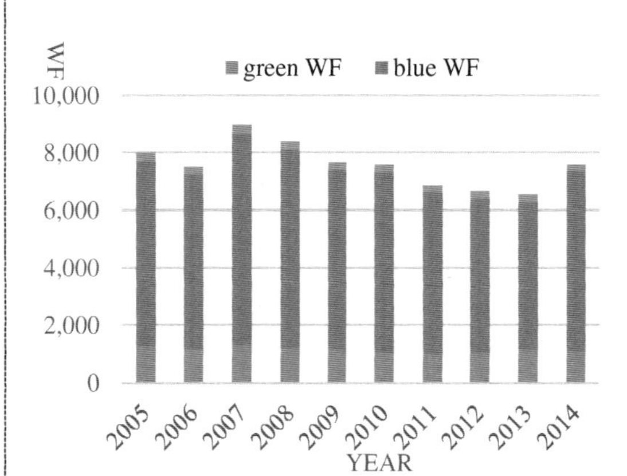

Figure 1. WFs from 2005 to 2014

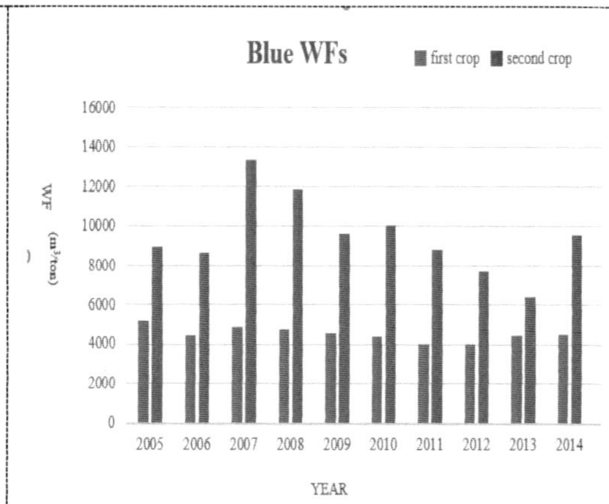

Figure 2. Blue WFs in different growth periods

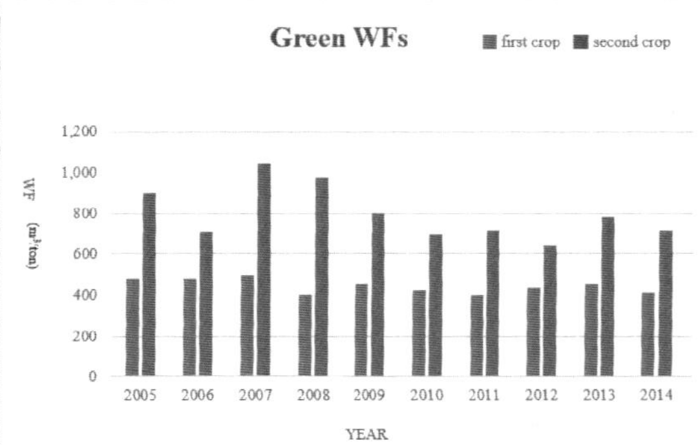

Figure 3. Green WFs in different growth periods

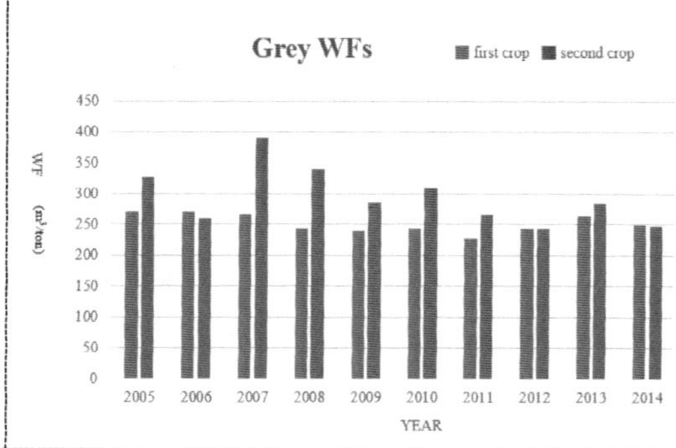

Figure 4. Grey WFs in different growth periods

the amount of water consumed for the second crop was higher than the amount consumed for the first crop. One reason for this difference is that the second crop is raised from July to November. During this period, the average annual amount of evapotranspiration for the years in question was 317 mm, and the effective rainfall was 8853 mm. Both of those values were higher than the corresponding values for the first crop season, which were 238 mm and 6620 mm, respectively. At the same time, the second crop is often affected by typhoons, heavy rain, cold damage, and other weather events, such that the amount of rice produced by the second crop is less than that produced by the first crop, even as the amount of irrigation water used for the second crop is not reduced.

The results of this study indicating the changes in the composition of green, blue, and grey water in Taiwan's first and second season crops are shown in the following figures. As can be seen from the figures, the main source of water resources for rice production in Taiwan is blue water, accounting for 80% of the total, a finding which demonstrates the importance of irrigation water. During the 2005~2008 period, the proportion of blue water in the overall WF was higher than the proportions of green and grey water for both the first and second crops. From 2005 to 2008, the blue WFs increased by 32%. Then, in 2009, the government activated the fallow farmland, causing the blue WFs to decrease. The green WF consistently accounted for a low proportion of the overall first crop and second crop WFs, and the proportion accounted for by the green WF should be affected by the climate and fallow system.

3.2. Comparison of WFs for rice in different regions

Figures 5-7 show the consumption of water resources in the northern region, central region, southern region, and eastern region of Taiwan during the 2005~2014 period. The average WF for the northern region, 35,415 m^3/ton, was the highest. The average WF for the eastern region was 19,786 m^3/ton. The average WF for the central region, 14,112 m^3/ton, was smaller than that for the eastern region, while that for the southern region, 8,311 m^3/ton, was the lowest of all. Therefore, the water use efficiency of the southern region was the best, while the water use for the northern region was the most inefficient, especially in terms of blue WFs.

3.3. Water-land-rice nexus: land, rice yield, and WF

Rice requires both water and land resources. The annual rates of change for land usage, rice production, and WFs during the 2005-2014 period are shown in Figure 8. The rates of change in cultivated land and rice production in Taiwan during the period decreased in a stable manner. However, the annual rates of change in the WFs for rice production during the period were unstable. Until 2009, when the active fallow land policy was implemented, the rice acreage and yield gradually increased. The annual rates of change in the WFs for rice production in 2010 ~ 2011 were reduced.

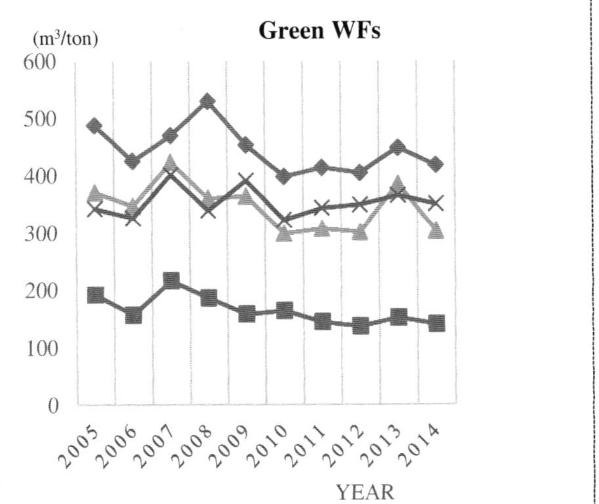

Figure 5. Green WFs in different regions

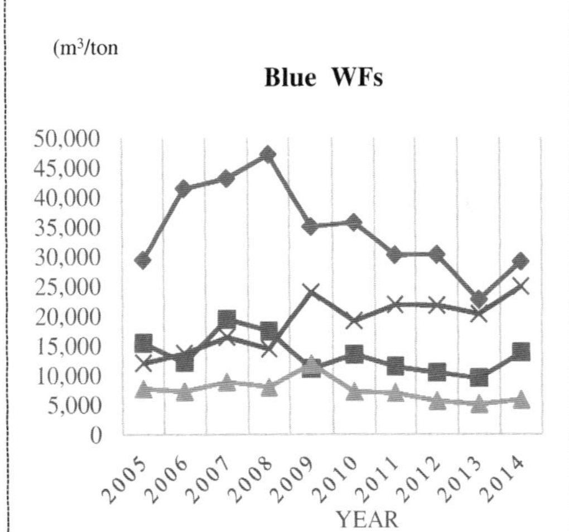

Figure 6. Blue WFs in different regions

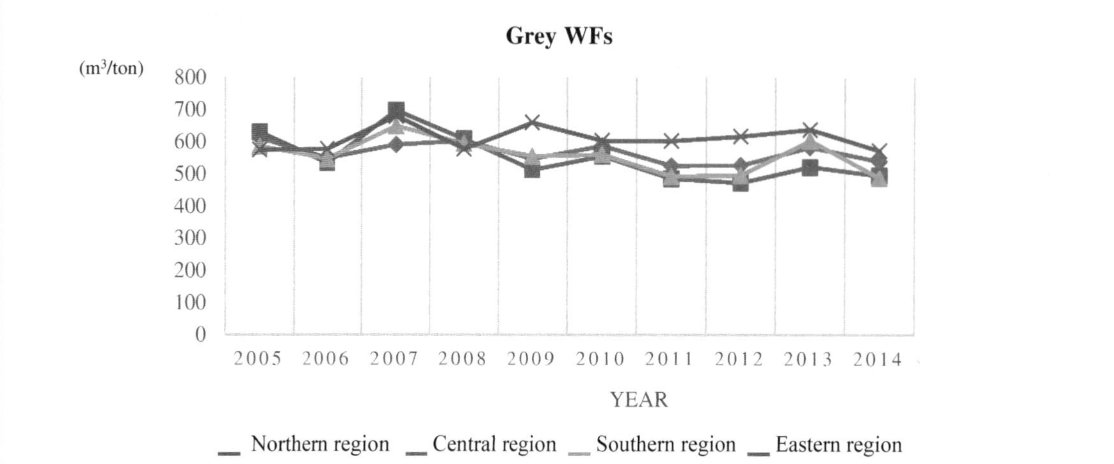

Figure 7. Grey WFs in different regions

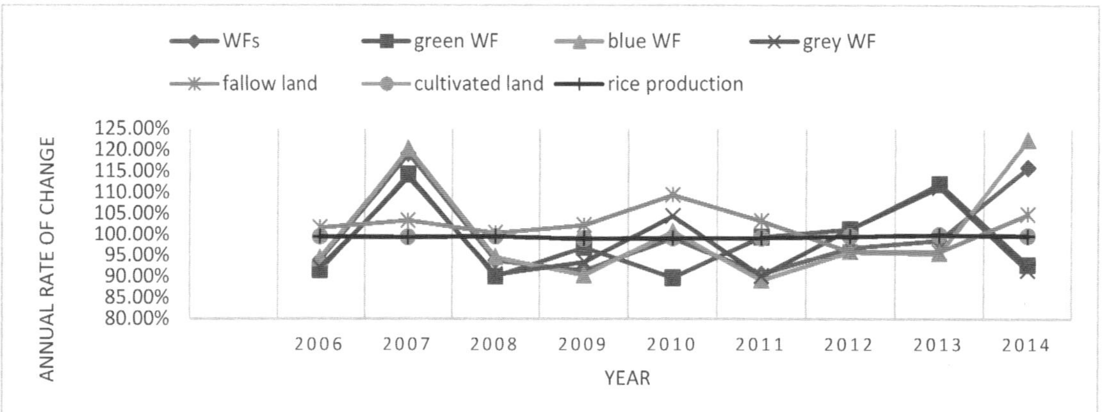

Figure 8. Annual rates of change for land, rice production, and WFs

4. Conclusion

Taiwan is listed as a country that suffers from water scarcity [28]. The results of this study showed that the average annual rice WF for the years 2005-2014 was about 7,580 m^3/ton, of which 80% was blue, 17% was green, and 3% was grey. This average annual footprint was about 5.7 times larger than the 1325 m^3/ton 2000-2004 average annual WF of rice for countries around the globe reported by [3], of which 48% was green, 44% was blue, and 8% was grey. The blue WF is the most important source of water for rice production in Taiwan. The water consumption of the second crop is higher than that of the first crop. The water use efficiency in the southern region is the best, while that of the northern region is the worst. The rates of change in cultivated land and rice production in Taiwan during the 2005-2014 period decreased in a stable manner. However, the annual rate of change in the rice WF was unstable. Until 2009, when the active fallow land policy was implemented, the rice acreage and yield gradually increased. Furthermore, land and water resources are central to agricultural development, and are intrinsically linked to challenges of food insecurity and poverty.

This research suggests that climate change will bring greater variation in weather events, including more frequent weather extremes. We will also face growing water scarcity, which will impact rural and urban livelihoods, food security, and economic activities. More specifically, water shortages will result in increasing competition, which will constrain agricultural production and affect the incomes and livelihood opportunities of many residents in both rural and urban areas. With increasing competition over water for agricultural purposes and in other sectors, governments will need to effectively communicate water scarcity conditions and use water wisely, thereby ensuring that water is allocated equitably and efficiently. However, in Taiwan, land and water institutions have not kept pace with the growing intensity of agricultural development and the increasing degree of interdependence and competition over land and water resources. As such, much more adaptable and collaborative institutions are needed to respond effectively to natural resource scarcity, particularly with regard to water and agriculture. Improvements will thus have to come from sustainable intensification that ensures the effective use of land and water resources, in addition to ensuring that such resources are not harmed [24]. Furthermore, the nexus of land, water, and food needs to be taken into consideration in order to protect water availability, maintain agricultural production, and avoid land degradation.

5. References

[1] Chapagain A K and Hoekstra a Y 2011 The blue green and grey water footprint of rice from production and consumption perspectives *Ecol. Econ* 70 **(4)** 749-58
[2] Lovarelli D, Bacenetti J and Fiala M 2016 Water footprint of crop productions: A review *Sci.of the Total Environ.* **548-549** 236-51
[3] Hoekstra A Y and Hung P Q 2002 A quantification of virtual water flows between nations in relation to international crop trade *Water Res.* **49** 203-9
[4] Hoekstra AY, Chapagain A K, 2007 Water footprints of nations: water use by people as a function of their consumption pattern *Water Resour. Manag.* **21 (1)** 35-48
[5] Hoekstra A Y, Chapagain A K, Aldaya M M and Mekonnen M M 2011 The water footprint assessment manual: setting the global standard Earthscan London UK
[6] Zhang Y, Huang K, Yu Y and Yang 2017 mapping of water footprint research: A bibliometric analysis during 2006-2015 *J. of Cleaner Produ.* **149** 70-9
[7] Chapagain A K, Hoekstra A Y, Savenije H G and Gautam R 2006 The water footprint of cotton consumption: an assessment of the impact of worldwide consumption of cotton products on the water resources in the cotton producing countries *Ecol. Econ.* **60 (1)** 186–203
[8] Chapagain A K and Hoekstra A Y 2010. The green, blue and grey water footprint of rice from both a production and consumption perspective Water Research Report Series No. 40. UNESCO-IHE, Delft, The Netherlands.
[9] Yoo S-H, Choi J-Y, Lee S-H and Kim T 2013. Estimating water footprint of paddy rice in Korea Paddy *Water Environ.* **12 (1)** 43–54 http://dx.doi.org/10.1007/s10333-013-0358-2
[10] Marano R P and Filippi R A 2015 Water Footprint in paddy rice systems Argentina *Ecol. Indic.* **56** 229–36
[11] Chapagain A K and Hoekstra A Y 2007 The water footprint of coffee and tea consumption in the Netherlands *Ecol. Econ.* **64 (1)** 109–18
[12] Chapagain A K and Orr S 2009 An improved water footprint methodology linking global consumption to local water resources: a case of Spanish tomatoes *J. Environ. Manag.* **90(2) 1219**–28
[13] Aldaya M M and Hoekstra AY 2010 The water needed for Italians to eat pasta and pizza *Agric. System* **103** 351-360
[14] Vanham D, Mekonnen M M and Hoekstra A Y 2013 The water footprint of the EU for different diets *Ecol. Indic.* **32** 1–8 http://dx.doi.org/10.1016/j.ecolind.2013.02.020.
[15] Hoekstra A Y and Mekonnen M M 2011 The water footprint of humanity *P. Natl. Acad. Sci U. S. A.* **109 (9)** 3232-37
[16] Chang Y H 2011 *Assessment of agricultural water footprint in Taiwan: case study of rice production* Master Thesis National Taiwan University
[17] Mekonnen M M and Hoekstra A Y 2012. A global assessment of the water footprint of farm animal products *Ecosystems* 15 **(3)** 401–15
[18] Gerbens-Leenes P W, Lienden A R, Hoekstra A Y and Meer T H 2012. Biofuel scenarios in a water perspective: the global blue and green water footprint of road transport in 2030 *Glob. Environ. Chang.* **22 (3)** 764–75
[19] Rulli M C et al. 2016 The water-land-food nexus of first-generation biofuels. Scientific Reports 6,22521; doi: 10.1038/srep22521

[20] Morera S, Corominas L, Poch M, Aldaya M M and Comas J 2016 Water footprint assessment in wastewater treatment plants *J. of Cleaner Produ.* **112** 4741-48

[21] Chiu C C, Shiang W, Lin C J, Wang C and Chang D 2015 Water footprint analysis of second-generation bioethanol in Taiwan *J. of Cleaner Produ.* **101** 271-7

[22] Duarte R, Pinilla V and Serrano A 2014 The water footprint of the Spanish agricultural sector: 1860–2010 *Ecol. Econ.* **108** 200-7

[23] Mekonnen M M and Hoekstra A Y 2011 The green blue and grey water footprint of crops and derived crop products *Hydro. Ear. Sys. Sci.* **15** 1577-600

[24] FAO 2011 The state of the world's land and water resources for food and agriculture (SOLAW) – Managing systems at risk Food and Agriculture Organization of the United Nations Rome and Earthscan London

[25] Shih C C, Cao I S, Hsu Y B and Kan J E 1982 The irrigation drainage principles of the third edition Central Book Publish Taipei Taiwan

[26] Shih C C and Chang B C 1993 Sorghum consumptive use study by drip irrigation *J. Chin. Agric. Eng.* **39** 13-25

[27] Shih C C and Hwang C D 1994 Study on the coefficient of evapotranspiration of corn *J. Chin. Agric. Eng.* **40** 1-15

[28] Su M H, Huang C H, Li W Y, Tso C T and Lur H S 2015 Water footprint analysis of bioethanol energy crops in Taiwan *J. of Cleaner Produ.* **88** 132-8

[29] Hoekstra A Y, Chapagain A K, Aldaya M M and Mekonnen M M 2009 Water footprint manual: state of the art 2009 Water Footprint Network Enschede the Netherlands

[30] Taiwan EPA (Taiwan Environmental Protection Agency) 2013 Groundwater Pollution Control Standards Available from: http://w3.epa.gov.tw

Acknowledgments

The author thanks Miss L W Huang, Miss Y C Zhan, Mr. J R Li and Mr. T R Zhang for helping with data collection and computer processing.

IOP Conf. Series: Earth and Environmental Science **150** (2018) 012002 doi:10.1088/1755-1315/150/1/012002

Modeling air concentration over macro roughness conditions by Artificial Intelligence techniques

T Roshni[1] and S Pagliara[2]
[1] Assistant Professor, National Institute of Technology Patna, Bihar, India-800005
[2] Professor, University of Pisa, Via Gabba 22, Pisa, Italy- 56122
Email: roshni@nitp.ac.in, s.pagliara@unipi.it

Abstract. Aeration is improved in rivers by the turbulence created in the flow over macro and intermediate roughness conditions. Macro and intermediate roughness flow conditions are generated by flows over block ramps or rock chutes. The measurements are taken in uniform flow region. Efficacy of soft computing methods in modeling hydraulic parameters are not common so far. In this study, modeling efficiencies of MPMR model and FFNN model are found for estimating the air concentration over block ramps under macro roughness conditions. The experimental data are used for training and testing phases. Potential capability of MPMR and FFNN model in estimating air concentration are proved through this study.

1. Introduction

Roughness occurs in nature in various situations such as flows in hilly streams, gravel beds, etc. The natural flows are very dynamic in nature showing a very large-scale variation in different time scales. Aeration happens due to the turbulent flow over macro and intermediate roughness conditions. Aerated flows or two-phase flows is recreated artificially in rivers by flows over block ramps or rock chutes [1]. Natural rocky rivers act as river training structures and the human intervention leads to regenerate a similar situation in the natural rivers as river training structure. These structures offer an alternative form of drop structures, fish ladders etc when the protection of the morphology and ecology of water course is required [2]. Air entrained in the flow leads to a positive effect in terms of flow reaeration and dissolved oxygen [3] representing a valid methodology to improve the fish habitat and to reduce the pollutant content of rivers or water bodies. Aeration studies over block ramps and stepped spillways have been carried out experimentally by [1], [2], [4], [5] etc. A close view of the aerated flows over macro roughness conditions is shown in Figure 1.

Figure 1. Aerated flow over ramps in River Maglioggio (Italy).[6].

Recent literature surveys show that a few investigators attempted to investigate the potential of artificial intelligent techniques in aeration studies [7]. This paper aims to develop models to estimate the air concentration over block ramps by Feedforward Neural Network (FFNN) and Minimax Probability Machine Regression (MPMR) models. Models are developed based on the experimental results of [1], [2]. This paper also emphasizes the applicability of regression models in the estimation of hydraulic parameter.

2. Theory of Models

2.1. Feedforward Neural Network (FFNN)

Artificial Neural Network acts as a powerful computational model and can applied to all fields of engineering. Previous literature work proves the efficacy of conventional model FFNN with Back Propagation algorithm. In FFNN, two phases involved are training and testing phase. Correspondingly, 70% and 30% of the data are divided for training and testing. In the three-layer system, 1st layer denotes the input layer and nodes in the input layer corresponds to number of inputs. Second layer is the hidden layer and the selection of nodes are by trial and error method and third is the output layer and the node corresponds to the output required. Model development proceeds until the error generated between the output and the target is within the desirable range. Transfer function used in this study is sigmoid function [8].

2.2. Minimax Probability Machine Regression (MPMR)

In recent studies only, the potential applicability of MPMR is noted and it acts a good alternative for Support Vector Machine (SVM) [9]. Related work of SVM can be seen in [10]. MPMR is introduced by Strohmann and Grudic [11]. Recent literature work of MPMR in prediction of seismic ultrasonic attenuation [12], prediction of fast fading channel [13] forecasting evaporative loss [9] and a few more reveals the potentiality of this regression model.
MPMR follows a regression model for y as

$$y = [\sum_{i=1}^{N} \beta_i K(x_i, x) + b] \pm \varepsilon \quad (1)$$

In eqn. 1, x and y represents the inputs and corresponding outputs, $K(x_i, x)$ is kernel function and generally, Radial Basis Function (RBF) is used as the kernel function. The other variables β, b are outputs obtained by MPMR algorithm and ε shows the limits of error fluctuations. The MPMR algorithm is established in MATLAB 2010.

3. Experimental Facilities and Data Collection

The source of data for the modeling analysis is taken from the study of [1], [2], [5], [14]. The experiment is conducted in a 7m long tilting flume with an effective width of 0.3m. Bed surface is filled with irregular boulders to simulate the conditions of roughness. Inflow to the flume is measured by digital flowmeter, point gauge to measure the depth of flow and single tip conductivity probe to measure the air concentration. Both the point gauge and the conductivity probe are connected to the trolley assembly, such that measurements can be taken at all possible locations. The single tip conductivity probe is connected to a data logger. The experiments vary for different slopes, discharges and different bed materials [1], [2]. The measurements were taken only in the uniform flow region. The experimental range of the data are shown in Table 1.

Table 1. Experimental data range

	d_e/D_{84}	Slope	Q (m^3/s)	k_c/D_{84}	C_{mean}
Minimum	0.52	0.136	0.006	0.64	0.13
Maximum	1.81	0.46	0.03	2.12	0.42

Where d_e is the effective depth of flow, D_{84} is for the bed material used, Q is the discharge, k_c is the critical flow depth, C_{mean} is the average air concentration in the uniform flow region.

4. Performance Measures

As a performance measure, different indicators are used in this study. The indicators are Root Mean Square Error (RMSE), Coefficient of Correlation (R), Nash–Sutcliffe model efficiency (NSE), and Standard Deviation (STD).

$$RMSE = \sqrt{\left(\frac{1}{N}\sum_{i=1}^{N}(h_{oi} - h_{ci})^2\right)} \quad (2)$$

$$NSE = 1 - \frac{\sum_{i=1}^{N}((h_{oi} - h_{ci})^2}{\sum_{i=1}^{N}((h_{oi} - \overline{h_{oi}})^2} \quad (3)$$

$$R = \frac{\sum(h_{oi} \cdot h_{ci})}{\sqrt{\sum h_{oi}^2 \sum h_{ci}^2}} \quad (4)$$

Where, h_{oi} = observed value and h_{ci} = predicted value, and N = number of observations, P is the input quantity and $\overline{h_{oi}}$ and $\overline{h_{ci}}$ are the mean of the observed values and the calculated values [1].

5. Results and Discussion

The experimental results were modelled with FFNN model and MPMR model. The input parameters considered in this study for modeling are d_e/D_{84}, Slope, Q and k_c/D_{84}. The output obtained is air concentration. For the model development, 70% of the data is taken for training and 30% of the data for testing. For FFNN model, different architectures are developed with different hidden nodes and the best architecture is selected for lowest RMSE value of the model. Similarly, MPMR model is also developed by varying the internal parameters till optimum model is developed. Visual performance of the observed experimental data and the MPMR model and FFNN model data is shown in Figure 2a and Figure 3a. A plot of confidence interval (+95% to -95%) band and prediction interval band (+95% to -95%) is also plotted for MPMR and FFNN model results to show the performance of the developed models (Figure. 2b and 3b). Figures (2-3) display a very good correlation for both the models and it is difficult to infer a best model out of two. Overall performance of training and testing phase is verified by plotting Regression Error Characteristics (REC) Curve (shown in Figure 4). REC curve gives a plot of accuracy in terms of absolute deviation. In this curve also, both models show potential ability of prediction with a slightly more weightage to FFNN.

A novel representation of performance measure indicator is Taylor diagram. In this, Standard deviation, Correlation coefficient and RMSE (blue dotted line) (shown in Figure 5) values are considered together to find the best model. Based on the training result, FFNN is showing better result, for testing model MPMR is more towards the observed values. Overall performance shows that, there was a good agreement between the experimental and the model results for FFNN model. The performance indicators for training and testing phase is shown separately in Table 2. This is also in line with the above results.

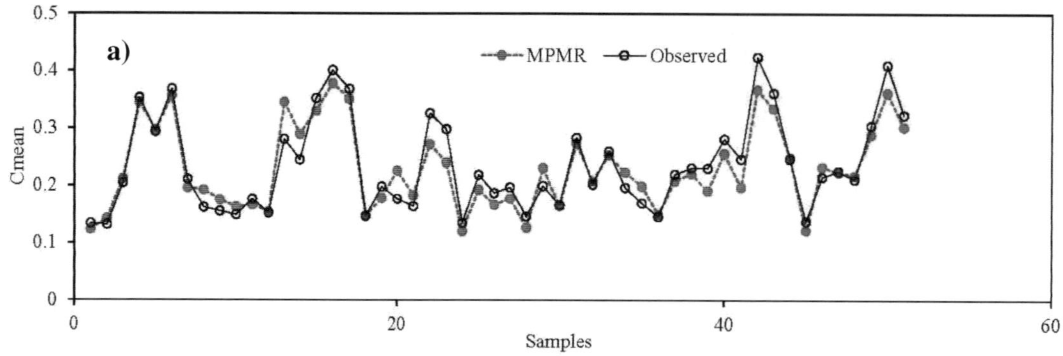

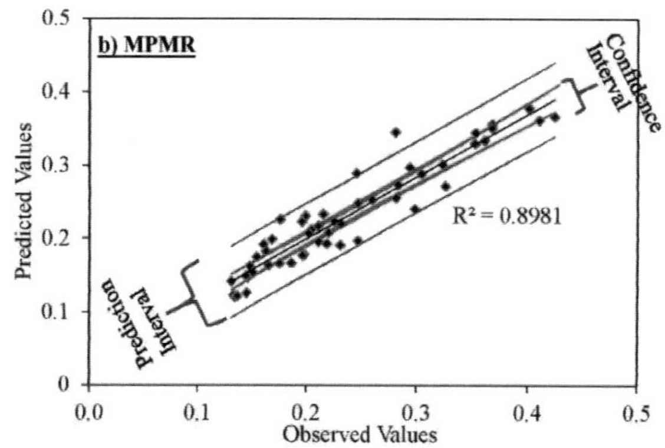

Figure 2. (a) Visual comparison of MPMR model result with that of observed data (b) Data within the confidence interval band and prediction interval band.

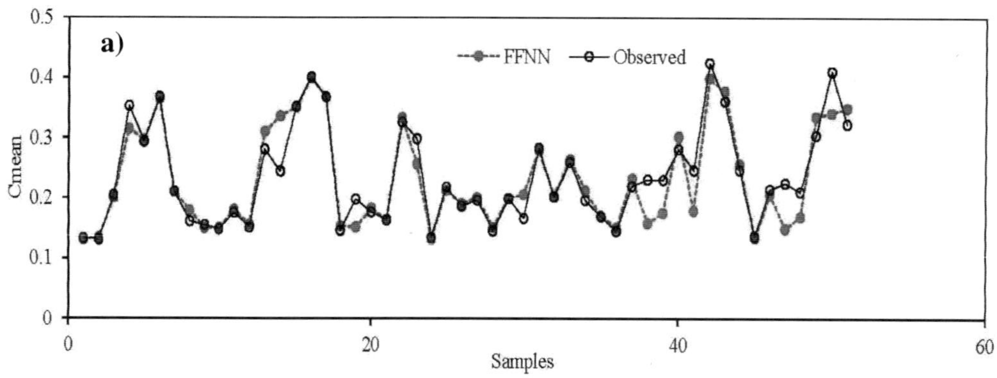

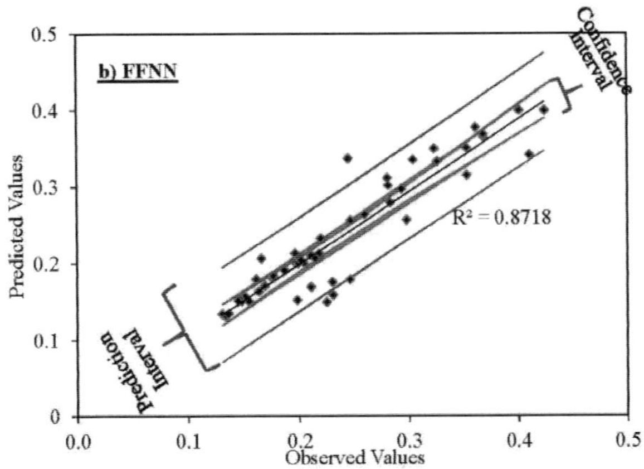

Figure 3. (a) Visual comparison of FFNN model result with that of observed data (b) Data within the confidence interval band and prediction interval band.

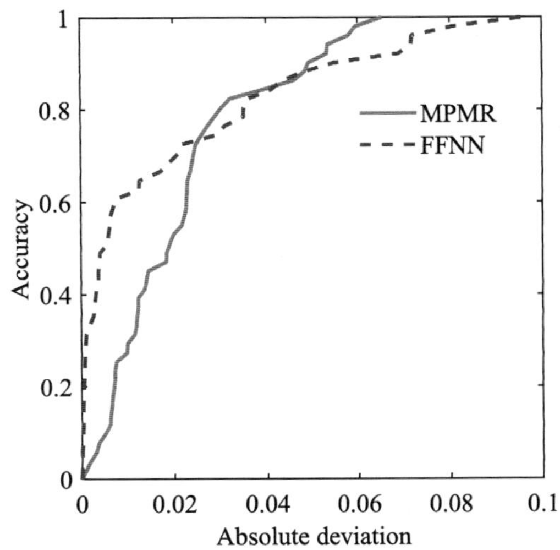

Figure 4. REC Curve for MPMR and FFNN model result

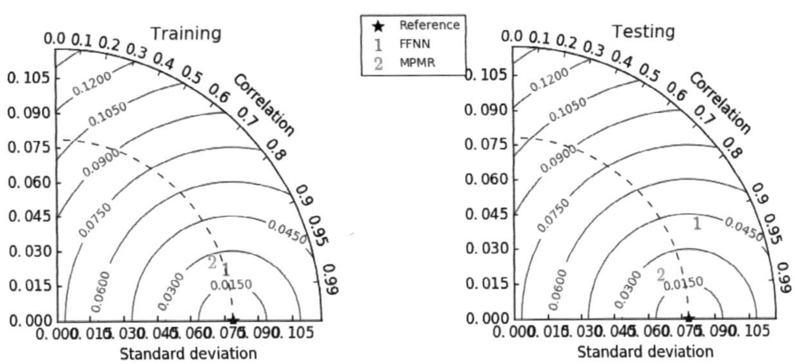

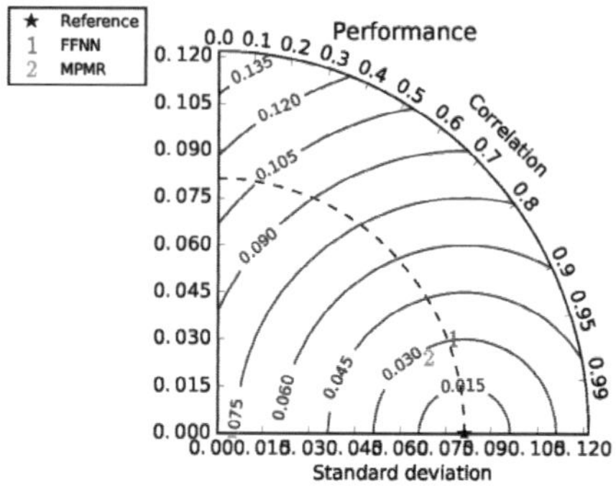

Figure 5. Taylor diagram plotted for MPMR and FFNN Models.

Table 2. Model Performances Indices for FFNN and MPMR Models

		FFNN	MPMR		FFNN	MPMR
Training	R	0.9595	0.94	Testing	0.8942	0.96
	RMSE	0.022	0.0267		0.0445	0.02
	NSE	0.9178	0.88		0.66	0.85
	STD	0.0787	0.0743		0.0918	0.0681

6. Conclusions

Aeration concentration varies with the macro roughness and the slopes. Block ramps or rock chutes are used in the experimental study to recreate the macroroughness conditions. Through this work, the potential efficacy of the soft computing techniques in the field of hydraulic engineering is studied. For that, MPMR and FFNN models are used in estimation of air concentration over block ramps. Both the models show equal potential in modeling efficacy. A slight better performance was noticed for FFNN model than MPMR model. Overall, both MPMR and FFNN models proved their capability in modeling aeration concentration over block ramps.

Further investigations in this field can be extended to the application of mathematical models in other fields of hydraulic engineering. Similar type of work can be used to study the effect of aeration over different roughness conditions in improving dissolved oxygen concentration, effect on pollutant loads, etc.

7. References

[1] Pagliara S, Carnacina I and Roshni T 2010 Air-water flows in presence of staggered and row boulders under macro-roughness condition *Water Resources Research* 46(8),11pp
[2] Pagliara S and Chiavaccini P 2006 Energy dissipation on block ramps *J. of Hyd. Engg*, 132(1), 41-48.
[3] Moog D B and Jirka G H 1999 Stream Reaeration in Nonuniform Flow: Macroroughness Enhancement *J. of Hyd. Engg.*, 125(1), 11-16.
[4] Chanson H 1995 *Air bubble entrainment in free-surface turbulent flows. Report CH46/95, Department of Civil Engineering, The University of Queensland, Brisbane.*
[5] Pagliara S, Carnacina I and Roshni T 2010 Self-aeration and friction over rock chutes in uniform flow conditions. *J. of Hyd. Engg.* 136(11), 959-964.
[6] Roshni T 2011 Two phase flows over macro and intermediate roughness. *PhD Thesis.* University of Pisa, Italy
[7] Khalifa A, Bayoumi S and Monayeri O EI 2011 Mathematical modeling of aeration efficiency and dissolved oxygen provided by stepped cascade aeration *Water Science and Technology.* 1-9.

[8] Sahoo S and Jha M K 2013 Groundwater-level prediction using multiple linear regression and artificial neural network techniques: a comparative assessment *Hydrogeol. J.* 21, 1865–1887.

[9] Ravinesh C D and Samui P 2017 Forecasting Evaporative Loss by Least-Square Support-Vector Regression and Evaluation with Genetic Programming, Gaussian Process, and Minimax Probability Machine Regression: Case Study of Brisbane *J. Hydrol. Eng.*, 22(6), 1-15.

[10] Nurcihan C 2014 Application of support vector machines and relevance vector machines in predicting uniaxial compressive strength of volcanic rocks *J. of African Earth Sciences.* 100, 634-644

[11] Strohmann T and Grudic G Z 2002 A formulation for minimax probability machine regression *Proc., Advances in Neural Information Processing System,* MIT Press, Cambridge, MA, 769–776.

[12] Manoj K, Mittal M and Samui P 2013 Performance assessment of genetic programming and minimax probability machine regression for prediction of seismic ultrasonic attenuation *Earthq. Sci.*, 26(2), 147-150.

[13] Liming Y, Wang L, Sun Y and Zhang R 2010 Simultaneous feature selection and classification via Minimax Probability Machine *Int J Comput Intell Syst.* 3(6), 754-760.

[14] Pagliara S, Carnacina I and Roshni T 2011 Inception point and air entrainment on flows under macro-roughness conditions *J. of Environ. Engg.*137(7), 629-638.

River Aquifer Interaction in Lower Gandak Command Area in Bihar, India

V Singh[1], C Raj[2] and B Chakraborty[3]

1 Professor, Dept. of Civil Engg., National Institute of Technology Patna, Patna, Bihar (India) 800005
2 Research Scholar, Dept. of Civil Engg., National Institute of Technology Patna, Patna, Bihar (India) 800005.
3 Scientist G & Head, National Institute of Hydrology Patna, Bihar, India
E-mail: vsingh@nitp.ac.in

Abstract. In this paper, the quantification of the interaction of aquifer and river during various stages of river Gandak has been carried out. A groundwater flow model, MODFLOW, has been used to quantify the interaction of river Gandak and the adjacent aquifer in all months of the year. The model has been calibrated with the observed water level data of the existing wells. Calibrated model was used to simulate the groundwater flow in the study area and the flows from river Gandak to adjacent aquifer and vice-versa. Results show that the river Gandak is recharging adjacent aquifer during the months May to October and again during January to February. In the months March, April, November and December aquifer is contributing water to Gandak River. It was found that the quantity of inflow/outflow from river Gandak is proportional to the difference of water level in river and aquifer. Further, a relationship for the interaction of Gandak River and adjacent aquifer has also been developed to find the aquifer recharge with rise in river stage in the study area.

1. Introduction

Multilayered aquifers occur in a sedimentary groundwater basin which is separated by confining layers. These confining layers have very less permeability. Aquifer-stream interaction occurs at multiple scales in both time and space. A solution for river aquifer interactions problems is important as groundwater and surface water are not independent resources [1]. A river in such basin may be penetrating either partially or fully in the upper aquifer. The river stage varies from month to month. In monsoon, the stage of the river is high and in non-monsoon months the river stage goes down. The rise in river stages above the aquifer water table in the vicinity of the river leads to induced recharge to groundwater. Due to rise in river stage, the upper aquifer is recharged directly through the bed and banks of the river and the lower aquifers are recharged from the top aquifer through the intervening aquitard. In the process of interaction either water flow from aquifer to river or from river to aquifer. Recharge takes place if water level in a river is above the water table in the aquifer having permeable or semi-permeable material. This type of groundwater recharge may be temporary, seasonal, continuous, natural or induced by man through groundwater extraction.

The recharge is proportional to the difference in the water level of river and aquifer. This proportionality constant is known as reach transmissivity. The reach transmissivity depends on the stream bed characteristics and the shape of the stream cross-section. River aquifer exchange also depends upon structural geology [2] and dynamics pressure waves [3]. The water level in the aquifer depends on all the withdrawal and recharges including recharge from river. Several studied have been

carried out on single river aquifer interaction problem analytically. Three different basic analytical solutions currently exist for flow besides stream. First solution is based on the assumption that the stream and aquifer are perfectly connected, second solution is on the assumption that there is a vertical semi permeable boundary between aquifer and stream bank. Riverbed and bank permeability are determined by laboratories and field modelling. Third assumption is that stream slightly penetrates top layer of the aquifer and there is a semi permeable bottom of negligible width, and aquifer is extended to infinity in all direction horizontally [4]. However, accuracy of the flow calculation by application of classical horizontal flow model is checked by [5] and found that the velocity approach model is more accurate. For a partially penetrating river the exchange flow rates are reduced significantly in comparison to fully penetrating rivers. Interaction relationship for aquifer and the river is generally linear, however different for different morphology of river [6]. With the advancement in computational resources numerical methods have increased. Diffusion equation with sources and sinks terms solved numerically to obtain the interactions between groundwater and surface water [7]. Several studies have been carried out for the interaction of river and aquifer ([8]-[11]). [12] Produced an over view of the methods that are currently applied for river aquifer studies.
In this study, interaction of River Gandak and adjacent aquifer has been carried out during various stages of river in different months of the year. A groundwater flow model, MODFLOW under transient state has been used. [13] Examined the accuracy of simulation through MODFLOW for connected and disconnected losing stream. The model has been used to compute flow to/from Gandak River to adjacent aquifer at different locations along the river. River aquifer interaction relationship has also been developed to find the recharge with rise in river stage.

2. Study Area

Study area lies in lower Gandak basin in Vaishali and Muzaffarpur districts of Bihar, India having 84°57' to 85°08' E longitudes and 25°56' to 26°05' N latitudes. It is bounded by River Gandak in the western and Vaishali Branch Canal (VBC) in eastern sides. Habibpur sub-distributory (HSD) takes off at 138.2 RD of Vaishali Branch Canal and flows in the middle of the study area as shown in Figure 1(a) and the discretized image is shown in Figure 1b. River stage measurements were taken on the left bank of River Gandak at locations A, B, C, D, E, F, G and H as shown in Figure 1(b). The area comprises of an extensive plain formed by the alluvium brought by the river Gandak. It is mostly plain although few shallow depressions called 'Chaurs' in local language are present. Estimated area is 181.6 km^2. The elevation of the ground varies from 56.27 m to 50.90 m above MSL. The river Gandak forming western boundary of the study area is perennial in nature flowing towards south eastern direction. The average bed slope of river varies from 1:15000 to 1:20000. Stage data of river Gandak is measured at Rewaghat and Lalganj gauging sites which come in the study area.

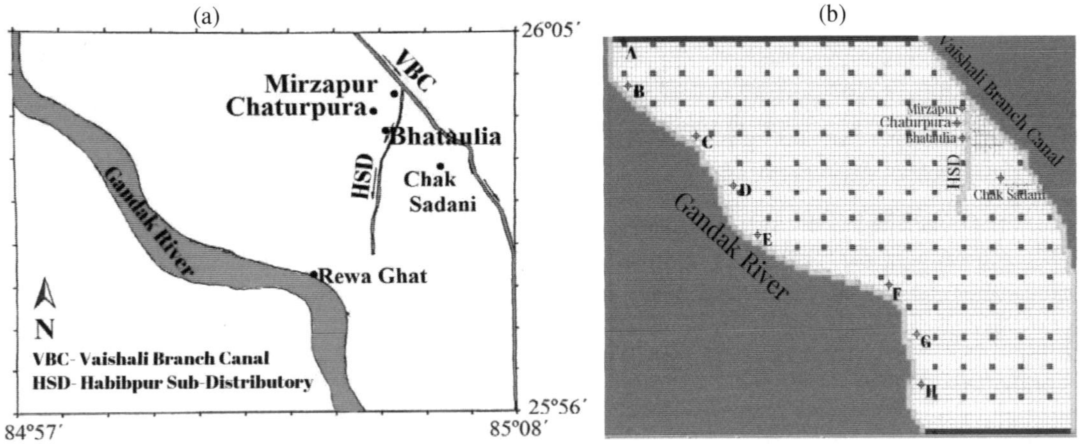

Figure 1. (a) Map of Study Area (b) Discretized image of study area.

Tirhut main canal is one of the three main canals receive water from Gandak main barrage at Balmikinagar. VBC takes off from Tirhut Main Canal at 554 RD (Reduced Distance = 1000 feet) forming eastern boundary of the study area. It is an unlined canal having an average bed slope of 1:7000 with side slopes 1:1.5. The bottom width of canal is 17.5m. HSD provides irrigation facility through 18 outlets of 0.25 cumec average discharge each. There are 5 state tube wells and more than 900 private tube wells in the command area. The subsurface formations a thick pile of unconsolidated alluvial sediments having different texture, along with their admixtures. The study area lies in monsoon sub-tropical zone. The average annual precipitation is 1046 mm [14].

3. Methodology

Three dimensional groundwater flow model, MODFLOW [15] has been used in this study. MODFLOW consist various modules and module 'River Package' is used to simulate interactions of river and groundwater. Details of the MODFLOW are not presented here as it is well known and generally used model to simulate the groundwater flow.

3.1. Model Formulation

This area has been discretized into 100 columns and 80 rows comprising of 8000 cells of dimension 200 × 200 m and three layers of average thicknesses 15 m, 15 m, and 50 m. The values of parameters like hydraulic conductivity (K) and specific storage (S_s), and porosity for different layers has been estimated by Pump test done in year 1993. Transmissibility, T = 400-700 m^2/day, Specific yield = 12-20% and Coefficient of storage = 0.015 have been taken from 'Vaishali district report' [16]. Water table elevation in the month of May, 1993 has been taken as initial condition. For convenience, all water table data have been raised by 100 m above MSL. West and east boundary of the study area is formed by Gandak river and VBC respectively whereas north and south boundary is open boundary termed as General Head Boundary (GHB). Wells and evapotranspiration have been considered as discharge component. According to census of Bihar, 957 shallow private tube wells were present in the study area. In modeling, 800 tube wells are distributed in first and third layer with discharge = 25 m^3/day. These wells are operational in Garma (summer) and Rabi crop seasons only. Monthly Evapotranspiration ET_0 is taken at Patna and Muzaffarpur [17]. In cropped region, evapotranspiration is considered and in the region where crop is not grown only evaporation is considered. In the model, ET in non-cropped region is taken 50% of value that is in cropped region. Rainfall and irrigation return flows are considered as recharge. 20 % of the monthly rainfall has been considered as the recharge in the months of March to June and from November to February, whereas from July to October it is considered 70%. There are 18 outlets of the HSD. Each outlet has average discharge of 0.025 cumec irrigating approximately 10 ha of land. In model, 70% of discharge through outlet is taken as recharge. The canal is operational from July to October, January and 15 days in February in a year. Stream bed hydraulic conductance for Gandak river, VBC and HSD are 3000 m^2/day, 350 m^2/day and 150 m^2/day respectively.

3.2. Model Calibration

The results from model with reliable input parameters and stresses period generally come in close to observed field data. In case of mismatch either input parameters are unreliable or the stresses. If total number of operating wells in area is not known exactly, this cause underestimation of pumpage which further results overestimation of water table and vice versa. Similarly, for potential evapotranspiration value, [17] considered only solar radiation and ambient temperature as parameters. This may be a cause of not actual estimation of evapotranspiration. This will also be a cause of incorrect parameter resulting wrong estimation of water table. Thus, calibration of model parameter is required to check uncertainty part. In order to calibrate, model was simulated with the input values and stresses. The output water table elevation is then compared with observed water table elevation data of existing observation wells and according to mismatch, the parameters are modified. Calibrated values of different parameters hydraulic conductivity for all three layers are calibrated and presented in Table 1, whereas the calibrated rainfall recharge and evapotranspiration have been presented in Table 2.

One year daily observed water table data during 10.05.1991 to 12.05.1992 were available for four locations namely Mirzapur, Chaturpura, Bhataulia and Chak Sadani (Figure 1b). In addition to these, nine wells are assumed adjacent to river Gandak whose water table is taken same as stage of the river.

The observed water table elevation is compared with the computed water table of the calibrated model. Figure 2 present the comparison of observed and computed water table after calibration. In this figure, all the data have been plotted on a straight line at 45° to show the match between the observed and computed results. It is found that computed values of water table data are well matched with that of observed water table data. Thus, the process of calibration is terminated at this stage, and the model is ready for simulation.

Table 1. Calibrated values of Hydraulic Conductivity

Layers	Hydraulic conductivity (m/day)	
	Initial Value	Calibrated value
Layer 1	10	5
Layer 2	1	1
Layer 3	25	20

Table 2. Calibrated values of Rainfall Recharge and Evapotranspiration

	Rainfall Recharge		Evapotranspiration			
Stress periods	Initial Values (mm/day)	Calibrated values (mm/day)	Initial value (mm/day)		Calibrated value (mm/day)	
			Cropped area	Uncropped area	Cropped area	Uncropped area
Mar	0.06	0.03	3.94	1.97	6.32	4.15
Apr	0.05	0.03	4.85	2.43	8.73	3.88
May	0.18	0.18	8.09	4.05	8.09	4.04
Jun	1.01	1.01	6.34	3.17	5.07	2.53
Jul	5.95	5.95	1.84	0.92	0.83	0.27
Aug	6.42	6.42	3.26	1.63	3.26	0.63
Sep	4.68	4.68	5.93	2.96	5.93	1.96
Oct	1.17	1.17	5.40	2.70	5.40	1.70
Nov	0.07	0.07	1.42	0.71	2.42	0.71
Dec	0.02	0.02	2.36	1.18	2.36	1.18
Jan	0.14	0.14	3.60	1.80	3.60	1.80
Feb	0.11	0.11	4.51	2.26	4.51	2.26

4. Results and Discussions

The ground water model has been developed and calibrated with the water table data at certain existing observation wells in the study area for all the months of the year. Calibrated model has been used to study the river aquifer interaction in the study area. For this purpose, nine different locations are selected based on the morphology of the river such as after the bends, in straight portions and near the bends. For instance, locations from the northern entry Point of Gandak River to the distance of 0.1, 4.4 and 7.2 km in downstream of the river are straight portion; at distance 2.0 and 19.4 km are after bends and locations at distances 9.2, 14.2, 17.2 and 21.4 km are near bends. In order to analyze the nature/direction of flow through Vaishali Branch Canal and Habibpur Sub-Distributory, five such more locations in Vaishali Branch canal and two in Habibpur Sub-Distributory are selected. Two series of simulations and analysis have been carried out in this study. In the first series of simulation, monthly analysis of inflow to and from aquifer at these locations has been carried out. Positive value of flow shows that water is flowing from river to the aquifer and negative value means flow is from aquifer to the river. Second analysis is the effect of incremental rise of river stage on the river aquifer interaction and a relationship has also been developed for this. Figures 3(a-c) present the nature/direction and magnitude of the flow between River Gandak and aquifer at different selected locations for all the months. Figure 3(a) shows results at distances 0.1, 2.0 and 4.4 km from the entry point of river Gandak. Similarly Figures 3(b) and 3(c) shows the interaction at the distance 7.2, 9.2 and 14.2 km and 17.2, 19.4 and 21.4 km, respectively along the river Gandak. Positive value of flow shows that water is going from river to the aquifer and negative value means flow is from aquifer to the river.

From figures it can be seen that aquifer is contributing water to river in almost all the months except May, June and July. Nature of flow at all the reaches of river is same but there is variation in magnitude.

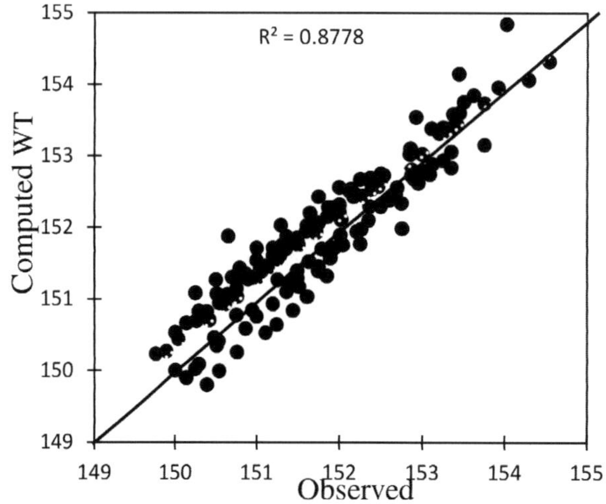

Figure 2. Comparison of observed and computed water table after calibration of model.

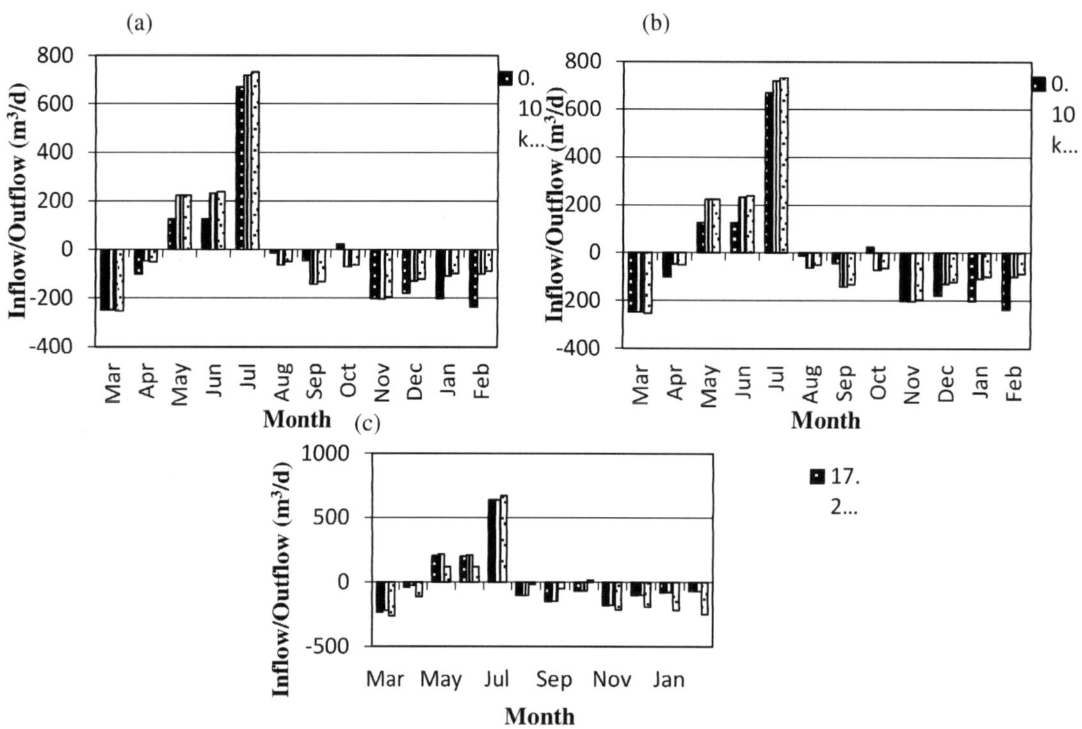

Figure 3. Inflow/Outflow to Aquifer at distances (a) 0.1, 2.0 and 4.4 kms (b) 7.2, 9.2 and 14.2 kms (c) 17.2, 19.4 and 21.4 kms from river Gandak.

Figures 4(a, b) show the nature of flow between VBC, HSD and aquifer. These figures show that VBC and HSD always contributing water to aquifer when these are operational. Figure 5 shows the nature and quantity of flow between aquifer of command area and River Gandak, VBC and HSD taken as a whole. It is found that the flow is from river to aquifer during the months May to October and in January and February with maximum quantity of flow is in the month of July. In the remaining months direction of flow is from aquifer to river with maximum value in month of March. Morphology of the river, water table condition and other stresses may be the reason to these variations in relations.

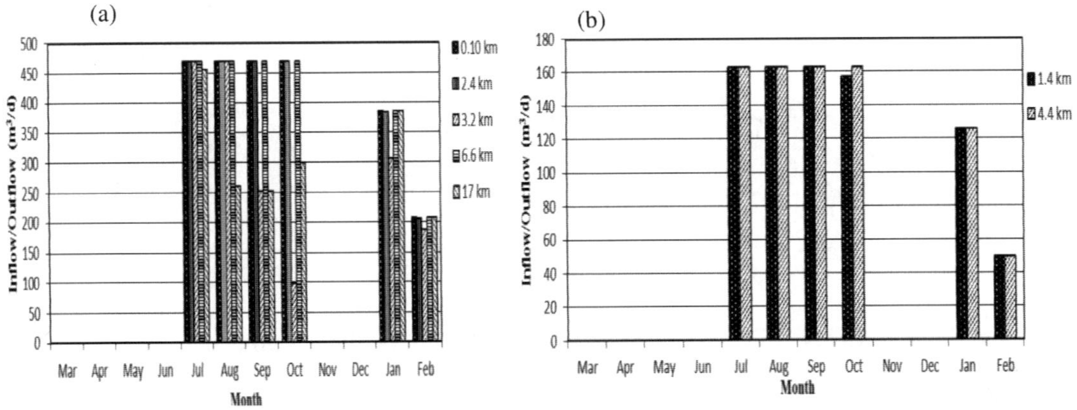

Figure 4. Inflow/Outflow to Aquifer at (a) 0.1, 2.4, 3.2, 6.6 and 17.0 kms from VBC (b) at 1.4 and 4.4 kms from HSD.

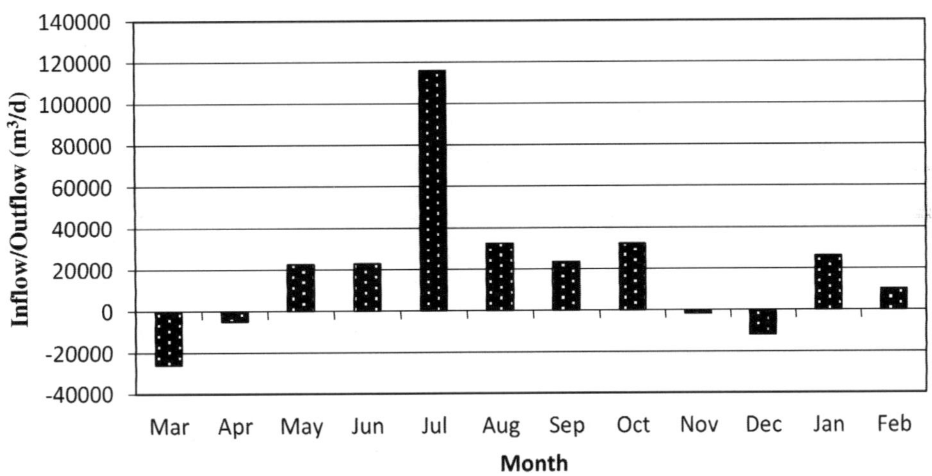

Figure 5. Inflow/Outflow to Aquifer from whole Gandak command area.

For second series of simulation, discharge is noted at each location after every unit increase of river stage. This incremental rise in stage versus discharge gives us a linear relationship. A generalized relation for whole study area is obtained which gives quantum of discharge between river and adjacent aquifer. The obtained linear equation is y=14986x-2^6, shown in Figure 6

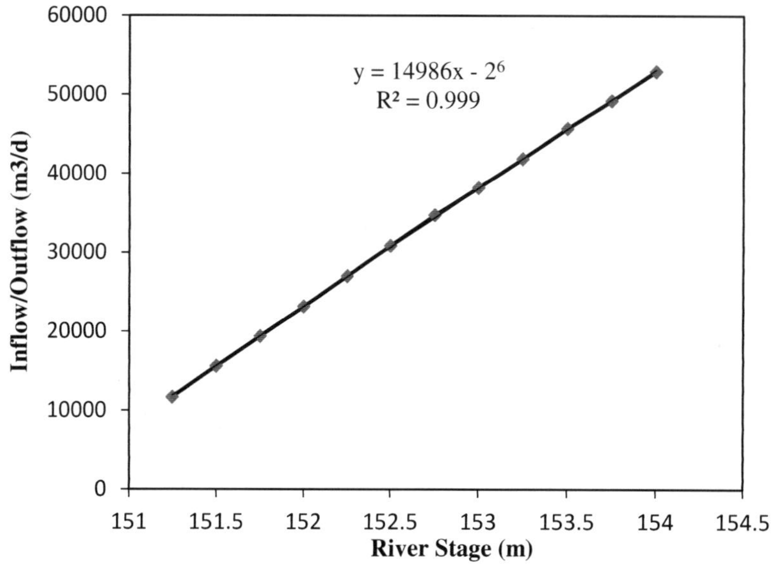

Figure 6. River aquifer interaction relationship for Gandak command area.

5. Conclusions

River aquifer interaction in the command area of Gandak in Bihar, India has been simulated using module 'Rivar Package' of ground water model, MODFLOW. Model has been developed and calibrated with the water table data at certain existing observation wells in the study area for all the months of the year. Calibrated model has been used to simulate the groundwater flow for one complete year. Results obtained as the water level data at all the grids and in term of the flow direction and magnitude either from river Gandak to adjacent aquifer or vice-versa. It is found that the flow is from river to aquifer during the months May to October and in January and February with maximum quantity of flow is in the month of July. In the remaining months direction of flow is from aquifer to river with maximum value in month of March. This shows that flow of water between Gandak and the aquifer depends upon stage of river, morphology of the river, and water table condition in the aquifer adjacent to river. The relationship of river aquifer interaction is linear in general and different for different reaches. River aquifer interaction relationship ($y = 14986\ x - 2^6$) developed for incremental rise in river stage gives the quantity of flow due to stage of river in study area as a whole.

6. References

[1] Mishra G C and Fahimuddin M, 2005 'Stream Multiaquifer Well Interactions', Journal of Irrigation and Drainage Engineering, Vol. 131, No. 5: pp 433-439.

[2] Fleckenstien J H, Niswonger R G and Fogg G E, 2006 River-Aquifer Interactions, Geological Heterogeneity and Low-Flow Management. Vol.44, No.6- Groundwater- November- December 2006, p-837-852.

[3] Calderon H and Uhelenbrook S 2014 Investigation of seasonal river- aquifer interactions in a tropical coastal area controlled by tidal sand ridges. Hydrology and Earth System Sciences. Discussion. 11, pp 9759-9790.

[4] Hunt B, Weir J and Clausen B, 2001 A stream depletion field experiment. Ground Water 39(2) pp 283-289.

[5] Nawalany M and Grodzka M, 2014 'River- aquifer interaction- analytical solution for groungwater horizontal flow model vs. velocity oriented approach in 3D'. Revista Latino-Americana de Hidrogeologia,vol. 9, No. 1, pp. 131-136.

[6] Chakraborty B and Pandey N G, 2004 Flood and its effect on River Aquifer Interaction.

[7] Anderson M P and Woessner W W, 2002 Applied groundwater modeling: simulation of flow and advective transport. Academic Press, London.

[8] Constantz J, 2008 Heat as a tracer to determine streambed water exchanges, Water Resour. Res., 44, W00D10.

[9] Ruehl C, Fisher A T, Hatch C, Huertos M L, Stemler G and Shennan C, 2006 Differential gauging and tracer tests resolve seepage fluxes in a strongly-losing stream, J. Hydrol., 330(1-2), pp 235–248.

[10] Hatch C E, Fisher A T, Revenaugh J S, Constantz J and Ruehl C, 2006 Quantifying surface water-groundwater interactions using time series analysis of streambed thermal records: Method development, Water Resour. Res., 42, W10410.

[11] McCallum A M, Anderson M S, Rau G C, Larsen J R and Acworth R L, 2014 'River-aquifer interactions in a semiarid environment investigated using point and reach measurements. Water Resour. Res., 50, pp 2815–2829.

[12] Kalbus E, Reinstorf F and Schirmer M, 2006 Measuring methods for ground water and surface water interactions: a review. Hydrology and Earth System Sciences Discussions, European Geosciences Union, 2006, 10 (6), pp 873-887.

[13] Brunner P, Simmons C T, Cook P G and Therrien R, 2010 Modelling Surface water Groundwater interaction with MODFLOW: Some considerations. National Groundwater Association.Vol.48, No.2- GROUND WATER- March-April 2010 pp 174-180.

[14] Bihar Irrigation Commission Report (1994) Vol-II, pp 53-60.

[15] Barlow P M and Harbough A, 2006 USGS directions in MODFLOW development. *Groundwater* 44(6) pp 771-774.

[16] CGWB 1993 Hydrology and Groundwater resources of Vaishali district, Bihar Technical report no. 103, series 'D'.

[17] Hargreaves G N, Prasad U K and Sumani Z A 1985 Acrop water evaluation manual for India., The Irrigation centre, Utah State University, USA, pp 85-86.

IOP Conf. Series: Earth and Environmental Science **150** (2018) 012004 doi:10.1088/1755-1315/150/1/012004

Detection of Leaks in Water Distribution System using Non-Destructive Techniques

H Aslam[1], M Kaur[1], S Sasi[1], Md M Mortula[2], S Yehia[3] and T Ali[2]

1 Student, American University of Sharjah, Sharjah, UAE
2 Associate Professor, American University of Sharjah, Sharjah, UAE
3 Professor, American University of Sharjah, Sharjah, UAE

Abstract. Water is scarce and needs to be conserved. A considerable amount of water which flows in the water distribution systems was found to be lost due to pipe leaks. Consequently, innovations in methods of pipe leakage detections for early recognition and repair of these leaks is vital to ensure minimum wastage of water in distribution systems. A major component of detection of pipe leaks is the ability to accurately locate the leak location in pipes through minimum invasion. Therefore, this paper studies the leak detection abilities of the three NDT's: Ground Penetration Radar (GPR) and spectrometer and aims at determining whether these instruments are effective in identifying the leak. An experimental setup was constructed to simulate the underground conditions of water distribution systems. After analysing the experimental data, it was concluded that both the GPR and the spectrometer were effective in detecting leaks in the pipes. However, the results obtained from the spectrometer were not very differentiating in terms of observing the leaks in comparison to the results obtained from the GPR. In addition to this, it was concluded that both instruments could not be used if the water from the leaks had reached on the surface, resulting in surface ponding.

1. Introduction

A water distribution system is a reliable set of infrastructure used for transportation of safe, potable and good quality water to the consumers. It generally comprises of pipes, pumps, valves, storage tanks, reservoirs, meters, fittings, and other hydraulic accessories that provides a link between treatment plants or the source of the water to houses of the consumers. The loss of water during the distribution process can be quite significant for any utilities. Water loss due to leakage contributes the most significant part of the water loss. Water leakage is defined as loss of water from an opening in the supply system. The leakage could occur in different parts of the distribution system such as, transmission pipes, distribution pipes, service connection pipes, joints, valves [1]. The amount of water that is leaked at a given time in a distribution network depends on the structure of the network, material of the network, the pipe flows, the age of the network and the length of the work cycle [1]. There are several factors that contribute to the leakage of the pipes. Pipe material, surrounding soil properties, and water quality in the distribution system are some of these factors. Pipe material plays a significant role in the leakage behaviour of pipes since pipes of different materials, due to the material properties, will fail in certain characteristic ways. In addition to this, the pressure exerted on the pipes by the surrounding soil and corrosivity of the soil that surrounds the pipe are contributing factors which could potentially lead to leaks in pipes. Additionally, the water flowing in the pipes might contain high chlorine content or other chemicals which could corrode the pipes leading to pipe leakages. Other factors which could lead to leaks in the water distribution system are: faulty

installation, excessive water pressure, material defects, water hammer, ground movement due to drought or freezing, excessive loads and vibration from road traffic and stray electric current [1].
Different methods are used for the detection of leaks in pipes in the water distribution system. Non-Destructive Techniques (NDT) have certain advantages to detect leaks because they allow to evaluate the condition of the leak without changing the structure of the water distribution network, excavations of the soil above the water pipes, or disrupting the serviceability of the system. This not only helps in maintaining the services provided by the distribution system, but also save a lot of costs associated with the leak detection.
Ground Penetrating Radar (GPR) is non-invasive, geophysical method, which employs short electromagnetic pulses in the microwave range to produce an image of the subsurface. In a study conducted a laboratory on Plastic and Metallic pipes surrounded by silt and clay, it was concluded that the GPR technique are good at detecting the location of the leak; however, these results are negatively influenced by the presence of foreign buried objects [2]. Additionally, a study conducted in a laboratory on Polyvinyl chloride (PVC) pipes surrounded by dry sand concluded that GPR probing technique might be applicable to detect leaks in real soil [3].
The basic function of a spectrometer is to disperse the light emitted by an object into its spectral components and digitize the signal as a function of wavelength; thus, the physical and chemical properties of the object emitting the light can be inferred by analyzing its spectral components. The spectral reflectance obtained by soil is affected by the moisture content of the soil. In event of a leak in the distribution system, the moisture content of the soil surrounding the pipes increases. Therefore, the soil moisture content is the soil property of interest when dealing with spectrometers. Studies have successfully proven that as that moisture content of soil increases the reflectance obtained from the sand decreases and vice-versa [4], [5]. However, literature shows that the reflectance decreased non-linearly with the increase in moisture content of soils and this effect was not noticed in saturated soils [6]. Also, studies have shown that the decrease in the reflectance with the increase in moisture content is more noticeable in the infrared bands [7]. Even though soil moisture content is measured using spectrometer, it has never been used for detection of leaks in water distribution system.
This research intends to determine and compare the abilities of the chosen NDT's (GPR and Spectrometer) in detecting leaks in water distribution systems. In this research, laboratory based experiments were conducted to assess the effectiveness of these NDTs.

2. Research Methodology

To achieve the objective of this study, laboratory scale experimental approach was taken. An experimental setup was designed to replicate the existing conditions of the underground leaking pipes in water distribution systems.

2.1. Experimental Setup

The experimental setup designed for experimentation is a soil filled wooden box with four pipes placed at a depth of two-third from the top edge of the box as shown in Figure 1. The box is made of wood because it has minimum signal reflection, which makes it suitable for using GPR. Plastic and concrete were not chosen as the material for the box because these materials give strong signal reflections whenever the signal strikes the material surface. The bottom surface of the wooden box will have a layer of metal to calibrate the GPR. The designed dimensions of the box (L x B x H) in inches are 66 x 40 x 24. The pipes were fitted at a height 16 inches from the top surface of the wooden box According to the user's manual published by the manufacturer, the minimum distance between the pipes should be greater than 3 inches for a 1.6 GHz GPR to avoid any interferences in the images generated by the GPR. Hence, the distance between the pipes was chosen to be 12 inches.

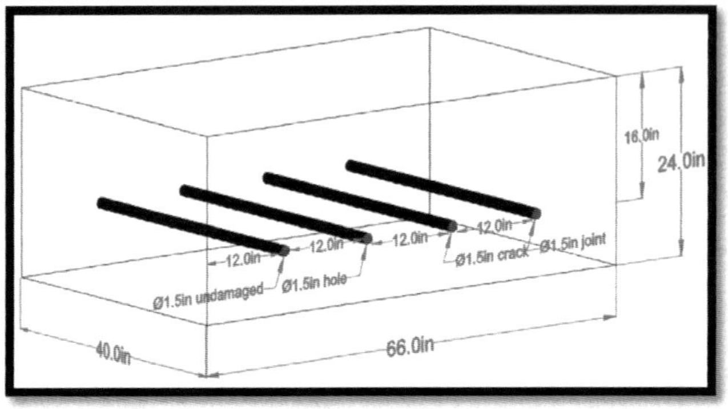

Figure 1. Dimensions of the Experimental Setup.

2.2. Pipes Used

The pipe materials used in our experiments is Polyvinyl Chloride (PVC). The diameter of these pipes is 1.5 inches. Also, the experimental setup consists of four pipes in which three pipes are damaged to simulate a leak and one pipe is an undamaged pipe. The first pipe in the experimental setup is a reference pipe with no damage to simulate no water leak scenario. In the second pipe, a hole was drilled to damage the pipe and stimulate a water leak due to defects in the pipe. In the third pipe, a crack (vertical cut) was created to damage the pipe. The last pipe consisted of a faulty joint to stimulate a leak. The sequence and type of damages stimulated in the pipes have been shown in Figure 1.

2.3. Fill Material for the Experimental Setup

Soil characteristics as low-electrical- loss materials, low relative dielectric constant and low absorption coefficient are suitable soils to obtain ideal GPR readings. Hence, taking all these factors into consideration the fill material chosen for experimentation is dune sand. In addition to this, the dune sand was compacted uniformly to imitate the real conditions of underground pipes. If compaction was not performed the damages stimulated in the pipes would leak faster because of lack of pressure from the surrounding soil. In addition to this, the water would spread faster because when the degree of compaction is less, the pore space is more and water can seep through these pores in the soil affecting the readings from other pipes. In this research, a compaction level of 85% was achieved to replicate the compaction level achieved in the road and pavement constructions as these pipes are usually under the pavements. A moisture content of approximately 5% was achieved.

2.4. Description of Instruments Used

The GPR used for experimentation is the Geophysical Survey Systems, Inc. (GSSI) Structure Scan Mini GPR with an antennae frequency of 1.6 GHz and penetration depth of about 20 inches. RADAN software was used to process and analyse the scans obtained from the GPR. Whereas the Spectrometer used in during experimentation was the Exemplar LS Spectrometer. The Exemplar LS Spectrometer displays real-time data and requires the use of BW Spec software to display the data obtained by the spectrometer.

2.5. Locations for Data Collection

During the experiments, the data was collected along the entire length of the pipe and not only at the location of damage because this experiment aims to observe how effective the instruments are at identifying the leak.

2.5.1. Locations of Data Collection points for the GPR

A grid was drawn on plywood to take readings for the GPR. For each pipe, three readings were taken which are: 6 inches right of the pipe, centre of the pipe and 6 inches left of the pipe.

IOP Conf. Series: Earth and Environmental Science **150** (2018) 012004 doi:10.1088/1755-1315/150/1/012004

2.5.2. Locations of Data Collection Points for the Spectrometer
Spectrometer readings were taken at three locations along the pipes which are the right side of pipe, middle of the pipe and left side of the pipe.

3. Results and Discussions
The GPR and the spectrometer were used on each of the four pipes to detect the leaks. The readings were taken at intervals of 30 minutes and a total of four readings were obtained. The total experimental time was 2 hours.

3.1. Results obtained for GPR
A dielectric constant of 6.4 was used to detect the depth of the calibration sheet. The calibration sheet is shown in a bright white tone in the images. The results obtained for pipe 3 (which is the pipe with crack) for the PVC setup are shown in Table 1. It can be seen from the images that there has been a significant drop in the depth of the calibration sheet with time as the moisture content level of the sand increased due to the leaks produced from the cracks in pipe 3. In addition to this, it is observed that the readings taken by running the GPR along the center of the pipe are not clear and the calibration sheet is not visible in the images. This is because the radiation emitted by the GPR reflects to the detector of the GPR on hitting the PVC pipe. Hence, the radiation does not reach the calibration sheet because of which it is not visible in the images obtained by running the GPR along the center of the pipes.

Table 1. Data Obtained on Pipe 3 Using the GPR

Time	0 mins	30 mins	60 mins	90 mins
Left	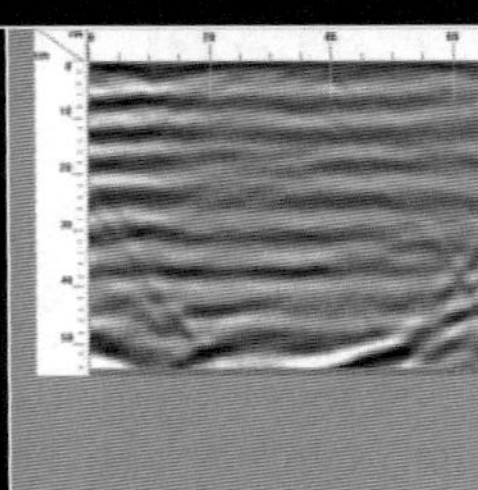	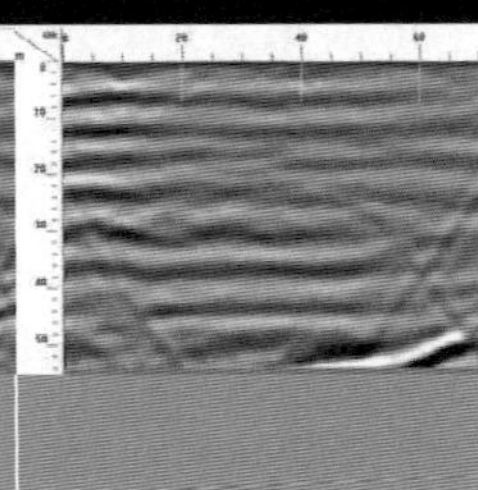	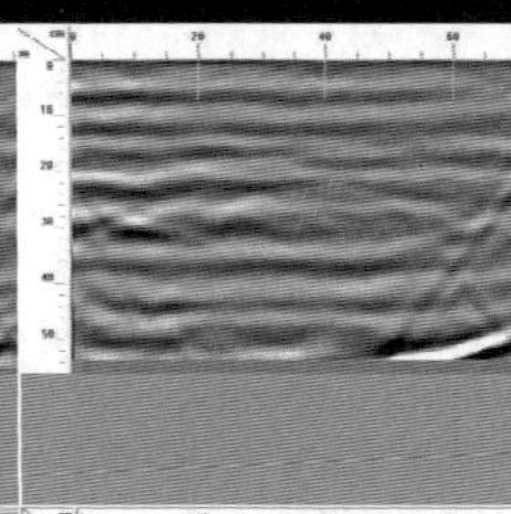	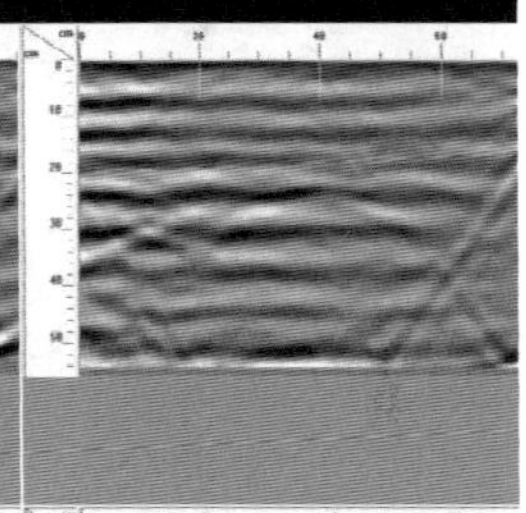
Middle	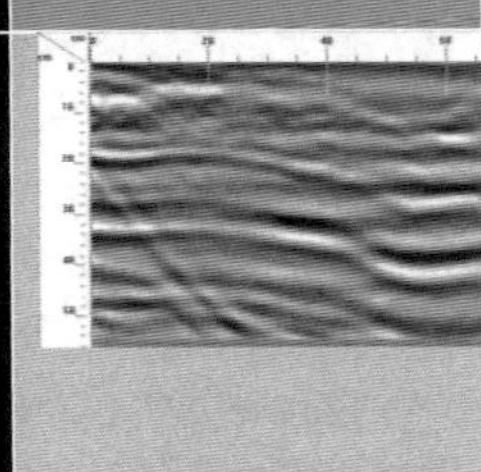	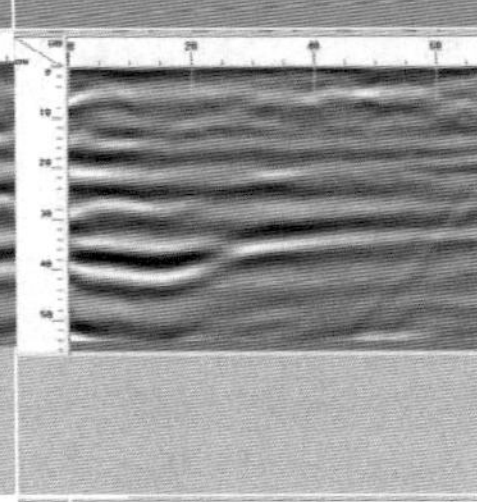	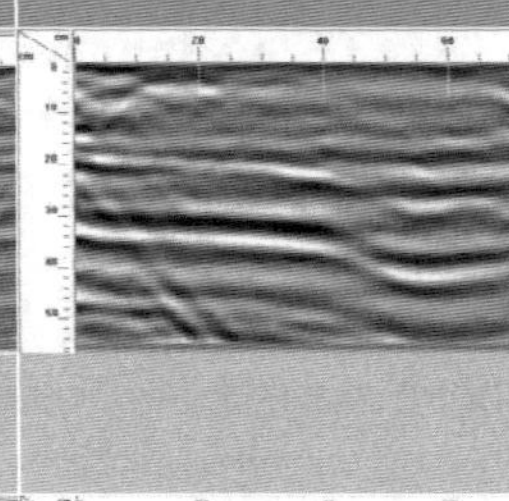	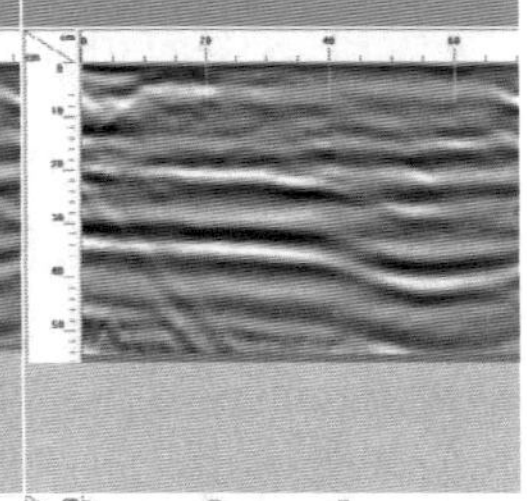
Right	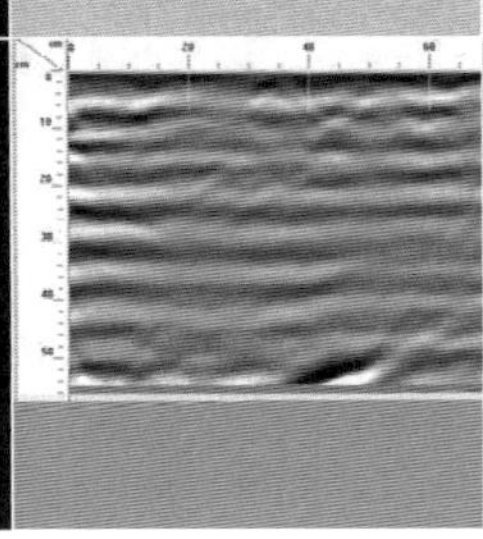	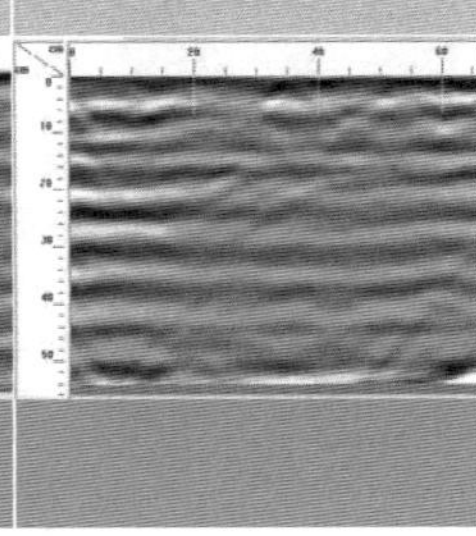	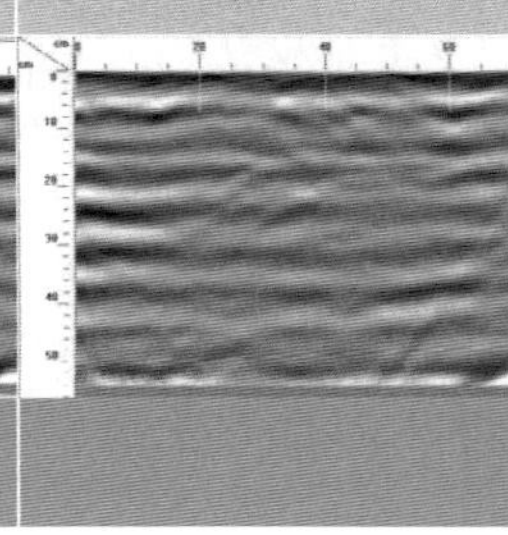	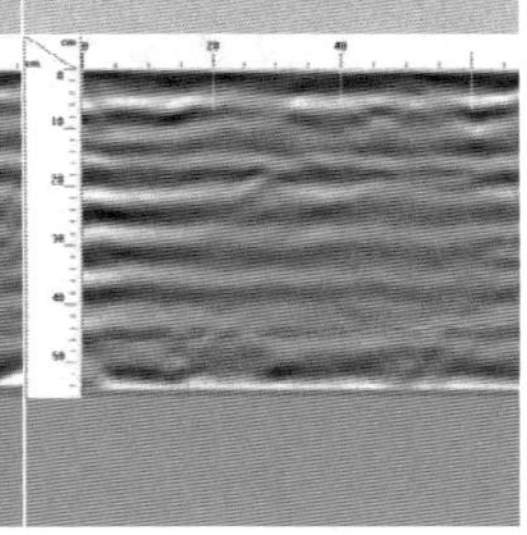

3.2. Results obtained for Exemplar Spectrometer

As observed in Table 2, the reflectance at all the three locations has slightly reduced at reading 3 (60 mins) when compared to reading 1 (0 mins). Furthermore, the reflectance reduces even further at the last reading, reading 4 (90 mins). Thus, this indicates an increase in the moisture content of the soil, which is caused by the leaks in the pipe 3 (i.e. pipe with a crack). However, the readings obtained after 30 minutes show an unusual pattern in comparison to the other three readings. Hence, this reading was not considered when analyzing the data. Also, the reduction in spectral reflectance is the highest in the readings obtained in the left side of the pipe, indicating that the leak was in the right side of the pipe.

Table 2. Data Obtained on Pipe 3 Using the Spectrometer.

Time	0 mins	30 mins	60 mins	90 mins
Right	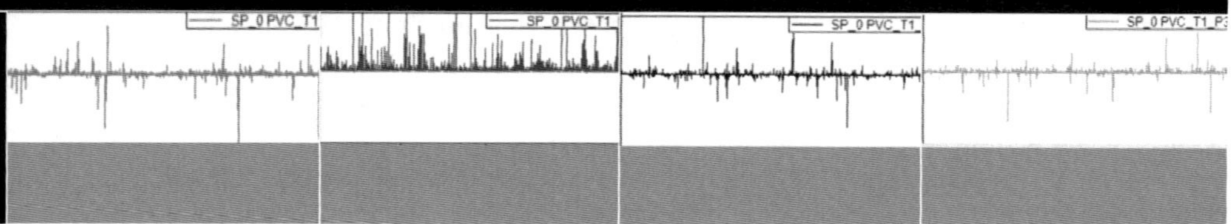			
Middle	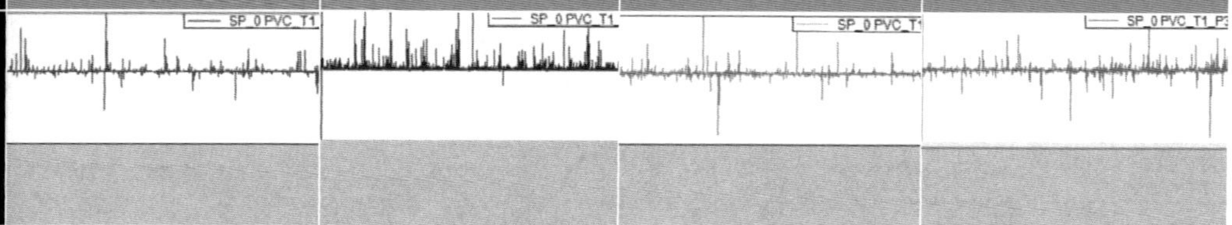			
Left	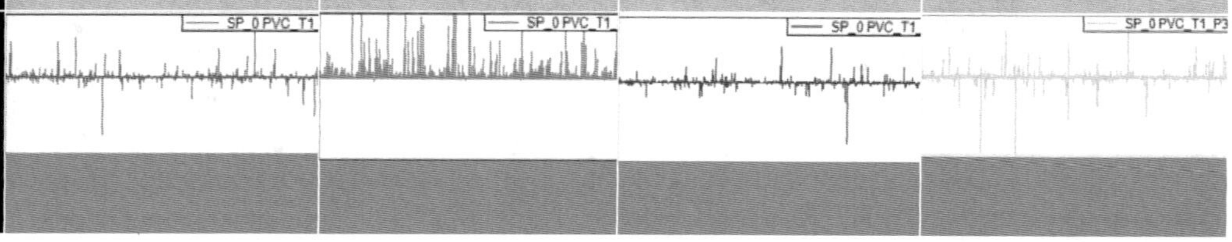			

After conducting the experiments for all four pipes using the two devices and analyzing the data, the observations have been summarized in Table 3. Both the instruments could identify leaks in pipe 2 and pipe 3 which are the pipe with the crack and the pipe with the hole respectively. However, both instruments were unable to detect the leak on the fourth pipe which is the pipe with the joint due to the accumulation of water on the surface of the soil in 30 minutes which resulted in ponding, as a result it was not possible to use both the instruments. This is coherent with the findings in literature which suggested that the spectral reflectance did not show any variation in saturated soils [6]. Nevertheless, the leak in the fourth pipe could be detected by just visual inspection as the water from the faulty joint had reached the surface.

Table 3. Summary of Leak Detection Potential of NDT's.

	GPR	Exemplar LS Spectrometer
Crack	Able to identify	Able to identify
Hole	Able to identify	Able to identify
Faulty Joint	Not able to identify	Not able to identify

4. Conclusion

Both the instruments were able to detect leakage in the pipes with the hole and the crack. In these pipes the water from the leaks did not cause ponding on the surface of the soil. However, in the fourth pipe which consisted of a faulty joint, water from the surface caused ponding on the surface of the soil because of which the GPR and the spectrometer could not be used. Nonetheless, inft case of surface ponding, these instruments will not be required as visual inspection itself will be enough to understand that there is a leak at that location. The results obtained from the GPR were easier to analyze. However, the initial calibration of the GPR was a time-consuming process which is necessary to determine the dielectric constant for the soil. On the other hand, the results obtained from the spectrometer showed very little deviation from the initial reading; making data analysis less accurate. However, the instrument was very easy to use during experimentation. This is valid for the soil type which is used in the current investigation. It is recommended to investigate other type of soils to determine the effectiveness of both interments.

5. References

[1] Hunaidi O 2000 Detecting Leaks in Water Distribution Pipes *Construction Technology Update* **40**

[2] Liu Z and Kleiner Y 2013 State of the art review of inspection technologies for condition assessment of water pipes *Elsevier* **46** pp 1-15

[3] Eyuboglu S, Al-Mahdi H and Al-Shukri H 2003 Detection of Water Leaks using Ground Penetrating Radar *Geophysics 2003: Conference on Applied Geophysics,* ed Anderson N pp 8-12

[4] Younis S M Z and Iqbal J 2015 Estimation of Soil Moisture Using Multispectral and FTIR techniques *The Egyptian Journal of Remote Sensing and Space Science* **18** pp 151-161

[5] Demattê J A M, Sousa A A, Alves M C, Nanni M R, Fiorio P R and Campos R C 2006 Determining Soil Water Status and Other Soil Characteristics by Spectral Proximal Sensing *Geoderma* **135** pp 179-2006

[6] Nolet C, Poortinga A, Roosjen P, Bartholomeus H and Ruessink G 2014 Measuring and Modeling the Effect of Surface Moisture on the Spectral Reflectance of Coastal Beach Sand PLoS *ONE* **9** p. e112151 http://dx.doi.org/10.1371/journal.pone.0112151

[7] Fabre S, Briottet, X, and Lesaignoux, A 2015 Estimation of Soil Moisture Content from the Spectral Reflectance of Bare Soils in the 0.4–2.5 μm Domain *Sensors* **15** pp 3262-3281

Progresses in Polystyrene Biodegradation and Prospects for Solutions to Plastic Waste Pollution

S S Yang[1,2], A M Brandon[2], D F Xing[1], J Yang[3], J W Pang[1], C S Criddle[2], N Q Ren[1,*] and W M Wu[2,*]

1 Harbin Institute of Technology, Harbin 150090, PR China.
2 Stanford University, Stanford, CA 94305-4020, USA.
3 School of Chemistry, Beihang University, Beijing 100191, China
* E-mail address: rnq@hit.edu.cn; billwu@stanford.edu

Abstract. Petroleum-based plastic pollution has been a global environmental concern for decades. The obvious contrast between the remarkable durability of the plastics and their short service time leads to the increasing accumulation of plastic wastes in the environment. A cost-effective, sustainable strategy to solve the problem should focus on source control and clean up. Polystyrene (PS) wastes, a recalcitrant plastic polymer, are among the wide spread man-made plastic pollutants. Destruction of PS wastes can be achieved using various abiotic methods such as incineration but such methods release potential air pollution and generation of hazardous by-products. Biodegradation and bioremediation has been proposed for years. Since the 1970's, the microbial biodegradation of plastics, including PS, has been evaluated with mixed and isolated cultures from different sources such as activated sludge, trash, soil, and manure. To date, PS biodegradation by these microbial cultures is still quite slow. Recently, the larvae of yellow mealworms (*Tenebrio molitor* Linnaeus) have demonstrated promising PS biodegradation performance. Mealworms have demonstrated the ability to chew and ingest PS foam as food and are capable of degrading and mineralizing PS into CO_2 via microbe-dependent activities within the gut in less than the 12-15 hrs gut retention time. These research results have revealed a potential for microbial biodegradation and bioremediation of plastic pollutants.

1. Plastic pollution status

Since 1950s, synthetic plastic products derived from petroleum have been widely used in our daily lives from electronic products, water supply and drainage pipelines, coating and wiring, to packaging and containers. In 2013, global plastic production was 299 million tons, which is a nearly 200 fold increase over the annual production of 1.5 million tons in 1950 [1]. The annual produced plastic production has been and will continue increasing in the foreseeable future [2]. According to USEPA [3], plastics wastes (PSW), which are a rapidly growing segment of municipal solid waste (MSW), are found in all major MSW categories, such as the containers and packaging category (e.g. bags, sacks, and wraps, other packaging, PET bottles, jars and HDPE natural bottles, and other containers). Compared with plastics wastes (390,000 tons) generated in 1960, the quantity of plastics wastes generation has increased relatively steadily to 31.8 million tons in 2011 [3]. With regard to percentage of MSW generation, plastics wastes generation was increased from less than 1% in 1960 to 12.7% in 2011. A recent study indicates that up to 6,300 million metric tons of plastic waste have been generated to date [4]. However, among the generated plastic wastes, less than half of it was confined to landfills or recycled. Of the remaining plastic wastes, a large proportion litters continents, oceans, and every corner of our world to make the world a "Plastic World" [5]. In our daily life, the most widely used plastics are thermoplastics, the most common of which are: polyethylene (PE) 29.6%, polypropylene (PP) 18.9%, polyvinyl chloride (PVC) 10.4%, polyurethane (PUR) 7.4%, polystyrene (PS) 7.1%, and polyethylene terephthalate (PET) 6.9% [2] (Figure 1). These plastic materials are also major sources of microplastic pollution in ocean [2]. Of particular concern is that, several of these plastic materials, especially PS, PVC, PUR as well as polycarbonate [6]-[8], making up approximately 26% of the plastics production, are the most hazardous to environment and human health [5]. There is a growing scientific consensus demonstrating that plastic wastes are a major environmental concern of increasing global significance:

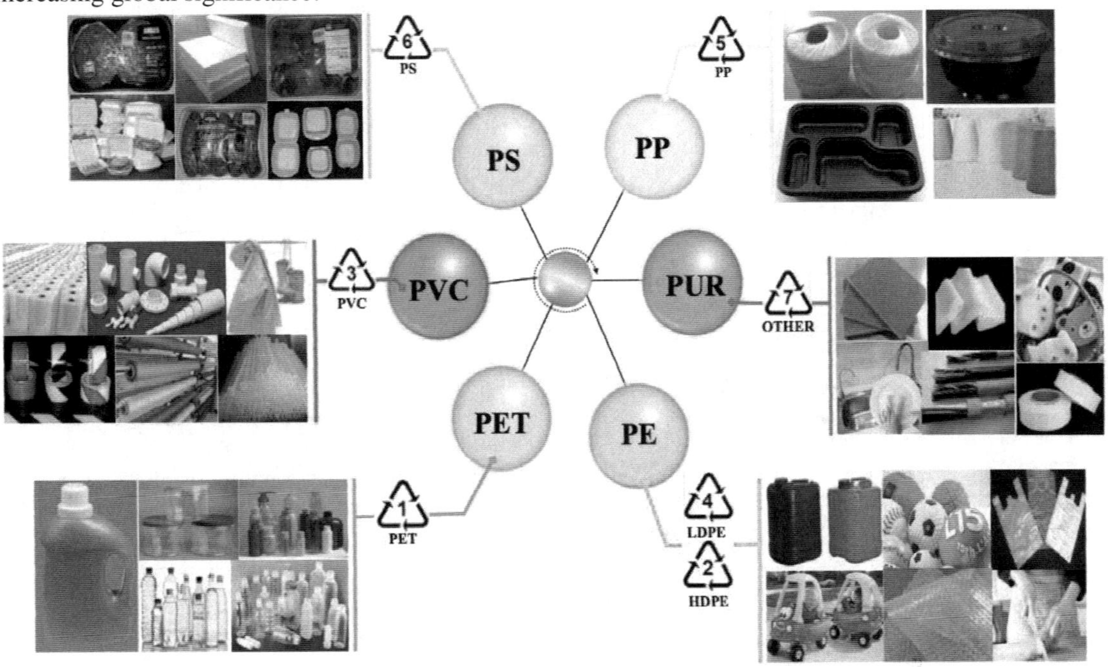

Figure 1. The most widely used plastics on the market

- Soil contamination by plastic wastes causes severe environmental pollution, and could lead to the decline of crop production. The plastics wastes in landfill occupy a great area of land (Landfill disposal of 10,000 tons plastic wastes occupy 0.067 hm^2 land), and also inevitably release large amounts of chemicals including oligomers, catalyst remnants, polymerization solvents, and a wide range of plastic additives [7]. These hazardous chemicals could leech from the plastic wastes and impact the soil quality [9], as well as neighboring groundwater.
- Water contamination by plastic wastes has been a major concern because they harms large ocean animals, deteriorate fishing industry, and damage aquatic environment. A recent report indicates that approximately 0.48-1.27 million tons of plastic waste enters ocean annually and the introduction of plastic into the ocean is increasing at an astonishing rate, with an estimated doubling time of 10 years [10], [11]. Studies have shown that the plastics wastes occupied 60-80% of all marine debris are widespread in the marine environment, at the sea surface, on shorelines and on the seabed [12]. Abiotic degradation, by wind, rain and sun, could cause harmful materials and chemicals to leech from plastic wastes, causing a secondary pollution problem within surface water, groundwater, and marine water bodies.
- Air pollution is another serious environmental concern. Referring to the disposal and management of solid wastes, PSW is co-incinerated with MSW. The combustion of PSW could result in the emission of air pollutants such as volatile organic compounds (VOCs), heavy metals, polycyclic aromatic hydrocarbons (PAHs), polychlorinated dibenzofurans (PCDFs), dioxins, NO_x, SO_x, etc. if off gases are not managed properly. At worst, some carcinogenic substances such as PAHs, nitro-PAHs, dioxins, etc., could be released as airborne pollutants during the

incineration of plastic wastes containing PVC, PET, PS and PE [13].

- Plastic pollution has the potential to poison animals and pose a serious threat to human health. According to a hazard-ranking model based on the United Nations' Globally Harmonized System of Classification and Labeling of Chemicals, the chemical ingredients of more than 50% of plastics are hazardous [7]. These harmful chemicals leeched from the plastic wastes or in the form of small or micro plastic debris are more likely to infiltrate food webs [14], and potentially impact ecologically important species including mussels, salt-marsh grasses and corals [14], [15]. Humans and mussels that ingested the chemicals from plastics and small or micro plastic debris could accumulate in the body, and harm the cells and other tissues [14], [16].

2. Plastic wastes control and management strategies

Except for the risks and hazards to the human health and the environment safety, the improper management and disposal of plastic wastes impact the aesthetic of the environment, badly affect the beauty of urban and remote environments. The generation of plastic wastes can be mainly classified into two categories (Figure 2): (1) preconsumer or industrial plastic wastes; and (2) postconsumer plastic wastes. Preconsumer or industrial plastic wastes referred to the discarded plastics, which are produced during the plastic production and product fabrication processes. These kinds of plastic wastes can be directly recycled and re-used by the processing and manufacturing departments.

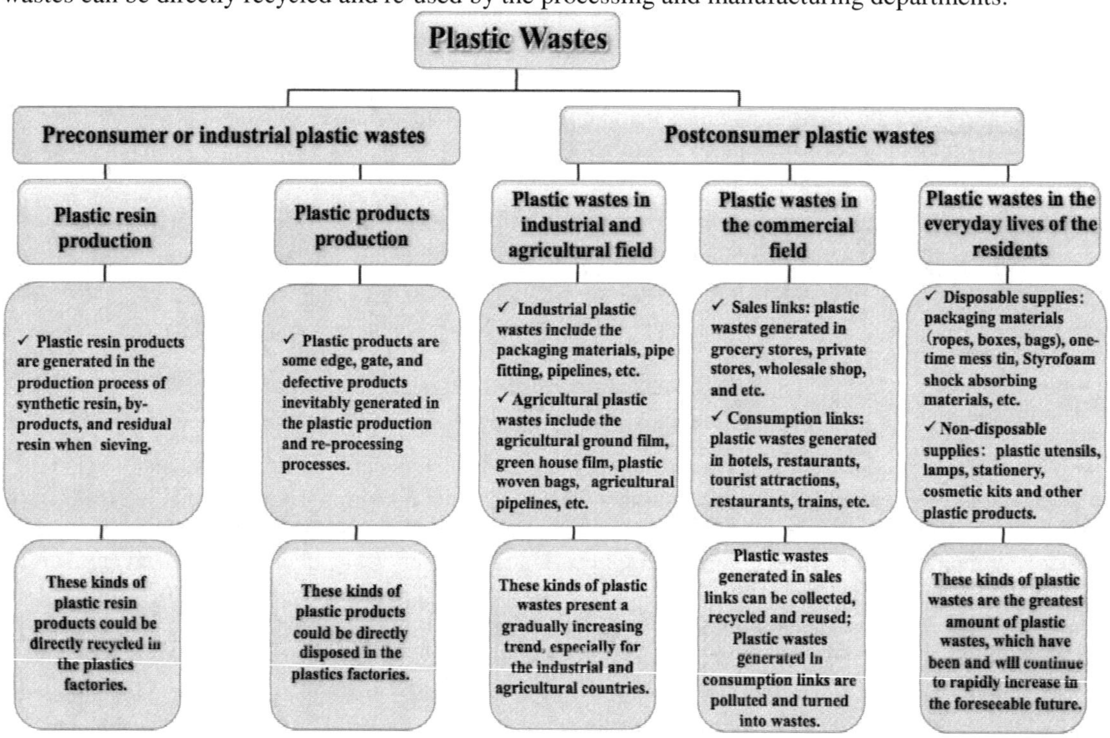

Figure 2. Sources of plastic wastes and strategies for the management of plastic wastes

Sustainable plastic wastes management solutions should apply methods or technologies that meet the requirements for efficient recycling of resources without generation of harmful substances (toxic and hazardous by-products or end-products) to humans or the environment. As we described elsewhere [2], the overall strategy for controlling plastics pollution should primarily focus on source reduction, increased use of biodegradable materials, improved reuse, recycle and recovery of plastic via improved separation efficiency from waste streams, and subsequently the development of cost-effective clean-up and bioremediation technologies. We will use biodegradation of polystyrene (PS) as example to evaluate the feasibility of biodegradation.

3. Biodegradation of polystyrene wastes

Industrial production of PS began around 1930 [17]. Today, PS products are widely used in our daily lives. The application of PS products includes use of expanded PS (EPS), trade name Styrofoam, in building insulation and packing, extruded PS (XPS) in containers (coffee cups and food trays), and high density PS materials as liquid containers. Most PS are used as foam products (Figure 1). With the rapid development of economy and society, the demand for PS materials and products has been continually increasing. By 2013, annual global production of PS was around 21 million tons, accounting for about 7.1% of the total global plastic production [18].

3.1. Feasibility of Biodegradation of PS

The term of "biodegradation" represents the predominance of biological activity in the degradation of plastic polymers into lower molecular weight compounds and mineralization into CO_2 and water [19]. The plastic polymers are too large to pass through cellular membranes. During biodegradation, the polymers are first converted into small fragments, then further degraded and mineralized within microbial cells [20]-[23]. Attempts to investigate direct biodegradation of PS has been performed since 1970s. PS products have been considered extremely durable and recalcitrant against microbial attack.

In 1980s styrene, a monomer used for synthesis of polystyrene [24], was assumed as an intermediate of PS breaking down product. Late professor D. Grbić-Galić at Stanford University and others conducted comprehensive research on its biodegradation. In earlier studies [25]-[27], mixed microbial cultures and *Pseudomonas* were found to grow on styrene as sole carbon source. Based on previous research, a strain of *Pseudomonas fluorescents*, which was capable of growing on styrene as the sole carbon source, was isolated from enrichment cultures (styrene, phenyl acetic and *o*-hydroxyl phenyl acetic acids) [28]. Styrene-degrading bacterium, *Xanthobacter* sp. strain 124X, and 14 strains of aerobic bacteria and two strains of fungi were isolated from a sewage culture with the styrene as the carbon source [28], [29]. Additionally, a novel flavin adenine dinucleotide-requiring styrene monooxygenase activity was further detected (oxidized styrene to styrene oxide (phenyloxirane)) [30]. The metabolic pathways of bacteria for styrene biodegradation under anaerobic pathways was also investigated under strictly anaerobic conditions [31] and a pure culture of *Enterohacfer cloacae* DG-6 (ATCC 35929) performed an almost stoichiometric conversion of styrene to CO_2 and low concentrations of aromatic and alicyclic compounds. To date, however, it has not been confirmed that styrene is the intermediate produced during PS biodegradation.

Since the early 1970s, researchers have studied the biodegradation of PS and other plastic wastes by microorganisms. Guillet et al. (1974) tested biodegradation of two photodegraded polystyrene-5% vinyl ketone copolymers which were laboratory-synthesized using ^{14}C styrene. The polymers were pretreated by a UV accelerator, resulting in decreasing viscosity molecular weight from 795,300 and 450,000 to 32,000 and 10,000. After pretreatment, bio-oxidation was observed based on the release of $^{14}CO_2$ with added activated sludge. Approximate 0.7% of total ^{14}C was released as $^{14}CO_2$ after 75 days. The $^{14}CO_2$ release rate from undegraded polymers was very low and then appeared level off [32]. Kaplan et al. [33] investigated the PS biodegradation capabilities of 17 species of fungi in axenic cultures, five groups of soil invertebrates, and a variety of mixed microbial communities including sludge, soils, manures, trash, and decaying plastics. After 35 days, only 0 to 0.24% PS appeared to be biodegraded by the fungi in axenic cultures. The PS biodegradation mixed microbial communities (including silt loam, cow manure, activated sludge, decaying plastics, etc.) ranged from 0.04 to 0.57% after 11 weeks. They concluded that the PS biodegradation by the fungi in axenic cultures and mixed microbial communities was performed extremely slowly. Sielicki et al (1978) studied the biodegradation of [β-^{14}C] polystyrene in the soil and liquid enrichment cultures by measuring $^{14}CO_2$ released. Biodegradation of polystyrene in soil was varied from 1.5 to 3.0% over a 4 month period based on the release of $^{14}CO_2$ [34]. The PS removal was still limited but appeared a 15 to 30 times of those reported by Kaplan et al. [33]. Mor and Sivan also reported the isolation of biofilm-producing strain (C208) of the actinomycete *Rhodococcus ruber* which was could degrade PS [35]. The strain C208 was incubated with PS flakes as sole carbon source, resulting in 0.5% and 0.8% of weight reduction in PS flakes within 4 and 8 weeks. Atiq et al. (2010) reported that *Paenibacillus urinalis* strain NA26, *Bacillus* sp NB6, and *Psueudomons aeruginosa* NB26 were isolated from buried PS film in soil after 8 months [36]. These bacterial strains colonized and used EPS as a sole carbon source. Some intermediate-like compounds (1-phenyl-1, 2 ethandiol and 2-phenylethanol) were detected in the medium using HPLC but no significant surface changes were observed using FTIR analysis. The PS biodegrading ability of these strains has not yet been convinced.

In addition, Nakamiya et al. (1997) showed that hydroquinone peroxidase of a lignin decolorizing bacterium, *Azotobacter beijerinckii* HM21 degraded polystyrene in a dichloromethane-water system [37]. However, no follow up studies of biodegradation of PS by strain HM21 have been reported.

Previous studies on PS biodegradation, as seen above, have shown promising signs although the degradation is limited (Table 1). More research is still needed to understand if any microorganisms capable of rapid-degrading PS are present in environment and their living conditions.

Table 1. Reported microbial cultures degrading polystyrene

Authors	Culture source	PS removal	Test Period
Sielicki et al. [34]	The soil and liquid enrichment cultures	1.5%–3.0%	4 months
Kaplan et al. [33]	The fungi in axenic cultures	0%–0.24%	35 days
	Mixed microbial communities	0.04%–0.57%	5-11 weeks
Mor & Sivan [35]	*Rhodococcus ruber* C208	0.5% or 0.8% weigh reduction	4 or 8 weeks
Atiq et al. [36]	*Paenibacillus urinalis* NA26 *Bacillus* sp NB6 *Psueudomons aeruginosa* NB26	No weight reduction reported but found intermediates	8 weeks
Yang et al. [38]	*Exiguobacterium* sp. strain YT2 from mealworms	7.4 ± 0.4% of weight reduction (2500 mg PS/L in 10^8 cells/mL)	60 days

3.2. Biodegradation of PS Wastes by PS-Eating Mealworms

Yellow mealworms are the larvae of *Tenebrio molitor* Linnaeus which is a holometabolic insect, passing through the stages of egg, larva, pupa, and beetle. They are pest insects for storage and also are commercial animal feed available in pet markets and stores. Mealworms eating Styrofoam and isolation of bacteria from mealworms' gut were reported by several middle and high school students for science fairs via public media (newspapers and internet) in 2000s. However, these early studies did not get academic attention as they did not result in peer-reviewed publications or archiving of isolates.

Mealworms have chewing and eating behaviour for Styrofoam (Figures 3) and other plastic foam. The confirmation results of the biodegradation and mineralization of PS in the larval gut of the *Tenebrio molitor* Linnaeus were published by Yang et al. (2015) [39]. The mealworms were purchased from a pet market in Beijing, China. After fed with Styrofoam (5.8 g) as the sole diet for 30 days, the mealworms (500 in an incubator) consumed 31.0±1.7% of Styrofoam. A test for the determination of the survival rate (SR) over a 1 month period showed that the difference between of the SR of Styrofoam-feeding mealworms and the SR of conventional diet (bran)-feeding mealworms was not significant (average 85%). Similar results have been observed using different batch of mealworms from Harbin, China and Compton, California, the USA. The SR was approximate 81.3 ± 2.5% and 86.7 ± 3.3% for the Styrofoam-feeding mealworms from Harbin and California, respectively when Styrofoam was fed as a sole diet over one month. For one incubator containing 120 mealworms and Styrofoam was consumed by 29.9 ± 3.84% and 23.6 ± 0.1% over one month period, respectively. The consumption of Styrofoam and high SG indicated that Styrofoam-feeding could be digested and utilized as energy source to maintain their life activity.

The retention time of ingested PS or Styrofoam in mealworm gut is relatively short, estimated to be 12-15 hrs. Characterization of Styrofoam (or PS) degradation by mealworms and other insects is primarily based on the changes in chemical and physical properties of egested residues (in the frass)

from the PS-feeding insects. PS depolymerization and biodegradation were characterized by gel permeation chromatography (GPC), Fourier transform infrared spectroscopy (FTIR), solid-state ^{13}C cross-polarization/magic angle spinning nuclear magnetic resonance (^{13}C-CP/MS NMR) and liquid-state ^{1}H nuclear magnetic resonance (^{1}H-NMR) analysis as well as other methods [38]-[40]. GPC analysis provides information on three key indicators of depolymerization and degradation of plastic materials: the number-average molecular weight (M_n), the weight-average molecular weight (M_w), and molecular weight distribution (MWD). The M_n and M_w of the residual polymer in frass decreased significantly from those of the original Styrofoam material, indicating depolymerisation is occurring. The MWD also shifts towards to lower molecular weight [38]-[40]. FTIR, ^{1}H-NMR or ^{13}C-CP/MS NMR analysis were conducted to identify functional groups indicative of depolymerization and oxidation [38]-[40]. In addition, the release ^{13}C-labeded CO_2 from the mealworms fed with α ^{13}C-labeled and β ^{13}C-labeled PS further confirmed the biodegradation and mineralization of PS. Our recent studies indicated that supplementation of nitrogen-containing feed (e.g. bran) can enhance PS consumption and degradation by mealworms [40]. At 25 °C, the PS consumption rate can be doubled with co-feeding bran [40].

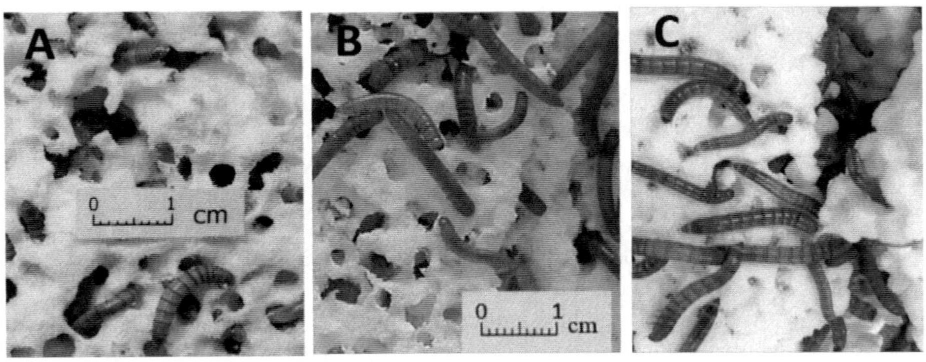

Figure 3. Styrofoam-eating mealworms from various sources (A) Beijing, China, (B) Harbin, China; and (C) Compton, California, USA. All these mealworms are capable of PS-degradation.

To better understand the mechanism of PS-degrading by the larvae, antibiotics, such as gentamicin, were fed to mealworms to understand the activity of gut bacteria and the effect on depolymerization of the PS [38]. The mealworms fed with gentamicin for 10 days showed significantly less depolymerization and mineralization after the activities of the gut bacterial were [38], suggesting that PS degradation is likely gut-microbe-dependent. A PS-degrading bacterial strain, *Exiguobacterium* sp. strain YT2 was isolated from the guts of the mealworms tested in Beijing [38]. The single culture of strain YT2 removed PS by 7.4 ± 0.4% in liquid medium containing PS pieces (2500 mg/L) during a 60 day period. GPC analysis of residual PS also confirmed PS depolymerisation by strain YT2. This has proved that PS-degrading microorganisms are present in mealworms' gut.

3.3. Future research and perspectives of microbial degradation of plastics

Mealworms degraded ingested PS in their gut up to 40-50% within the relatively short 12-15 hrs gut retention time. Further studies with antibiotics have suggested that the PS biodegradation is likely gut microbe-dependent. To date, however, no microorganisms isolated, including strain YT2 isolated from the mealworm gut, can degrade PS at the same rate as the mealworms. More research is needed to understand the mechanisms of the PS biodegradation in mealworm gut and interactions between the mealworm host and gut microbes. The mealworms' ability to biodegrade PS may expand to other plastic materials such as PE, PP, PET, and PUR and will be tested. The ubiquity of PS biodegradation by mealworms around world in relation to their gut microbiota should be investigated in order to further understand the mechanisms and select highly efficient mealworm strains and microorganisms.
Future microbiological work can focuses on isolation and characterization of more PS-degrading single cultures and define high efficiency mixed cultures (if present) from various PS-degrading mealworms.

Biodegradation of PS as well as other plastics requires breaking down or depolymerisation of plastic polymers. Further research on biochemistry can involve in the separation and identification of enzymes, especially extracellular enzymes. This could lead to production of plastic-depolymerizing enzymes for future large-scale use.

The research on PS-degradation by mealworms is just beginning to open a door for future scientific studies and development of potential biotechnologies to solve plastic pollution problem.

4. References

[1] Panda A K, Singh R K and Mishra D K 2010 Thermolysis of waste plastics to liquid fuel: A suitable method for plastic waste management and manufacture of value added products–A world prospective. *Renew. Sustain. Energy Rev.* 14 (1) 233-248.

[2] Wu W M, Yang J and Criddle S C 2017 Microplastics pollution and reduction strategies. *Front. Environ. Sci. Eng.* 11(1), 6-DOI 10.1007/s11783-017-0897-7.

[3] USEPA 2011 http://www.epa.gov/osw/nonhaz/municipal/msw99.htm.

[4] Geyer R, Jambeck J R and Law K L 2017 Production, use, and fate of all plastics ever made. Sci. Adv. 3: e1700782, DOI: 10.1126/sciadv.1700782.

[5] Rochman C M, Browne M A, Halpern B S, Hentschel B T, Hoh E, Karapanagioti H K, Rios-Mendoza L M, Takada H, Teh S and Thompson R C 2013 Policy: Classify plastic waste as hazardous. *Nature* 494 (7436) 169-171.

[6] Miller A 1994 Industry Invests in Reusing Plastics. *Environ. Sci. Technol.* 28, 16A.

[7] Lithner D, Larsson A and Dave G 2011 Environmental and Health Hazard Ranking and Assessment of Plastic Polymers Based on Chemical Composition. *Sci. Total. Environ.* 409, 3309-3324.

[8] Teuten E L, Saquing J M and Knappe D R U, et al. 2009 Transport and release of chemicals from plastics to the environment and to wildlife. *Phil. Trans. R. Soc. B* 364 2027-2045..

[9] Crompton T R 2007 Additive migration from plastics into foods. A Guide for the Analytical Chemist. Shrewsbury: *Smithers Rapra Technology Limited Publishing* (Shawbury, Shrewsbury, Shropshire, SY4 4NR, UK).

[10] Jambeck J R, Geyer R, Wilcox C, Siegler T R, Perryman M, Andrady A, Narayan R and Law K L 2015 Marine pollution. Plastic waste inputs from land into the ocean. *Science*, 347 (6223), 768-771.

[11] Cózar A, Echevarría F, González-Gordillo J I, Irigoien X, Úbeda B, Hernández-León S, Palma Á T, Navarro S, García-de-Lomas J, Ruiz A, Fernández-de-Puelles M L and Duarte C M 2014 Plastic debris in the open ocean. *Proceedings of the National Academy of Sciences of the United States of America*, 111 (28), 10239-10244.

[12] Lusher A L, McHugh M and Thompson R C 2013 Occurrence of microplastics in the gastrointestinal tract of pelagic and demersal fish from the English Channel. *Marine Pollution Bulletin* 67 94-99.

[13] Al-Salem S M, Lettieri P and Baeyens J 2009 Recycling and recovery routes of plastic solid waste (PSW): A review. *Waste Management* 29 (10) 2625-2643.

[14] Browne M A, Dissanayake A, Galloway TS, Lowe D M and Thompson R C 2008 Ingested Microscopic Plastic Translocates to the Circulatory System of the Mussel, Mytilus edulis (L.). *Environ. Sci. Technol.* 42 5026-5031.

[15] Uhrin A V and Schellinger J 2011 Marine debris impacts to a tidal fringing-marsh in North Carolina. *Mar. Pollut. Bull.* 62 2605-2610.

[16] Pauly J L, Stegmeier S J, Allaart H A, Cheney R T, Zhang P J, Mayer A G and Streck R J 1998 *Cancer Epidem. Biomarkers Prev.* 7 419-428.

[17] Krueger M C, Harms H and Schlosser D 2015 Prospects for microbiological solutions to environmental pollution with plastics. *Appl. Microbiol. Biotechnol.* 99 8857-8874.

[18] Plastics Europe. Plastics-The facts 2014/2015 http://www.plasticseurope.org/Document/plastics-the-facts-20142015.aspx?Page=DOCUMENT%26FolID=2, 2015.

[19] Gautam R, Bassi A S and Yanful E K 2007 A review of biodegradation of synthetic plastic and foams. *Appl. Biochem. Biotechnol.* 141 85-108.

[20] Griffin G J L 1980 Synthetic polymers and the living environment. *Pure Appl. Chem.* 52 399-407.

[21] Kamal M R and Huang B 1992 Natural and artificial weathering of polymers. In: Hamid SH, Ami MB, Maadhan AG, editors. In: Hamid S H, Ami M B, Maadhan A G, editors. *Handbook of Polymer Degradation.* (New York, NY: Marcel Dekker) 127-168.

[22] Swift G 1997 Non-medical biodegradable polymers: environmentally degradable polymers. In: Domb A J, Kost J, Wiseman D M, editors. *Handbook of Biodegradable Polymers.* (Amsterdam: Harwood Academic) 473-511.

[23] Shah A A, Hasan F, Hameed A and Ahmed S 2008 Biological degradation of plastics: A comprehensive review. *Biotechnol. Adv.* 26 246-265.

[24] Hartmans S, Smits J P, Van der werf M J, Volkering F and De Bont T J A M 1989 Metabolism of Styrene Oxide and 2-Phenylethanol in the Styrene-Degrading *Xanthobacter* Strain 124X. *App. Environ. Microbiol.* 55 2850-2855.

[25] Kyrikou I and Briassoulis D 2007 Biodegradation of Agricultural Plastic Films: A Critical Review. *J. Polym. Environ.* 15 125-150.

[26] Sielicki M, Focht D D and Martin J P 1978 Microbial transformations of styrene and [^{14}C] styrene in soil and enrichment cultures. *App. Environ. Microbiol.* 35 124-128.

[27] Shirai K and Hisatsuka K 1979 Production of ~-phenethyl alcohol from styrene by Pseudomonas 305-STR-1-4. *Agr. Biol. Chern.* 43 1399-1406.

[28] Baggi G, Boga M M, Catelani D, Galli E and Treccani V 1983 Styrene Catabolism by a Strain of Pseudomonas fluorescens. *System. Appl, Microbiol.* 4 141-147.

[29] van den Tweel W J J, Janssens R J J and de Bont J A M 1986 Degradation of 4-hydroxyphenylacetate by Xanthobacter 124X. *Antoine van Leeuwenhoek* 52 309- 318.

[30] Hartmans S, Van der werf M J and De Bont J A 1990 Bacterial Degradation of Styrene Involving a Novel Flavin Adenine Dinucleotide-Dependent Styrene Monooxygenase. *Appl. Environ. Microbiol.* 56 1347-1351.

[31] Grbić-Galić D, Churchman-Eisel N and Mraković I 1990 Microbial transformation of styrene by anaerobic consortia. *J. Appl. Bacferiol.* 69 247-260.

[32] Guillet J E, Regulski T W and McAneney T B 1974. Biodegradability of photodegraded polymers II: tracer studies of biooxidation of Ecolyte PS polystyrene. *Environ. Sci. Technol.* 8, 923-925.

[33] Kaplan D L, Hartenstein R and Sutter J 1979. Biodegradation of polystyrene, poly(methyl methacrylate), and phenol formaldehyde. *App. Environ. Microbiol.* 38 551-553.

[34] Sielicki M, Focht D D and Martin J P 1978. Microbial degradation of [^{14}C] polystyrene and 1, 3-diphenylbutane. *Can. J. Microbiol.* 24 798-803.

[35] Mor R and Sivan A 2008 Biofilm formation and partial biodegradation of polystyrene by the actinomycete Rhodococcus ruber. *Biodegradation* 19 851-858.

[36] Atiq N, Ahmed S, Ali M I, Andleeb S, Ahmad B and Robson G 2010. Isolation and identification of polystyrene biodegrading bacteria from soil. *Afr. J. Microbiol Res.* 4 1537-1541.

[37] Nakamiya K, Sakasita G, Ooi T and Kinoshita S 1997 Enzymatic Degradation of Polystyrene by Hydroquinone Peroxidase of Azotobacter beijerinckii HM121. *J. Ferment Bioeng*, 84, 480-482.

[38] Yang Y, Yang J, Wu W M, Zhao J, Song Y L, Gao L C, Yang R F and Jiang L 2015 Biodegradation and Mineralization of Polystyrene by Plastic-Eating Mealworms: Part 1. Chemical and Physical Characterization and Isotopic Tests. *Environ. Sci. Technol.* 49 12080-12086.

[39] Yang Y, Yang J, Wu W M, Zhao J, Song Y L, Gao L C, Yang R F and Jiang L 2015 Biodegradation and Mineralization of Polystyrene by Plastic-Eating Mealworms: Part 2. Role of Gut Microorganisms. *Environ. Sci. Technol.* 49 12087-12093.

[40] Yang S S, Anja Malawi Brandon A M, Flanagan J C A, Yang J, Ning D L, Cai S Y, Fan H Q, Wang Z Y, Ren J, Benbow E, Ren N Q, Waymouth R M, Zhou J Z, Criddle C S and Wu W M 2017. Biodegradation of polystyrene wastes in yellow mealworms (larvae of Tenebrio molitor

IOP Conf. Series: Earth and Environmental Science **150** (2018) 012005 doi:10.1088/1755-1315/150/1/012005

Linnaeus): factors affecting biodegradation rates and the ability of polystyrene-fed larvae to complete their life cycle. Chemosphere (in press).

Acknowledgments
The authors gratefully acknowledge the financial support by the Woods Institute for Environment at Stanford University (award 1197667-10-WTAZB), partially by US National Science Foundation SBIR award 1648559, the National Nature Science Foundation of China (Grant No. 51708154), the Special Financial Grant from the China Postdoctoral Science Foundation (Grant No. 2016T90302), the class General Financial Grant from the China Postdoctoral Science Foundation (Grant No. 2016M601434), and the State Key Laboratory of Urban Water Resource and Environment (Grant No. HC201621-01).

IOP Conf. Series: Earth and Environmental Science **150** (2018) 012006 doi:10.1088/1755-1315/150/1/012006

Thermal Conductivity in Soil: Theoretical Approach by 3D Infinite Resistance Grid Model

A Changjan[1] and N Intaravicha[1]

[1] Department of Environmental Science and Technology, Faculty of Science and Technology, Pathumwan Institute of Technology, 833 Rama 1 Road, Wangmai, Pathumwan, Bangkok 10330, Thailand.
E-mail: otto_sinkronity@yahoo.com

Abstract. Thermal conductivity in soil was elementary characteristic of soil that conduct heat, measured in terms of Fourier's Law for heat conduction and useful application in many fields: such as Utilizing underground cable for transmission and distribution systems, the rate of cooling of the cable depends on the thermal properties of the soil surrounding the cable. In this paper, we investigated thermal conductivity in soil by infinite three dimensions (3D) electrical resistance circuit concept. Infinite resistance grid 3D was the grid of resistors that extends to infinity in all directions. Model of thermal conductivity in soil of this research was generated from this concept: comparison between electrical resistance and thermal resistance in soil. Finally, we investigated the analytical form of thermal conductivity in soil which helpful for engineering and science students that could exhibit education with a principle of physics that applied to real situations.

1. Introduction

The thermal conductivity of the soil was critical physical properties in the soil that describe soil's ability to conduct heat. Heat flow potentiality in soil could be applied in many fields such as agricultural, climatology, engineering and environmental science [1-9]. Thermal conductivity in soil depended on several parameters such as soil density, quantity of air and water in soil (macrospores and microspores contents) and soil particle (sand, silt and clay). Effect of particle size making the soil had the ability to retaining air and water in the soil. Cause of this reason, Kim et.al [10] and Devries [11] found thermal conductivity in soil depended by soil particles, soil moisture and soil organic matter. Markert et.al [12] measured thermal conductivity in full moisture soil by single heat probe measurements and evaporation method with three different soil (sand, sandy loam and silt loam). Results show that the accuracy of the thermal conductivity measurements depends on heating power, soil texture and soil moisture. Yu et.al [13] determined thermal conductivity in soil by numerical simulation of a heating test on a heat exchanger pile and compared with field and laboratory experiment. The numerical results close agreement with the laboratory test results but differed from the field experimental value. Jin et.al [14] studied on the calculation model of rock soil thermal conductivity for the design of the ground source heat pump system. Nagy [15] estimated thermal conductivity in soil by modified Johansen's method for higher sand containing samples and de Vries's method for high silt and clay containing sample. Lu et.al [16] developed a simple model for estimating soil thermal conductivity from soil texture, bulk density and water content. The new model had the potential for studied heat movement in soil and could be incorporated into numerical algorithms for describing coupled heat and mass transfer process. Zhang et.al [17] examined the effects of biochar amendment on the thermal conductivity, surface reflectance and temperature of soil. The effect of biochar amendment on soil temperature could be explained by the combined action of changes in soil thermal conductivity and reflectance. El Shamy et.al [18] studied on effect of shear-induced anisotropy on thermal conductivity of granular soils by discrete element method. Results of conducted simulations show that shear-induced anisotropy results in an anisotropic thermal conductivity tensor. Nikolaev et.al [19] reported of an experimental investigation of the soil thermal conductivity over a wide temperature range, for various water contents and two soil types (Ottawa sand and Richmond Hill fine sandy loam). Bansal et.al [20] studied the underground temperature characteristics of the soil surrounding the Earth Air Tunnel Heat Exchanger (EATHE) pipe and the effect of duration of operation of EATHE on its thermal performance. Thermal conductivity of soil was a critical parameter for design a Ground Heat Source Pump (GHSP) system: for estimated initial investment, energy saving effects and operating performance of the system. Perpar et.al [21] determined the effect of the soil thermal conductivity coefficient on the heat loss from pre-insulated pipes during operation. Jia et.al [22] analyzed soil thermal conductivity with different compositions and water contents. Ju et.al [23] investigated the influences of soil aggregation on soil thermal conductivity higher water contents by the thermo-time domain reflectometry (TDR) technique. The results showed that thermal conductivity was significantly higher in the non-aggregated soils than in the aggregated soils at intermediate water contents. Lu et.al [24] developed an improved model that describes the relationship between thermal conductivity and volumetric water content of soils. The new model could be estimated using soil bulk density, sand (or quartz) fraction and water content. Song et.al [25] developed model of heat exchanger with horizontal pipe of GSHP. Fundamental of physics can be illustrated with thermal conductivity in soil by the schematic similarity of the electrical circuit. However, the soil in natural has multiplex structure. Then 3-D model Fourier's Law can develop from infinite resistance concept [26-28]. In this research, we investigate the thermal conductivity of the soil in analytical form by 3-D infinite resistance grid model.

2. Model and Calculations

Heat transfer describes the exchange of thermal energy between physical systems. The fundamental modes of heat transfer are conduction, convection and radiation. In soil, the main process of heat transfer is by heat conduction [28]. Thermal conductivity is the properties of a material to conduct heat. It is evaluated in term of Fourier's law for heat conduction. Heat conduction occurs at a lower rate across materials of low thermal conductivity and higher rate across materials of high thermal conductivity. Materials of low thermal conductivity are used as thermal insulator and materials of high thermal conductivity are widely used in a heat sink. The thermal properties of soil have found important used in engineering, microclimate and agrophysics. In this research, we study thermal conductivity in soil by analogy heat resistance in soil with electrical resistance in infinite resistance grid 3-D and generate a simple analytical model of thermal conductivity in soil.

The equation that explains the rate of heat transfer can analogy with the ohm's law of the simple direct current circuit. Figure 1 show comparison between the model of heat transfer in soil and electric current in circuit.

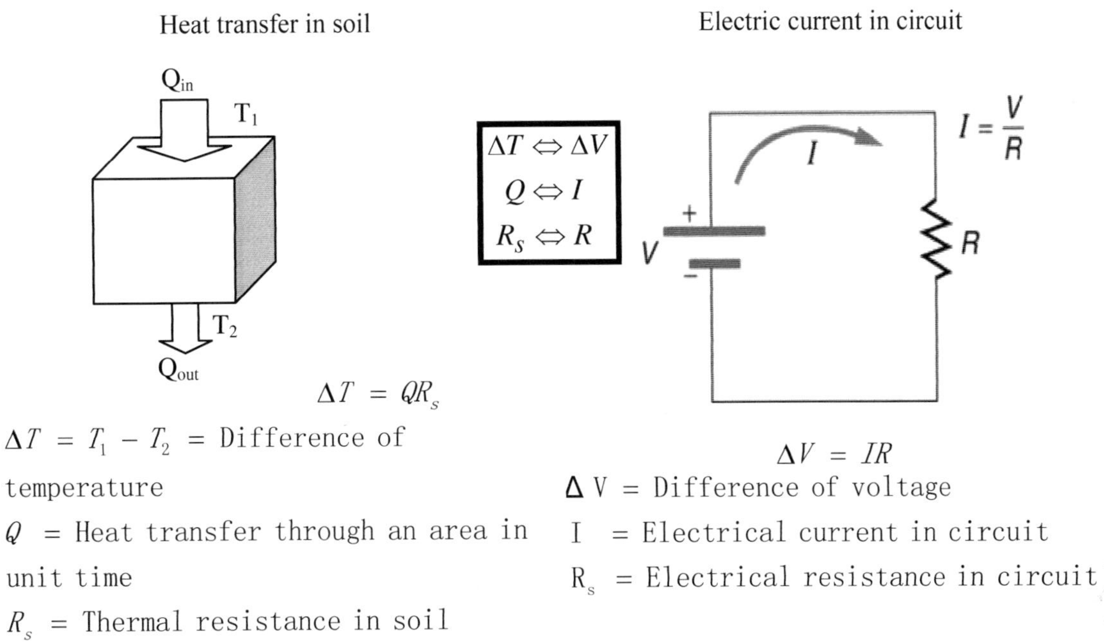

Figure 1. Comparison between model of heat transfer in soil and electric current in circuit.

From Fourier's law, the relationship according to which the heat flux density is proportional to the temperature gradient along the direction of the heat flow as show in Figure 2(a). Fourier's law can be express as

$$G = -\lambda \frac{\Delta T}{\Delta x} = \frac{\Delta Q}{\Delta t \cdot A}. \tag{1}$$

Then, we get

$$\frac{\Delta Q}{\Delta t} = \dot{Q} = -\lambda A \frac{\Delta T}{\Delta x}. \tag{2}$$

Here λ is thermal conductivity $\left[Wm^{-1}K^{-1}\right]$,

$\Delta x \equiv d$ is thickness of soil layer $[m]$,

ΔQ is amount of heat $[J]$ transferred through a cross section of unit area $A\left[m^2\right]$

Consider in an infinite uniform sphere $(r_{out} \to \infty)$ as show in Figure 2(b),

$$\dot{Q} = 4\pi\lambda \frac{\Delta T}{\frac{1}{r_{in}} - \frac{1}{r_{out}}} = 4\pi\lambda r_{in} \Delta T. \tag{3}$$

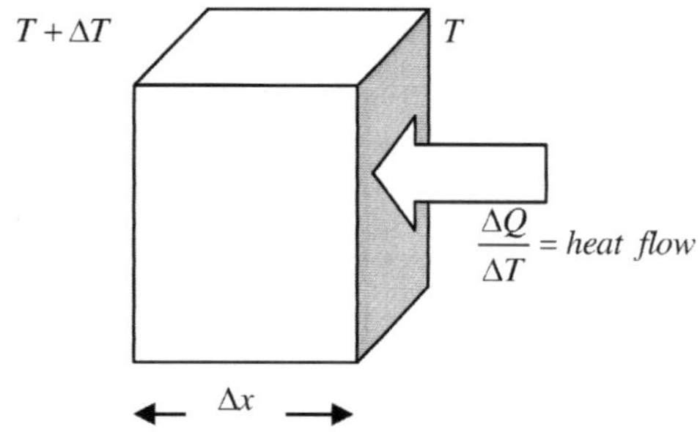

Figure 2. (a) Definition of Fourier's law

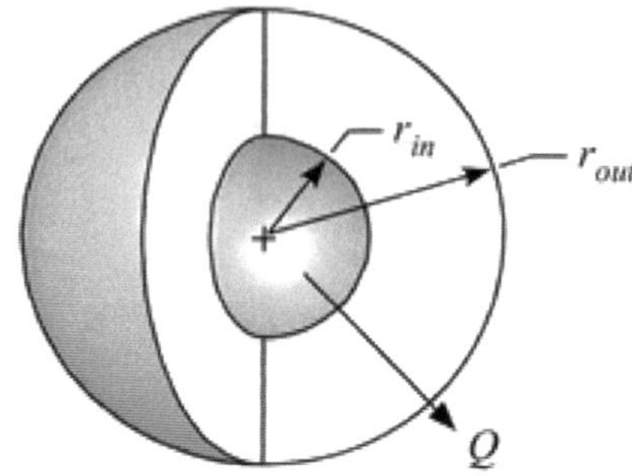

Figure 2. (b) Fourier's law in uniform sphere model

So that, thermal resistance in soil for one unit size (one particle) with the radius r_k in an infinite uniform medium can express by

$$R_S = \frac{1}{4\pi\lambda r_k}. \tag{4}$$

The volumetric unit of soil consisted of many units size of soil (Figure. 3). It is assumed that connecting between neighbor unit size will be represent by serial and parallel connection of thermal resistors as an infinite resistance grid 3-D.

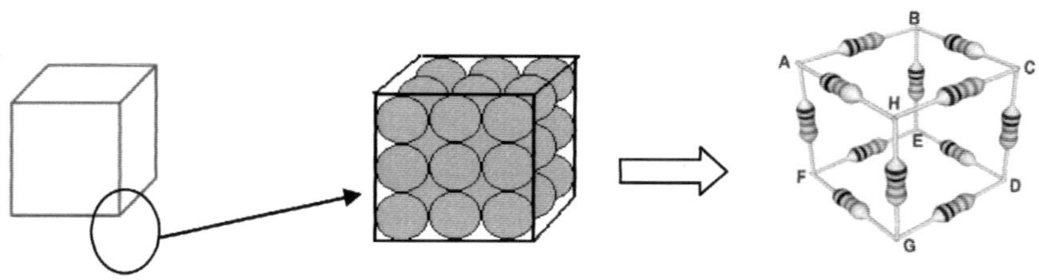

Figure 3. Schematic constructions of the thermal conductivity model of soil by resistance infinite grid 3-D

Applied the lattice Green's function for calculating the resistance of infinite networks of resistors [29], we get

$$R_{total} \approx (0.505462...)R_S. \tag{5}$$

Finally, thermal conductivity in soils as

$$\lambda = \frac{d}{(0.505462...)R_S A}. \tag{6}$$

3. Conclusion and Discussion

It is well known that the resistance between two adjacent grid points of an infinite square lattice in which all the edges represented as $R/2$ [30]. Cserti [29] presented an alternative approach using lattice Green's function to calculate the resistance between two points of an infinite square (2-D) and cube (3-D) lattice. In this research, we developed an analytical form of thermal conductivity in soils from Fourier's Law and infinite resistance grid 3-D concept. Thermal conductivity in soil investigated by infinite resistance grid 3-D model that useful for thermal conductivity prediction in almost uniform soil structure. This incidence similar to the electric circuit in case infinite resistance grid 3-D. Analytical form of thermal conductivity in soils could be useful in science and engineering education field by helpful for students that could be exhibited education with a principle of physics that applied to real situations and made inspirations for lecturer who followed with assumption for learning development.

4. Acknowledgements

The financial supports of Pathumwan Institute of Technology are acknowledged.

5. References

[1] Vasilev G, Parvanova S, Dineva P and Wuttke F 2015 *Soil. Dynam. Earthquake Eng.* 70 104
[2] Barry M D, Bouazza A, Singh R M, Wang B and Ranjith P G 2013 *Eng. Geol.* 164 131
[3] Perpar M, Rek Z, Bajric S and Zun I 2012 *Energy* 44 197
[4] W A. Van Wijk 1963 *Physics of Plant Environment* (North-Holland, Amsterdam)
[5] Astm D 5334-00, 2000. Standard test method for determination of thermal conductivity of soil and soft rock by thermal needle probe procedure. Vol. 04. 08, ASTM, 100 Barr-Harbor Dr., West Conshocken, PA 19428-2059.
[6] Smucker A J M, Park E J, Dorner J and Horn R 2012 *Vadose Zone J.* 6 282
[7] Abu-Hamdeh N H 2000 *Soil Tillage Res.* 56 145
[8] IEEE Std. 442-1981 IEEE Guide for soil thermal resistivity measurements.
[9] P. Laohachai 2012 *KMUTT R&D Journal* 4 433
[10] Kim D, Kim G. and Baek H 2015 *Int. J. Heat Mass Transfer* 84 1049
[11] DeVries D A 1956 *Nature* 178 1074
[12] Markert A, Peters A. and Wessolek G 2016 *Soil Sci. Soc. Am. J.* 80(2) 275
[13] Yu K L, Singh R M, Bouazza A and Bui H H 2015 *Geotech. and Geol. Eng.* 33(2) 239
[14] Jin H W A and Fang Q S b 2014 *Adv. Mater. Res.* 889-890 1347
[15] Nagy B 2014 *Adv. Mater. Res.* 1041 281
[16] Lu Y, Lu S, Horton R and Ren T 2014 *Soil Sci. Soc. Am. J.* 78(6) 1859
[17] Zhang Q, Wang Y, Wu Y, Wang X, Du Z, Liu X and Song J 2013 *Soil Sci. Soc. Am. J.* 77(5) 1478
[18] Shamy U E, Leon D O and Wells R 2013 *Int. J. Geomech.* 13(1) 57
[19] Nikolaev I V, Leong W H and Rosen M 2013 *Int. J. Thermophys.* 34(6) 1110
[20] Bansal V, Misra R, Agarwal G D and Mathur J 2013 *Appl. Energy* 103 1
[21] Perpar M, Rek Z, Bajric S and Zun I 2012 *Energy* 44(1) 197
[22] Jia D Y, Sui L Y, Fan L K and Wang J S 2012 *Appl. Mech. Mater.* 170-173 516
[23] Ju Z, Ren T A and Hu C 2011 *Soil Sci. Soc. Am. J* 75(1) 26
[24] Lu S, Ren T, Gong Y and Horton R 2007 *Soil Sci. Soc. Am. J* 71(1) 8
[25] Song Y, Yao Y and Na W 2007 *J. Harbin Institute Tech. (New Series)* 14(SUPPL.) 381
[26] Liska S and Colonius T 2014 *J. Comput. Phys.* 278 76
[27] Zhao C and Zang M 2014 *J. Terramechanics* 55 29
[28] Usowicz B 1992 *Pol. J. Soil Sci.* XXV/1 25
[29] Cserti J 2000 *Am. J. Phys.* 68 896
[30] Aitchison R E 1964 *Am. J. Phys.* 32 566

Diurnal Change of Soil Carbon Flux of Binhai New District

T F Wang[1], T Y Mao[2] and W Ye[3]

1 Assistant Engineer, Tianjin Research Institute for Water Transport Engineering, Ministry of Transport, Tianjin, China
2 Senior Engineer, Tianjin Research Institute for Water Transport Engineering, Ministry of Transport, Tianjin, China
3 Research Associate, Tianjin Research Institute for Water Transport Engineering, Ministry of Transport, Tianjin, China
E-mail: 295129016@qq.com

Abstract. In order to investigate the factors influencing diurnal change of soil carbon flux of Binhai New District. Field observation experiments were carried out by using LC pro-SD photosynthetic apparatus. The diurnal changes of soil carbon flux and its environmental factors such as atmosphere temperature and soil temperature were analysed. The results indicated that soil carbon flux appeared single diurnal pattern. The diurnal average of soil carbon flux ranked from 0.2761 to 2.3367μmol/m^2/s. Soil carbon flux varied significantly among different land use regimes($P<0.001$). Significant relationships were found between soil respiration rate and atmosphere temperature, which could he best described by exponential equations ($P<0.05$). The Q_{10} value was based on the exponential correlations. Its value of Tian Keyuan, ECO-city, Dagu-Outlet and Yongding-River was 8.331, 6.049, 2.651 and 1.391, respectively. There were quadratic correlations between soil carbon flux and soil temperature (10cm). And soil temperature could account for more than 32.27% of the soil carbon flux changes ($P<0.05$, R^2=0.3227-0.7465).

1. Introduction

At present the main reason for global warming is greenhouse gas emissions increase.CO_2 is the main greenhouse gas. It is also an important carbon cycle in the ecosystem. Soil respiration is the CO_2 flux on the soil surface. It is the main way to return to the atmosphere in CO_2 form when the plants are fixed carbon. Global annual carbon dioxide emissions from soil respiration to the atmosphere is about 75-120Pg [1], [2]. Therefore, soil respiration plays an important role in the global carbon cycle regulation [3]. The researches on soil respiration mainly focus on land, forest [4]-[6], grassland and other ecosystems. There are few studies on soil respiration in coastal salt marsh ecosystem. At the same time, coastal salt marsh has become an effective blue carbon sink to mitigate global warming due to its higher primary productivity and slower rate of organic matter degradation [7].The ports are concentrated in the coastal salt marshes. Therefore, it is of great significance to study the carbon sink mechanism in the coastal salt marshes.

Soil respiration is an important factor determining soil respiration. Researchers at home and abroad have carried out a series of studies on the factors affecting soil respiration. There was improved understanding of the relationship between soil respiration and factors. Raich believe that temperature and precipitation are the major factors affecting global soil respiration [8]. Ryan believes that soil respiration is affected by climatic factors and vegetation factors [9]. Soil fertility and human activity

also affect soil respiration [10]. Therefore, it is very important to study the effects of various factors on soil respiration in coastal salt marsh.
In this paper Tian Ke Yuan, ECO-city, DaGu-outlet and YongDing River as the research object. The diurnal variation characteristics of soil carbon flux were characterized by LCpro-SD photosynthetic apparatus. The main influencing factors of carbon flux in different types of soil were analyzed and relevant relationships were established. This will provide a reference for understanding the carbon sink mechanism in the port areas.

2. Materials and methods

2.1. Research area overview

The selected experimental observation sites are located in Tian Ke Yuan, ECO-city, DaGu-outlet and YongDing River. A total of four observation points (Table 1 and Figure 1). The area is located in the eastern margin of rift basin in North China and the west coast of Bohai basin. It is located at the northern end of Huanghua depression. Belongs to the warm temperate monsoon continental climate and has the marine climate characteristic. The annual average temperature is 13 ℃ , the high temperature extreme value is 40.9℃ and the extreme temperature is -18.3℃.The annual precipitation is average of 566 millimeters. The precipitation varies significantly with the season and less in winter and spring, and in summer.

Table 1. A survey of the measurement sites.

Sample	latitude	longitude	place name	altitude/m	Soil type
1	117°41' 53"N	39°0' 11"N	Tian Ke Yuan	2	Artificial greening soil
2	117°47' 43"N	39°2' 49"N	ECO-city	-2	Artificial greening soil
3	117°42' 55"N	38°58' 46"N	DaGu-outlet	3	Intertidal soil
4	117°44' 44"N	39°5' 02"N	YongDingRiver	0	Intertidal soil

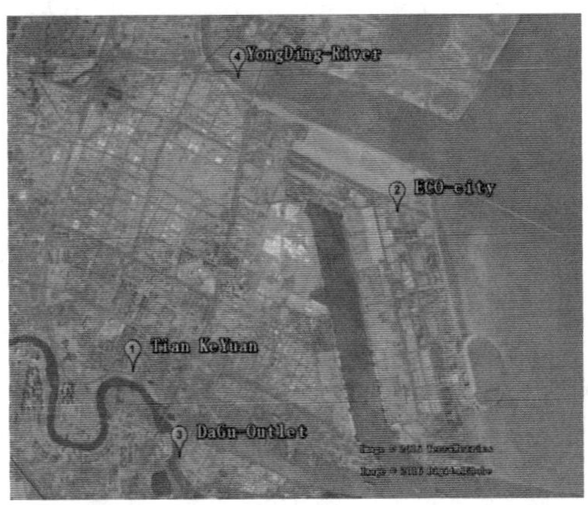

Figure 1. Spatial distribution of measurementsites No.1-4

The soil type of 1 point is artificial afforestation soil, and the main vegetation type is artificial lawn. The soil parent material of sample point 2 corresponds to fluvial alluvium and marine sediments. The status of land use land, farming, transportation, land and other residents living in salt-works. The soil types mainly include artificial greening soil, rock soil and marsh soil.The type of soil point 3 and point 4 corresponding to the intertidal zone. Mainly by Huang He, Hai He, Luan He, Jiyun He by ocean tides and waves of the backwater effect. Coastal land is piled up and the current land uses include salt, sea water and constructed wetlands.

2.2. Research methods

2.2.1. Soil respiration observation

A sample point was set at each observation site and observed from March 2016 to October 2016.Place the soil PVC ring (diameter 15cm) at each sample point for measurements. In setting up plots more flat areas should be chosen to reduce the influence of gradients on CO2 flux measurements. The plants and the dead in the PVC ring should be removed so as to avoid the disturbance of plant respiration to soil respiration measurement before measuring [11].
Soil respiration rate was measured by soil CO_2 flux automatic detection system (LCpro-SD). Soil temperature and atmospheric temperature at 10cm depth were measured by LCpro-SD soil temperature and atmospheric temperature probe respectively while measuring soil CO_2 flux.

2.2.2. Data analysis.
Soil carbon flux was analyzed by one-way ANOVA with different soil types as the influencing factors. Sampling multiple comparison method (LSD) was used to analyze the relationship between soil carbon flux groups in different land types. The relationship between soil carbon flux and atmospheric temperature at different observation sites was analyzed by nonlinear regression analysis. The exponential model is as follows after fitting [12], [13]: $y=a\,e^{bT}$(1). Y is the value of soil carbon flux, in unit mol/(m^2·s), T is the atmospheric temperature at the sample point(℃), A is the soil respiration rate when the temperature is 0℃ and B is the temperature response coefficient. A temperature sensitive coefficient of soil carbon flux using Q_{10} was used. The increase in soil respiration rate when the temperature is increased by 10℃ [14]. The formula is as follows: $Q_{10}=e^{10b}$ (2). The relationship between soil carbon flux and soil temperature at 10cm depth was analyzed by a two element equation. The formula is as follows: $y=a+b_1x+b_2x^2$ (3). a, b_1 and b_2 in the formula are the fitting parameters.

3. Results and analysis

3.1. Diurnal variation characteristics and differences of soil carbon flux

In April the diurnal change of soil carbon flux along the coast during the plant growing season was: it begins at 9:00 to 11:00. The maximum value appears between 11:00 and 12:30.After 12:30 to 15:00 gradually decline. The diurnal trends of soil carbon flux are similar to those already reported. It shows a unimodal form (Figure 2). The soil carbon flux of Tian Ke Yuan, ECO-city and DaGu-outlet reached maximum at around 11:00. The YongDing River reaches its maximum at 12:30. The soil carbon fluxes of Tian Ke Yuan, ECO-city, DaGu-outlet and YongDing River were between 0.01 to 0.67μmol/m^2/s, 0.01 to 0.65μmol/m^2/s, 0.30 to 0.58μmol/m^2/s and 1.49 to 2.82μmol/m^2/s respectively. The daily mean values of soil carbon flux were 0.3448μmol/m^2/s, 0.2761μmol/m^2/s, 0.4388μmol/m^2/s and 2.3367μmol/m^2/s respectively. The minimum and maximum values are found in ECO-city and the YongDing River respectively.

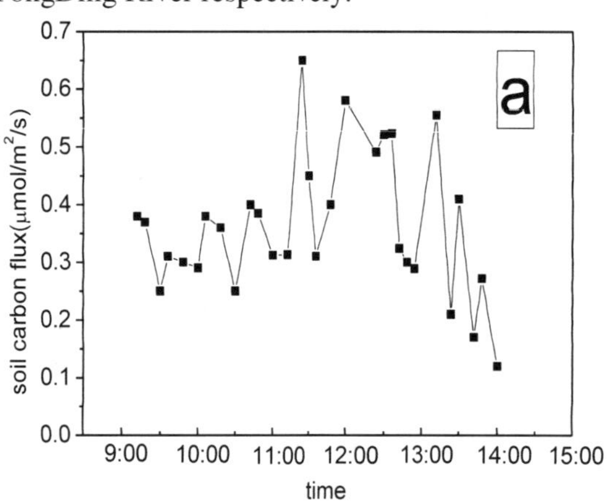

a) Tian Ke Yuan

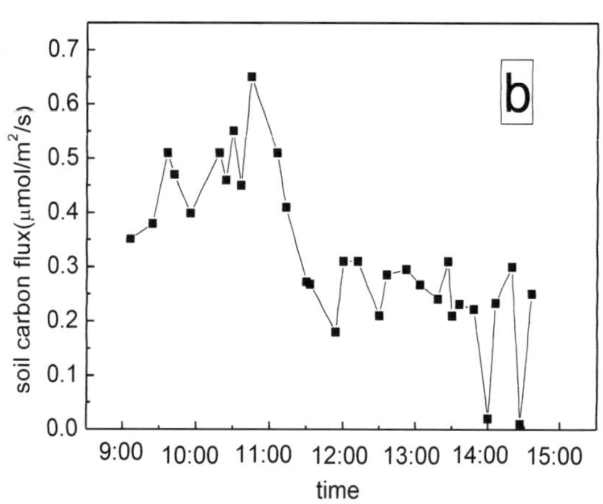

b) ECO-city

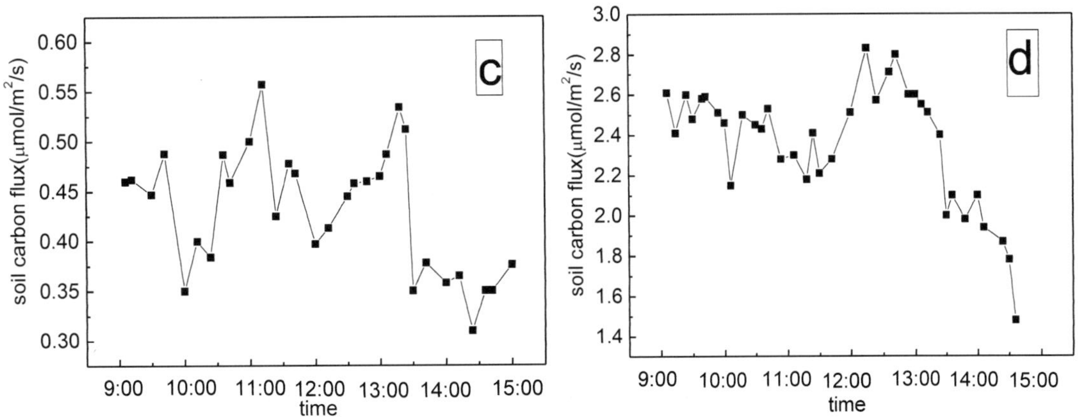

c) DaGu-outletd) d) YongDing River

Figure 2. Diurnal change of soil carbon flux.

The difference of soil carbon flux between the time and the daily mean indicates that the land type has a significant effect on the daily variation of soil carbon flux. In order to verify the effect of different land types on soil carbon flux, the single factor variance analysis of soil carbon flux was carried out with land type as influencing factors. The F value was 794.137, the significance P value was approximately 0($P<0.05$). It is further proved that the land type along the coastal area is the influencing factor of soil carbon flux.

Sampling Multiple Comparative Method (LSD) analyzes the relationship between soil carbon fluxes of different land types (Table 2). There was no significant difference in soil carbon flux between Tian Ke Yuan and ecological cities ($P>0.05$). However， there was significant difference between Dagu sewage outlet and Yongding River($P<0.05$). Often soil respiration is affected by vegetation root respiration, soil microbial respiration substrate supply, regional climate and soil environmental factors [15], [16].The soil types of the Tian Ke Yuan and the ecological cities are artificial green soils, covering the main vegetation types [17]. Soil type, environmental factors are small and covering vegetation type is a single reason for its no significant difference. The daily mean of soil carbon flux in DaGu-outlet and YongDing River was significantly higher than that of Tian Ke Yuan and ECO-city. The DaGu-outlet and YongDing River soil types are intertidal soils with high humidity. Soil respiration and humidity showed a significant positive correlation while soil moisture in a certain range [18], [19]. When the soil moisture content is low, hindering the diffusion of soluble organic carbon in soil solution, leading to microbial nutrient supply, soil respiration weakened. The plants covered by Tian Ke Yuan and ECO-city are mainly shrubs and the plants covered by DaGu-outlet and YongDing River are the coastal salt marsh plants. Studies on soil respiration for grassland and wetland have shown that there is a large difference in soil respiration among different biomes [20]. Therefore the biological type may also be the coastal area of soil respiration factors. This difference arises from differences in soil moisture and biological types as well as differences in atmospheric and soil temperatures.

Table 2. Multiple comparison of soil carbon fluxes and containing notes[a].

Significance(P value)	Tian Ke Yuan	ECO-city	DaGu-outlet	YongDing River
Tian Ke Yuan	—			
ECO-city	0.170	—		
DaGu -outlet	0.022	0.01	—	
YongDing River	0	0	0	—

[a]The significance level of the mean difference is 0.05.

3.2. The influence of atmospheric temperature on soil carbon flux

IOP Conf. Series: Earth and Environmental Science **150** (2018) 012007 doi:10.1088/1755-1315/150/1/012007

The relationship between atmospheric temperature and soil carbon flux in different research areas is different. Generally, the fitted equation is used to estimate soil respiration [21].The exponential equation $y=a\ e^{bT}$ and its sensitivity coefficient Q_{10} were used to estimate the relationship between atmospheric temperature and soil carbon flux in different regions. Among them Figure 3a for Tian Ke Yuan and Figure 3b for ECO-city, and Figure 3c for DaGu-outlet, Figure 3d for the YongDing River. The result is shown in Figure 3.

It can be seen that the atmospheric temperature and soil carbon flux in the four areas have obvious exponential relationship from the fitting results. The correlation was significant ($P<0.05$).Both increase with the increase of atmospheric temperature. The soil carbon flux of DaGu-outlet and YongDing River is better than the atmospheric temperature, and the coefficient of determination (R^2) is 0.517 and 0.5344 respectively. The decision values of Tian Ke Yuan and ECO-city were lower, 0.3269 and 0.289, respectively. According to the fitted exponential equation, the Q_{10} of Tian Ke Yuan, ECO-city, DaGu-outlet and YongDing River are 8.331, 6.049, 2.651 and 1.391, respectively, with an average value of 4.606. Atmospheric temperature acts as a function of the respiratory intensity of the ecosystem. That is the direct influence on the activity of root system and microorganism and the indirect influence on carbon distribution of photosynthesis and root system, affecting the net exchange of ecosystem CO_2 [22]. It has been reported that the respiration of wetland ecosystem is significantly affected by atmospheric temperature. The positive exponential relation is generally shown [23]-[25]. The relationship between ecosystem atmospheric temperature and soil carbon flux was consistent with that obtained in this study. And the index model described the relationship between the two is better ($R^2=0.289$ to 0.5344, $P<0.05$).

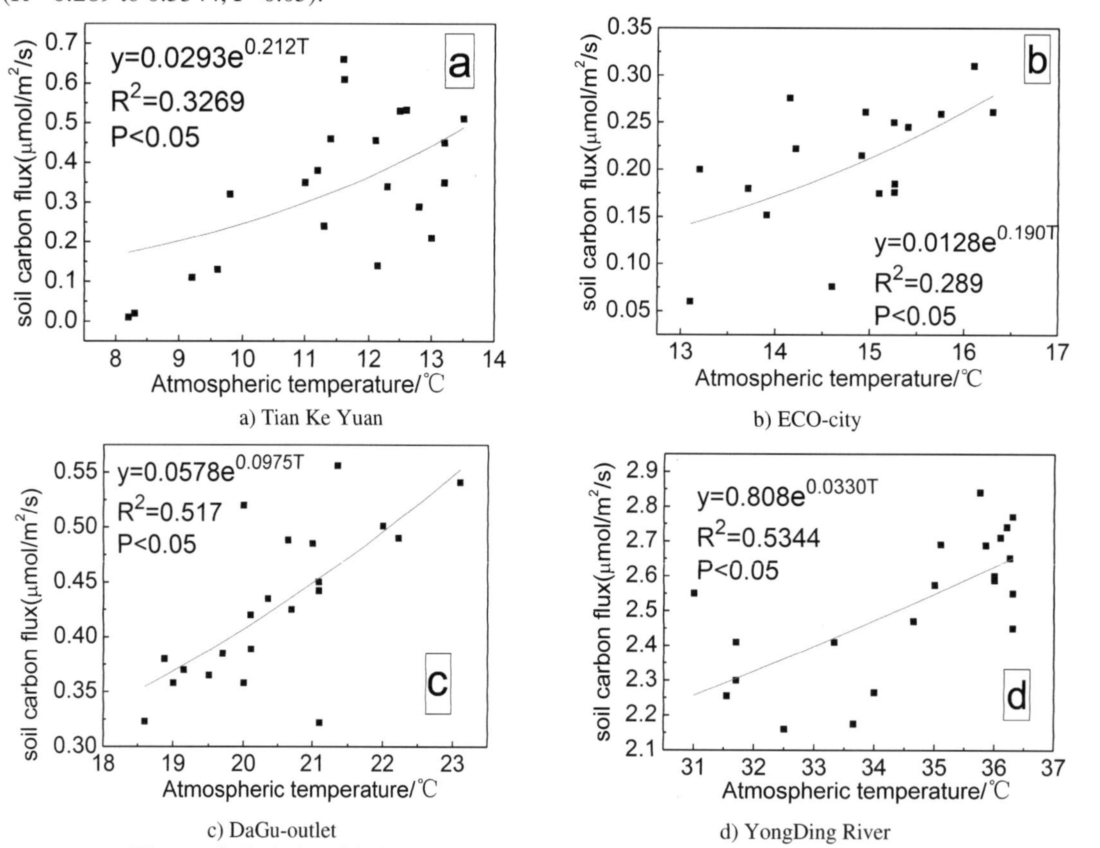

a) Tian Ke Yuan b) ECO-city

c) DaGu-outlet d) YongDing River

Figure 3. Relationship between atmospheric temperature and soil carbon flux.

3.3. The influence of soil temperature on soil carbon flux

The relationship between soil temperature (soil temperature in 10cm depth) and soil carbon flux was fitted by using the model two equation $y=a+b_1x+b_2x^2(a, b_1, b_2)$ as constant (Figure 4). From the level

IOP Conf. Series: Earth and Environmental Science **150** (2018) 012007 doi:10.1088/1755-1315/150/1/012007

of significance ($P<0.05$) and the coefficient of determination (R^2), the soil carbon flux and soil temperature in all regions had a good correlation (R^2 =0.3227-0.7465). It is indicated that soil temperature is an important factor affecting soil carbon flux and shows a trend of increasing at first and then decreasing. The relationship between soil temperature and carbon flux obtained in this study is inconsistent with the model adopted by other grassland and intertidal zones [26], [27].

Soil respiration reflects the intensity of soil microbial activity and soil metabolism in soil [28].Temperature and humidity are the major limiting factors for soil respiration. Temperature has significant effects on the activity of respiratory enzymes and microbial activity in plant roots [28]. Root system and microbial metabolism require higher soil moisture. The relationship between soil carbon flux and soil temperature in the experimental plots is no single. It is indicating that the factors of soil respiration in coastal areas are more complex. Soil temperature can explain the variation of soil carbon flux components of 32-75% without considering other factors. The fitting effect of the day hospital and eco city is better. This is because the day science and ecological city greening area belongs to the artificial soil and DaGu-outlet and YongDing River are located in the intertidal zone. The soil moisture variation from river water level significantly affected which is leading to poor effect of soil temperature and soil carbon flux.

Compared with the soil respiration in the salt marsh ecosystem in other areas, the higher soil moisture in the Binhai New District has the distinct characteristics of soil respiration. The correlation between vegetation carbon flux and thermal conditions in Binhai New District was investigated in the paper. However, the types and quantities of vegetation in the wetlands, the diversity of microorganisms in the soil and the soil organic carbon content still need further investigation. Therefore, in order to obtain a more complete understanding of the impact factors of soil carbon fluxes in the salt marsh, long-term measurements of soil carbon fluxes in the area need to be made and the correlation between other influencing factors and soil carbon flux. It provides the basis for comprehensive understanding of the mechanism of carbon sequestration in the salt marsh in Binhai New District.

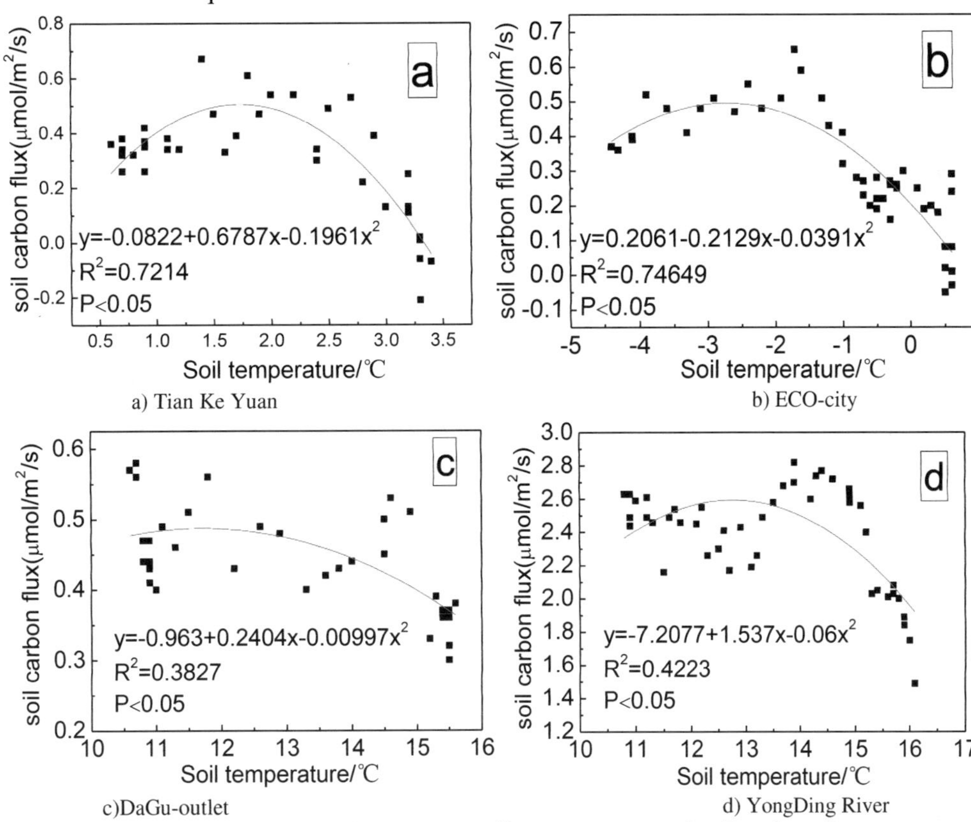

Figure 4. Relationship between soil temperature and soil carbon flux.

4. Conclusion

- Soil carbon fluxes in the days of Tian Ke Yuan, ECO-city, DaGu-outlet and YongDing River showed a single-peak curve in April 2016. Soil carbon fluxes began to increase in 9:00-11:00 and maximum occurs 11:00-12:30, then 12:30-15:00 decreased. Different types of land have a significant effect on soil carbon flux.
- The exponential model can better describe the effect of atmospheric temperature on soil carbon flux (R^2 is 0.289-0.5344, $P<0.05$). And the average Q_{10} was 4.606. The correlation between soil carbon flux and atmospheric temperature in DaGu-outlet and YongDing River is great. The result indicating that the atmospheric temperature on DaGu-outlet and YongDing River (intertidal soil) soil carbon flux is larger than Tian Ke Yuan, or ECO-city (artificial greening soil).
- The soil temperature and soil carbon flux at 10 cm were fitted with a quadratic equation $y=a+b_1x+b_2x^2$. The soil temperature could explain 32.27%-74.65% of soil carbon flux Changes.

5. References

[1] Schimel D S 1995 *J. Global change biology* **1** 77-91
[2] Hibbard K A, Law B E, Reichstein M and Sulzman J 2005 *J. Biogeochemistry* **73** 29-70
[3] Schlesinger W H and Andrews J A 2000 *J. Biogeochemistry* **48** 7-20
[4] Yang J and Wang C 2005 *J. Acta. Phytoecological. Sinica* **30** 286-94
[5] Yan J, Bing W, Yu-Ru W and Qing-Pei Y 2010 *J. Yingyong Shengtai Xuebao* **21** 1641-48
[6] Liu Y, Han S, Hu Y and Dai G 2005 *J. Ying yong sheng tai xue bao* **16** 1581-85
[7] Cao L, Song, Li X, Yuan H, Li N and Duan L 2013 *J. Acta. Ecol. Sin* **33** 5141-52
[8] Raich J W and Schlesinger W H 1992 *J. Tellus* B **44** 81-99
[9] Ryan M G and Law B E 2005 *J. Biogeochemistry* **73** 3-27
[10] O'neill K P, Kasischke E S and Richter D D 2002 *J. Canadian Journal of Forest Research* **32** 1525-41
[11] Feng C Y, Liu S H, Gao J X, Liu S H and Lin D 2008 *J. Journal of Beijing Forestry University* **30** 20-26
[12] Luo Y, Wan S, Hui D and Wallace L L 2001 *J. nature* **413** 622-25
[13] Grace J and Rayment M 2000 *J. Nature* **404** 819-20
[14] Guo H, Dong X and Jiang F 2010 *J. Scientia. Silvae. Sinicae* **46** 110-15
[15] Davidson E A, Janssens I A and Luo Y 2006 *J. Global Change Biology* **12** 154-64
[16] Wang G B, Tang Y F, Ruan H H, Shi Z, He R, Wang Y and Su G X 2009 *J. Acta. Ecologica. Sinica* **29** 966-75
[17] Li X C, Guo X P, Hu Y C and Liu W 2014 *J. Guangdong Agricultural Sciences* **16** 034
[18] Wang Y, Hu Y, Ji B, Liu G and Xue M 2003 *J. Advances in Atmospheric Sciences* **20** 119
[19] Hanson P J, Wullschleger S D, Bohlman S A and Todd D E 1993 *J. Tree physiology* **13** 1-15
[20] Cao L, Song J, Li X, Yuan H, Li N and Duan L 2013 *J. Acta. Ecol. Sin* **33** 5141-52
[21] Liu Y, Han S, Hu Y and Dai G 2005 *J. Ying yong sheng tai xue bao* **16** 1581-85
[22] Davidson E, Belk E and Boone R D 1998 *J. Global change biology* **4** 217-27
[23] Franzluebbers K, Franzluebbers A J and Jawson M D 2002 *J. Soil Science Society of America Journal* **66** 254-62
[24] Griffis T J 2004 *J. Agricultural and Forest meteorology* **125** 207-23
[25] Song C, Yan B, Wang Y, Wang Y, Lou Y and Zhao Z 2003 *J. Chinese Science Bulletin* **48** 2749-53
[26] Sun Q, Fang H L, Liang J, Qian X W, Liu M D, Zhang Q F and Hao G J 2009 *J. Chin. J. Ecol.* **28** 1572-78
[27] Xu L, Tang H, Yang G and Xin X 2010 *J. Acta. Botanica. Boreali-Occidentalia. Sinica* **30** 1882-86
[28] Andrews J A, Matamala R, Westover K M and Schlesinger W H 2000 *J. Soil Biology and Biochemistry* **32** 699-706

IOP Conf. Series: Earth and Environmental Science **150** (2018) 012008 doi:10.1088/1755-1315/150/1/012008

Possibility of forming artificial soil based on drilling waste and sewage sludge

J Kujawska[1], M Pawłowska[1] and H Wasag[2]

1 Institute of Renewable Energy Engineering, Faculty of Environmental Engineering, Lublin University of Technology, 40B Nadbystrzycka Street, Lublin 20-618, Poland
2 Institute of Environmental Protection Engineering, Faculty of Environmental Engineering, Lublin University of Technology, 40B Nadbystrzycka Street, Lublin 20-618, Poland
E-mail: j.kujawska@pollub.pl

Abstract. Land redevelopment is necessary due to the amount of a degraded area. Depositing waste on the small area of landfills is harmful for the environment. New methods of managing and utilizing waste are being sought in order to minimize the deposition of waste. In small amounts, many types of waste can be treated as a substrate or material improving physicochemical properties of soils, and hence can be used in reclamation of degraded lands. The study analysed the effect of different doses of sewage sludge (35%, 17.5%) with addition (2.5% and 5%) of drilling waste on the properties of degraded soils. The results show that created mixtures improve the sorption properties of soil. The mixtures contain the optimal the ratio of nutrient elements for growth of plants is N:P:K.

1. Introduction

Extraction and exploitation of minerals leads to forming extensive areas of completely degraded surface. According to the Institute of Environmental Protection and the Institute of Ecology of Industrial Areas, in Poland there are over 800,000 hectares of degraded land and 3950,000 hectares of the area threatened by industrial degradation [1]. In keeping with the idea of sustainable development, these sites should be redeveloped and reused. Reclamation of such areas becomes an urgent necessity [2].

Covering these areas with waste mixtures that will allow the production of vegetation is a simple solution to protect them. The idea of mixing rock crumbs with organic waste such as animal excrements, compost or sewage sludge to create a reclamation layer on degraded land was first used in 1977 by Laga. Haraldsen created a mixture of rock crumbs, sand and various organic waste: sewage sludge and compost. It was observed that the type and amount of organic matter had a significant effect on plant growth. A mixture with sewage sludge proved the best mixture for plant growth [3].

An increasing amount of unconventional and conventional hydrocarbon exploration generates large quantities of drilling waste. The mineralogical composition of drilling waste is similar to that of soil. Their physical and chemical properties indicate that they can be used for soil reclamation, especially the sandy ones. The use of this type of waste on degraded soils can improve soil properties, increase plant yield and optimize its biogenic components.

The purpose of this article is to use drilling waste to produce the waste component that would have the properties of a soil improvement agent. We propose using sludge to achieve this goal. The resulting mixture of drilling waste and sewage sludge would constitute a biologically active layer that could be used to recreate degraded soil.

Sewage sludge and drilling waste mixed together in appropriate proportions may contribute to improving the quality of physically or chemically degraded areas. The porosity, layout and water-air conditions of soil can be improved by increasing the level of organic matter. Increasing the pH value may lead to adsorption, complexation and precipitation of heavy metals, which will subsequently limit their mobility in groundwater and bioavailability for plants. The use of waste components instead of individual waste is much more advantageous because of the ability to adjust the composition and properties of a soil improvement agent.

2. Materials and Methods

The following components were used in the mixture:

- soil material was extracted from the field situated in Felin, near Lublin (Eastern Poland)
- granulated sewage sludge coming from the Hajdów Sewage Farm in Lublin
- drill cuttings that were taken from facility of mining waste in Luchow

The scope and analytical methods of determination of physical and chemical parameters of mixtures and their components:

- dry matter content by means of loss on drying method in the temperature of 105°C (PN EN ISO 11465:1999P soil quality. Determination of dry matter content of soil and water in soil per dry mass of soil. Gravimetric method),
- pH value by means of potentiometric method in water solution and in 1 M solution of potassium chloride (PN-ISO 10390: 1997P soil quality. Determination of pH),
- specific gravity (ρs), Le Chatelier flask method, in the temperature of 22°C,
- granulometric composition sieve analysis (PN-EN ISO 14688-2:2006),
- bulk density (ρb), Kopecky's cylinder method with standard volume of 100 mL,
- total porosity (n) was calculated on the basis of actual density and volumetric density, according to the following equation:

$$n=\frac{\rho_s - \varrho s}{\rho_s} 100\%$$

- hydrolytic acidity (Kh) by Kappena method, in 0.5 mol/dm^3 $(CH_3COO)_2Ca$,
- total exchangeable bases (S) by Pallmann method in 0.5 mol/dm^3 NH_4Cl at pH 8.2,
- cation exchange capacity (T) as a sum of S and K_h,
- carbonate content was determined with Scheibler's volumetric method,
- total organic carbon by means of TOC-5050A analyzer Schimadzu (PN-ISO 4335 standard),
- total nitrogen content by Kjeldhal method using 1002 Kjeltech distillation unit
- total phosphorous and heavy metals were determined by ICP-OES Ultrace 238 (Jobin Yvon-Horriba France) using direct calibration method after microwave digestion (Multiwave 3000, Anton Paar). The samples of homogenized soil (1 g) were digested in acid mixture of HNO3: HCl (5:2) and water samples (15 g) were digested in HNO3 (3 mL). The digestion process lasted 45 min at 180°C and at the pressure of 18 bars. Metal concentrations were determined at different wavelengths (213–395 nm). Detection limits for particular metals did not exceed 10 ppb.

3. Mixture composition

Granulated sewage sludge was obtained from Hajdów Sewage Farm in Lublin. The content of Cd, Cr, Cu, Pb, Zn and Ni in sediments is shown in Table 1. The concentration of metals is low and does not exceed the limits for the agricultural use of sewage sludge specified in the Regulation of the Polish Minister of Environment on Municipal Sewage Sludge (Journal of Laws, 2015, item 257) [4]. The drill cuttings were extracted from the waste layer deposited at a waste disposal facility in Luchow. They are strongly alkaline (pH 11). As indicated in Table 2, the concentrations of metals in drill cuttings did not exceed the limit values given in the Regulation of the Polish Minister of Environment on the Quality of Soil Quality Standards (Journal of Laws, 2016, item 1359) with the exception of barium [5].

Table 1. Concentrations of heavy metals in sewage sludge.

Elements	Concentration (ppm)	Limit values for concentrations of heavy metals in agriculture (ppm) [4]
Cd	0.2	20
Cu	915	1000
Ni	204	300
Pb	119	750
Zn	1770	2500
Hg	< g.o.	16
Cr	< g.o.	500

Table 2. Concentrations of heavy metals in drill cuttings.

Elements	Concentration (ppm)	Permissible concentration in soils (ppm) [5]
Ba	1820	1000
Cd	0.2	15
Co	0.08	200
Cr	62.63	500
Cu	99.23	600
Pb	27.21	600
Ni	39.92	300

3.1. Composition of waste components.

A series of ternary soil-like mixtures were prepared: acidic podzol soil with the following additions of drill cuttings and sewage sludge in the proportions shown in Table 3.

Table 3. Scheme of experiment.

Component P	Soil + 2.5% drill cuttings (v/v) + 35% sewage sludge (v/v)
Component R	Soil + 5% drill cuttings (v/v) + 35% sewage sludge (v/v)
Component M	Soil + 2.5% drill cuttings (v/v) + 17.5% sewage sludge (v/v)
Component O	Soil + 5% drill cuttings (v/v) + 17.5% sewage sludge (v/v)

3.2. Statistical analysis

Mean values with standard deviation were calculated for the results of this study. Then, the statistical analyses based on the one-way analysis of variance (ANOVA) and multiple T-Tukey tests with the significance level $\alpha = 0.05$ were carried out. T-Tukey's multiple comparative tests provided a detailed comparative analysis of mean values by isolating statistically homogeneous groups (homogeneous groups). The matrix of correlation coefficients between the variables was determined.

The calculations were made using the Statistica 13.0 (the license of the Faculty of Environmental Engineering, Lublin University of Technology).

4. Results and discussion.

Table 4 shows the properties of ternary mixtures which may have a significant influence on the direction and intensity of the processes occurring in soils.

The addition of the waste component increased the fraction of sand in the created mix.tures. All the mixtures produced are characterized by a fine grain size with a prevalence of PM fraction (0.05 ÷ 0.002 mm) that accounts for over 50% of their total mass. The sand fraction (0.05 ÷ 0.002 mm) represents 30 to 50%. The dominant PM fraction increases the water holding capacity of soils. The created mixtures can be classified as sandy soils [6].

The variable composition of the added waste components did not significantly affect the bulk and actual density and porosity of the light soil. The post hoc statistical analysis using Tukey's honest significance test did not show a significant statistical effect of the waste component on the density, volume and porosity of the soil either. The density of the created mixtures is characteristic for mineral-organic soils in which this parameter ranges from 1.8 to 2.60 g/cm^3. Besides, the bulk density of the mixtures was in the lower range of typical values for mineral soils, ranging from 0.75 to 1.30 g/cm^3. The porosity of individual mixtures is comparable to the porosity of peat soils that amounts to 55-80% [6].

Table 4. Physical properties of the soil and the mixtures (±standard deviation).

Mixtures	Granulometric composition (%)			Specific gravity (g/cm^3)	Bulk density (g/cm^3)	Total porosity (%)
	Sand fraction	Silt fraction	Clay fraction			
Soil	23	71	6	$2.57^a \pm 0.02$	1.14^a ($\pm$ 0.01)	55.45
P	45	50	5	$2.48^a \pm 0.01$	$0.91^a \pm 0.03$	63.31
R	38	56	6	$2.47^a \pm 0.001$	$0.92^a \pm 0.01$	62.80
M	30	65	5	$2.46^a \pm 0.002$	$1.07^a \pm 0.01$	59.39
O	31	64	5	$2.47^a \pm 0.06$	$1.04^a \pm 0.01$	56.64

The uppercase letter ^{a} next to the average content values in the Table (Tukey Homogeneous Groups) indicates statistically homogeneous groups. The presence indicator designates the lack of statistically significant difference between them.

Addition of the waste component increases the dry matter content in sandy soil (Table 5). The Tukey's test confirmed that the addition of the waste component constitutes a statistically significant increase in the organic matter content. No statistically significant differences in the organic matter content between the analysed soil mixtures P, R and M, O were observed. The resulting mixtures exhibit organic matter content similar to that of black soil, in the range of (2.22% – 46.5%) [6]. High content of organic matter and humus in soil is a stabilising component to its structure, prevents soil compaction and degradation on account of water and wind erosion. Soils of agricultural lands demonstrate 5-7% organic matter content [7], which pales into insignificance in view of the created mixtures' organic matter content. It is precisely the said high content of the organic matter that earmarks the waste component as a perfect solution for agriculture applications [8].

Carbonate content upon soil enrichment with the waste component is marked with a tenfold increase in comparison with standard soil. This growth may be largely contributed to the presence of drill cuttings and amounts to 1% in all created soil mixtures, which corresponds with the average content of carbonate in soils of agricultural lands in Poland. Statistical analysis confirms that the addition of the waste component has had a statistically significant effect on the carbon content in all soil mixtures in question.

From the viewpoint of biochemical changes in the soil environment, there is a marked significance of the carbon to nitrogen to phosphorus ratio (C:N:P). The right proportion of these elements guarantees the proper course of changes of organic to mineral matter (mineralisation), or to humus components (humification). The optimal C:N:P ratio for soil reclamation 10:1:100, which is precisely the case with the created soil mixtures [9].

The *post hoc* statistical analysis conducted by means of the Tukey test showed that the addition of the waste component was statistically significant to the increase of nitrogen, carbon and phosphorus content in soil. No such difference was noted for the carbon content in the mixtures. The phosphorus content in mixture R shows statistically significant difference with the other soil mixture variants. Finally, no statistically significant differences were observed in nitrogen content in P, R and M, O soils.

Table 5. Selected chemical properties of the soils and the mixtures (±standard deviation).

Mixtures	Carbonate content (%)	Dry matter content (%)	N (%)	C (%)	P (%)
Soil	0.02^c ±0.02	1.9^c ±0.05	0.06^c ±0.01	0.03^b ±0.01	2.53^c ± 0.98
P	1.87^a ± 0.29	23.2^a ± 0.65	1.61^a ± 0.02	13.53^a ± 1.39	149.66^b ± 52.68
R	1.01^b ± 0.08	22.3^a ± 0.73	1.59^a ±0.06	12.82^a ± 2.32	184.22^a ± 78.21
M	1.45^{ab} ± 0.42	11.7^b ± 0.07	0.77^b ± 0.07	6.32^a ± 0.73	110.06^b ± 56.44
O	0.8^d ± 0.001	12.1^b ± 0.67	0.85^b ± 0.061	7.02^a ± 0.49	142.31^b ± 68.21

The uppercase letters [a,b,c,d] next to the average content values in the Table (Tukey Homogeneous Groups) indicate statistically homogeneous groups. The presence of the same indicator designates the lack of statistically significant difference between them.

The content of mineral elements smaller than 0.02 mm in soil is regarded to be the key indicator of humus content in soil. Hence, the richer the soil is in mineral elements, the higher humus content it should exhibit. Humus content in soil is calculated by multiplying the organic carbon content (Corg) by 1.7 factor. The humus content in all mixtures is higher than 3%, which is indicative of highly fertile chernozem soils (with humus content in the range of 1.2 – 4.1%) [6].

Cutitin *et al.* show that the increase in the C:N content in slightly acidic soils, the elevated levels of calcium and alkaline reaction accelerate soil mineralisation and foster plant biomass growth [10]. Similar results were obtained in the presented tests.

Table 6. pH value of the soils and of the mixtures (±standard deviation).

Mixtures	pH in H_2O	pH in KCl
Soil	4.16^b ± 0.04	4.01^b± 0.01
P	6.41^a ± 0.08	6.41^a ± 0.01
R	6.39^a ± 0.01	6.33^a ± 0.02
M	6.38^a ± 0.02	6.47^a ± 0.04
O	6.34^a± 0.03	6.4^a ± 0.03

The uppercase letters [a,b] next to the average content values in the Table (Tukey Homogeneous Groups) indicate statistically homogeneous groups. The presence of the same indicator designates the lack of statistically significant difference between them.

The presence of the waste component contributed to the elevated level of soil pH, hence the soil mixtures approached the neutral reaction (Table 6). The pH levels of the analysed soil material indicate their suitability for the majority of agricultural plants. In biological reclamation the reaction of the soil is paramount to the growth, development and harvesting of plants. Furthermore, it is the range of soil reactions from acidic to neutral that proves the most beneficial for the majority of plants [11].

The *post hoc* statistical analysis with the Tukey test showed that the addition of the waste component was statistically significant to the increase of soil pH.

Basic wastes raise the pH of acidic soils, which was moreover confirmed by the studies of Ociepa *et al.* and Roig *et al.* [12], [13]. The work of Wesołowska *et al.* showed that the addition of waste component in the volume of 10 Mg/ha contributed to elevating the acidity of soil from 5.2 to 6.1, whereas in mixtures with sewage sludge (400 m^3/ha) and mineral wool (10 Mg/ha) the pH of soil amounted to 5.7 [14].

The presence of the waste component in soil highly increases the cation exchange capacity and the total exchangeable bases in sandy soil (Figure 1). The cation exchange capacity of the mixtures is above 20 cmol(+)/kg, which is comparable to non-degraded chernozem soils. What is decisive in such high cation exchange capacity of soils is the silt and humus content and the basic reaction wastes. High cation exchange capacity is beneficial to soil as the nutritional elements in soil are stopped from leaching deeper into the soil profile and into the underground water, but are retained in the rooting zone of plants and released from the cation exchange complex during the plant growth periods. What is more, in soils contaminated with excessive metal content (particularly cadmium or lead) high cation exchange capacity prevents leaching and transfer of metals to the food chain [6]. The cation exchange capacity in the created soil mixtures is very high compared to soils characteristic of Poland, and may be only compared with fertile chernozem soils [15].

The soil mixtures are characterised by high values for the sum of the total exchangeable bases (S). According to IUNG data (2012), the average sum of basic cations in arable soils in Poland is approx. 8 cmol(+)/kg, and the highest recorded values amount to 40 cmol(+)/kg [16].

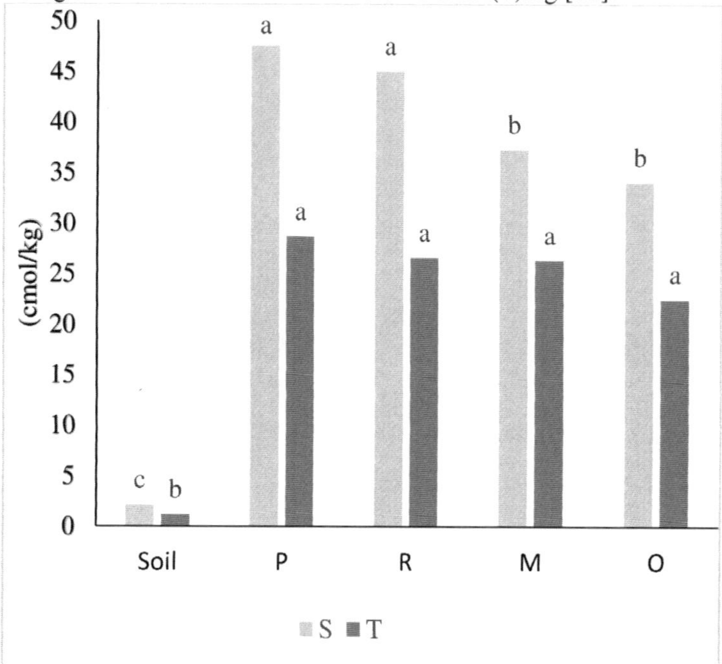

Figure 1. Effect of wastes on total exchangeable bases (S) and cation exchange capacity (T) in sandy soil. The uppercase letters [a,b] next to the average content values in the Table (Tukey Homogeneous Groups) indicate statistically homogeneous groups. The presence of the same indicator designates the lack of statistically significant difference between them.

High basic cation content in soil is of benefit to plant growth due to the fact that exchangeable ions in the cation exchange complex transferred into the soil solution constitute an important source of minerals for plants [11].

The *post hoc* statistical analysis with the Tukey test showed that the addition of the waste component was statistically significant to the increase in the cation exchange capacity and the sum of the total exchangeable bases in soil.

The increase in the exchangeable ion content, observed in the study, is the effect of drilling waste on sandy soil. The sewage sludge contains negligible amounts of calcium or magnesium, and it is only by adding the drilling waste, characterised by high content of exchangeable cations, that the soil becomes enriched in those cations [11]. Other researchers enrich the sewage sludge with ash and thus prepared composition add to soil with a view to reclaiming because, similarly to drilling waste, ash are rich in exchangeable cations and may therefore be used to improve soil fertility and growth of biomass of agricultural plants [17]-[19].

5. Conclusion

The results of the research confirm the possibility of using waste mixtures, i.e., drill cuttings and sewage sludge to improve the quality of degraded soils. The mixtures contain the optimal reaction and the ratio of nutrient elements to growth of plants N: P: K .The mixtures show very good physicochemical properties. The addition of the waste component increased the sorption capacity of the soil. The value of the sorption capacity of the created mixtures is comparable to that of chernozem soils.

6. References

[1] Nowak A and Schneider C 2017 Sci Total Environ. **590–591**
[2] Pawul M and Sobczuk W. 2011 Probl. Ekorozw. **6(1)**
[3] Haraldsen T K and Pedersen P A 2003 Urban. For Urban. Gree. **2(1)**
[4] Regulation of the Polish Minister of Environment on Municipal Sewage Sludge (Journal of Laws, 2015, item 257)
[5] Regulation of the Polish Minister of Environment on Soil Quality Standards (Journal of Laws, 2016, item 1359)
[6] Rattan L 2016 *Encyclopedia of Soil Science* (London: CRC Press Taylor & Francis)
[7] Krasowicz S, Oleszak W, Hrobik J, Debicki R, Jankowiak J, Stuczyński T and Jadczyszyn J 2011 Polish J. Agron. **7**
[8] Czechowska-Kosacka A, Cao Y and Pawłowski A 2015 Rocz. Ochr. Sr. **17(1)**
[9] Leys N M, Bastiaens L, Verstraete W and Springael D 2005 Appl. Microbiol. Biot. **66(6)**
[10] Curtin D, Campbell C A and Jalil A 1998 Soil. Biol. Biochem. **30**
[11] Menge K and Kirkby E A 2011 *Principles of plant nutrion* (Switzerland: Springer Science + Business Media B.V.)
[12] Ociepa A, Pruszek K, Lach J and Ociepa E 2008 Ecol. Chem. Eng. S. **15(1)**
[13] Roig N, Sierra J, Nadal M, Marti E, Navalon-Maadrigal P, Schumacher M and Domingo J L 2012 Sci Total Environ. **425**
[14] Wesołowska S, Baran S, Żukowska G, Myszura M, Bik–Małodzińska M, Pawlowska M and Pawłowski A 2017 An influence of municipal sewage sludge and mineral wool application on sorption properties of coarse–grained soil *Environmental Engineering V* ed M Pawłowska and L Pawłowski (London: CRC Press Taylor & Francis) pp 169–180
[15] Ahnert B and Schultz J 2005 *The Ecozones of the World: The Ecological Divisions of the Geosphere* (Switzerland: Springer Science & Business Media)
[16] Terelak H, Motowicka–Terelak T, Maliszewska-Kordybach B and Pietruch Cz 2002 *Monitoring of chemical properties of arable soils in Poland* (Warsaw: Environmental Monitoring Library)
[17] Sajwan K S, Paramasivam S, Alva A K, Adriano D C and Hooda P 2003 Adv. Environ. Res. 8
[18] Joel O F and Amajuoyi C A 2009 J. Appl. Sci. Environ. Manage. **13(2)**
[19] Jia–Qing X, Rui–Lian Y, Xiao–Yi D, Gong–Ren H, Xue–Song S, Qun W and Hu–Wei L 2012 J. Hazard. Mater. **217-218**

IOP Conf. Series: Earth and Environmental Science **150** (2018) 012009 doi:10.1088/1755-1315/150/1/012009

Sorption characteristics of cadmium in a clay soil of Mae Ku creek, Tak Province, Thailand

P Thunyawatcharakul[1,2] and S Chotpantarat[2,3,4,5*]

1 International Postgraduate Program in Hazardous Substance and Environmental Management, Graduate School, Chulalongkorn University, Bangkok 10330, Thailand
2 Center of Excellence on Hazardous Substance Management (HSM), Chulalongkorn University, Pathumwan, Bangkok, 10330, Thailand
3 Department of Geology, Faculty of Science, Chulalongkorn University, Bangkok, 10330, Thailand
4 Research Program of Toxic Substance Management in the Mining Industry, Center of Excellence on Hazardous Substance Management (HSM), Bangkok 10330, Thailand
5 Research Unit of Green Mining (GMM), Chulalongkorn University, Bangkok, 10330, Thailand
E-mail: csrilert@gmail.com

Abstract. Mae Sot is a district in Tak province, the northern part of Thailand where has encountered with cadmium (Cd) contaminated in soils. Exposure of Cd can lead to severe health effect, for examples, bone softening, osteoporosis, renal dysfunction, and Itai-Itai disease. This study aims at elucidating sorption behavior of Cd in the contaminated soil collected from Mae Ku creek, Mae Sot district, Thailand. Batch sorption experiment was conducted in order to investigate sorption characteristics of Cd onto the contaminated soil. The soil sample taken from the study area consists of 26% sand, 16% silt 58% clay, which categorized as a clay soil, based on USDA classification. Soil pH is slightly alkaline (pH~7.7) and organic matter in the soil is 2.93%. The initial concentration in the batch sorption experiment was in the range from 0- 200 ppm. The result from the batch sorption experiment showed that soil sample can adsorb Cd up to 173.5 ppm and the sorption behavior of the soil sample can be well described by Freundlich isotherm, indicating the multilayer sorption (R^2 = 0.9964), with Freundlich constants of 0.312 and 1.760 L g^{-1} for 1/n and K_f, respectively.

1. Introduction

Cadmium (Cd) is a heavy metal, which typically occurs with zinc (Zn), lead (Pb), and copper (Cu) mineral. Cadmium contamination in the environment can cause by fuel combustion, human activities such as mining and using fertilizers. Moreover, natural also causes Cd contamination, for examples, stream transportation or volcanic eruption. Exposure pathway of Cd can come from ingestion of contaminated food and water as well as inhalation of cadmium fume from industry or smoking cigarette can also lead to Cd exposure [1]-[4].

Mae Sot is a district in Tak province, the northern part of Thailand which has been encountered Cd contamination in soil. Simmons [5] studied Cd concentration in soil sample in paddy field in Ban Pha Te, Phatat Pha Daeng sub-district, Mae Sot, Tak Province, which Cd concentration in soil samples varied from 0.5 to 284 mg kg^{-1}, some samples excess the background concentration of Cd in Thailand (0.002- 0.14 mg kg^{-1}) [6]. Moreover, Simmons et al. [5] analyzed 15 soil profiles from 0-75 cm. depth

from the ground surface. As a result, it showed that distribution of Cd mostly high in 0-30 cm. depth, accounting for 88.84% of total Cd concentration in soil profile, indicated that the source of contamination may come from suspended sediment from Mae Tao creek where flows through Phadaeng zinc mine where people in the study area used as irrigation water for crops. Akkajit and Thongcumpou [7] investigated fractionation of Cd and other metals (Cu, Fe, Mn, Pb, and Zn) that related for bioavailability by using BCR sequential extraction. The result showed that Cd has highest proportion of exchangeable form (BCR1), compared with other metals as follows: Cd (29.9%) > Mn (27.7%) > Zn (25-27%) > Pb (2.6%) > Cu (less than 0.5%) >Fe (0.05%). Similarly, Kosolsakul [6] studied fractionation of Cd in Baan Mae Tao Mai, Phrathat Phadaeng sub-district, by using BCR and Tessier sequential extraction. The result showed that Cd in BCR 1 fraction had the highest proportion (approximately 67%-84%) compared with other fractions and higher BCR 1 was found in the high Cd concentration areas. The result from Tessier sequential extraction showed that Cd in the study area had Cd associated with the carbonate phase (T2) more than the exchangeable phase (T1). The higher Cd concentration contaminates, the higher proportion of T2 becomes, whereas the proportion of T1 does not change significantly.
Even though there are many studies, which investigated about leaching potential of Cd from contaminated soils in Mae Sot District, the study of the sorption behavior of such contaminated soils in the site are still lacking. The sorption plays an important role in controlling the migration of contaminant, influenced by many parameters, such as pH, clay content, and organic matter (OM) [8]. As a result, the batch sorption experiment was carried out in order to elucidate sorption behavior of the contaminated soil in the study area.

2. Methodology

2.1. Sample collection
The topographic map of study area was derived from the Royal Thai Survey Department (RTSD) in 2006. Cadmium concentration map [7], and types of soil map derived from the Land Development Department (LDD) in 1988 were together considered in order to assign the sampling point in the study area. After soil sampling spot was selected, field observation was done to collect the soil sample. Before collecting soil samples, top soil was removed to the desired depth. Then, based on LDD soil sampling method, the soil sample was collected by mixing 1 kg of 5 sub-sampling spots within 10 m of the radius in sampling spot in order to get soil sample as the representative soil in the interesting spot. all of the sampling spots were collected at particular depth around 15 cm depth from the ground surface. After that, this sample was dried at 60° C and sieved through 2 mm sieve.

2.2. Soil properties
The proportion of sand, silt, and clay of the soil sample in the study area was analyzed by using hydrometer. Based on ASTM 422 method, soil particle analysis was carried out. 50 g of soil sample of < 2 mm grain size, was put into 1,000 cm^3 deionized water with the dispersal agent (4% Sodium Hexametaphosphate) in the plastic cylinder. The solution was shaken to make soil suspension well mixed, and then placed the 152H hydrometer in to the solution. The measurement of hydrometer was recorded at specific time periods as follows: 8, 15, 30 min and 1, 2, 4, 8, 16, 24, 36, 50 hr in order to analyze for soil particles size [9].
For organic matter measurement, the Walkley and Black procedure was done. One gram of soil sample was mixed with 10 ml of 0.1 N Potassium Dichromate ($K_2Cr_2O_7$) and 15 ml of concentrated sulfuric acid, and leave it for 30 min After that, the solution was filled with 50 ml deionized water and cooled down the solution temperature. Five drops of O-phenantholine were added into the solution, then tritrate the solution with ammonium iron (II) sulfate hexahydrate ($Fe(NH_4)_2(SO_4) \cdot 6H_2O$; FAS) to finally determine organic matter [10].

2.3. Batch sorption experiment
In this study, the batch experiment was performed in order to investigate adsorption coefficients, which can describe sorption behavior of the soil sample. One gram of soil sample was put into a

polypropylene bottle with 20 ml of Cd solution, which varied the Cd concentration from 0 to 200 ppm (0, 25, 50, 75, 100, 150, 200 ppm), based on our preliminary of this soil type. Moreover, the pH value of soil solution was adjusted to 7 by using NaOH and HNO_3 and ionic strength was adjusted to 1-5 mM by adding $CaCl_2$ in order to adjust solution condition as same as surface water [11]. Then polypropylene bottles were shaken by orbital shaker with 200 rpm at room temperature for 24 hours to let Cd sorption reaches an equilibrium condition. After that, soil solution was filtered by using Whatman no. 42 filtered paper and analyzed for residue Cd concentration from filtered solution by using AAS (Analytic Jena).

In order to explain sorption behavior of soils, these sorption isotherms were applied to describe sorption behavior as follows: Linear isotherm, Freundlich isotherm, and Langmuir isotherm. The calculation of these isotherms is showed in Eqs. 1-3, respectively.

$$Q_e = K_d C \tag{1}$$

$$Q_e = K_f C_e^{1/n} \tag{2}$$

$$Q_e = \frac{Q_m K_L C_e}{1 + K_L C_e} \tag{3}$$

Where Q_e is the concentration that sorbed in sorbent (mg kg^{-1}), Q_m is maximum sorption capacity (mg g^{-1}), C_e is the concentration of Cd in solution at equilibrium (mg l^{-1}), K_f is Freundlich sorption coefficient (L g^{-1}), K_L is Langmuir sorption coefficient (L g^{-1}), K_d is linear coefficient (L g^{-1}), n is Freundlich exponent (-).

Sorption behavior of contaminated soil can be defined by the fitting graph with the sorption isotherm. If the graph is fitted well with Freundlich isotherm, the sorption behavior of contaminated soil can be defined as multilayer sorption of ions on a heterogeneous surface. If the graph can be fitted well with Langmuir isotherm, sorption behavior of contaminated soil can be described as monolayer sorption on mono layer surface [12]-[14]. Moreover, the linear sorption isotherm is Freundlich sorption isotherm that Freundlich exponent (n) equivalent to 1 [15].

(a)

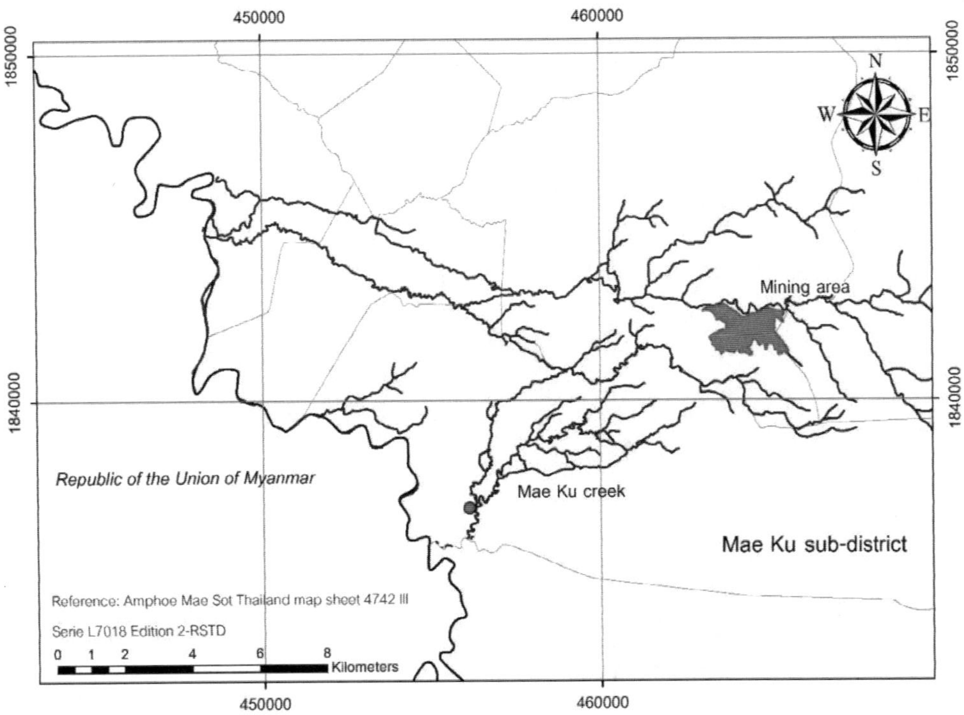

(b)

Figure 1. Soil sampling point in Mae Ku Creek, Tak Province, Thailand

3. Result and discussion

3.1. Study area

The soil sample was taken in Mae Ku Creek, Tak province within a range of UTM eastings of 455000 to 460000 and a range of UTM northings of 1835000 to 1840000 as shown in Figure 1. The soil type in the sampling point was classified as Mae Rim series by LDD.

3.2. Soil properties

Physico-chemical properties of the soil from the sampling spot are shown in Table 1.

Table 1. Physico-chemical properties of the soil

Sand (%)	Silt (%)	Clay (%)	OM (%)	pH
26	16	58	2.93	7.7

The soil collected can be classified as a clay soil by USDA soil triangle. The soil pH is 7.7, which is slightly alkaline and the percentage of organic matter in the soil sample is 2.93.

3.3. Batch sorption experiment

Sorption of Cd concentration in the sampling soil is shown in Table 2.

Table 2. Sorption and percent sorption of soil sample

Initial Cd concentration (mg L^{-1})	Cd Adsorption (mg g^{-1})	Adsorption (%)
nd.[a]	nd.	-
21.43	0.43	99.95

44.42	0.89	99.72
68.05	1.35	99.49
88.82	1.76	98.95
128.50	2.53	98.36
173.50	3.42	98.42

[a] nd: not detected

Cd sorption in soil (Q_e) and residue Cd in solution (C_e) from the experiment were plotted and fitted by Freundlich isotherm, Langmuir isotherm, and Linear isotherm in order to get sorption behavior of the soil. The result from the experiment was shown in Figure 2 and Table 2.

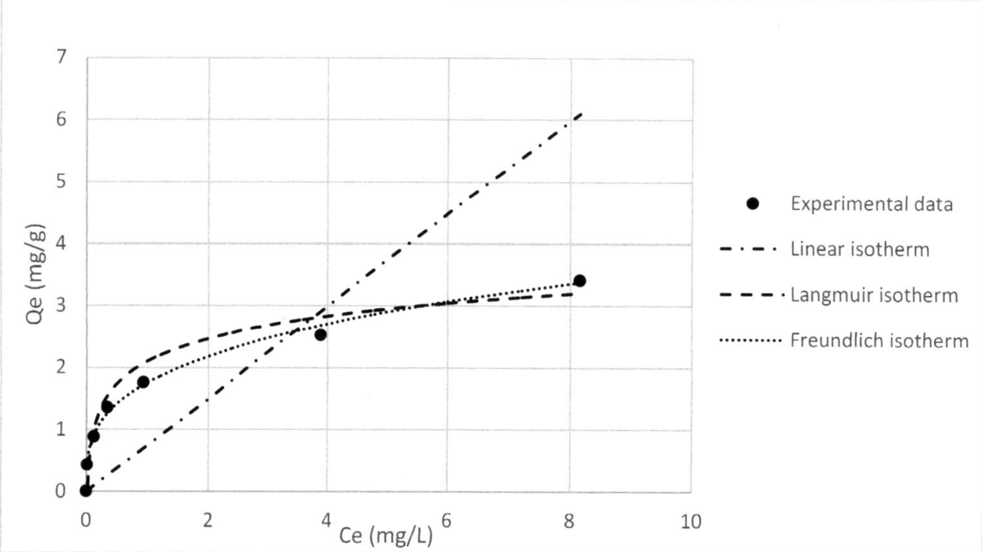

Figure 2. showing experimental data fitted with sorption isotherms

Table 3. Fitted parameters of sorption isotherms

Linear isotherm	Langmuir isotherm	Freundlich isotherm
K_d = 0.745 (L g^{-1})	K_L = 1.730 (L g^{-1})	K_f = 1.760 (L g^{-1})
	Q_m = 3.470 (mg g^{-1})	1/n = 0.312 (-)
R^2 = 0.2245	R^2 = 0.9747	R^2 = 0.9964

As shown in Table 3, it shows sorption of the soil sample in Mae Ku Creek is well fitted with Freundlich isotherm (R^2 = 9964.0), which can be described that Cd ions sorb on the soil surface as multilayer sorption [16].

3.4. Discussion

The result showed that the soil sample has high capability to sorb Cd as shown in Table 2. The sorption ability of the soil sample still exceeds 98 percent, even the initial concentration of Cd is 173.5 mg L^{-1}. The sorption behavior can be described by Freundlich sorption isotherm which is multilayer sorption [17]. Moreover, the sorption behavior of the soil sample has the same sorption behavior as Waleeithikul [17] who also study the sorption behavior of soil in this study area. As compared to Waleeithikul [17], in this study the sample has the percentage of clay higher than that of the previous study, consisting of 64.5% sand, 23.2% silt, and 12.3% clay, classified as a sandy loam soil by USDA. As a result, K_f of the soil sample was 3 times higher than that of the sandy loam soil (0.571 L g^{-1}), while 1/n is lower than that of the sample, indicating that the more preferential sorption of Cd^{2+} onto the clay soil. However, the values of 1/n of both soils were lower than 1, implying that sorption was favorable.

As compared to Karak et al. [18] who studied Cd sorption in a clay soil that has proportion of 18% sand, 34% silt, and 48% clay. K_f of the sample is 7.264 with 1/n equal to 0.466 for Freundlich isotherm (R^2 = 0.965), and K_L equal to 0.027 and Q_m is 84.03 mg g^{-1} (R^2 = 0.844). The results from Freundlich isotherm has higher K_f than the soil sample in the study area (see Table 3). Similarly, both studies showed that sorption behavior can be well fitted by Freundlich sorption isotherm.

4. Conclusion

The study area is located in Mae Ku Creek, Tak Province, that has Cd contamination problem. The soil sample in the study area has 26% sand, 16% silt, and 58% clay with 2.73% OM and soil pH of 7.7. The batch experiment was performed in order to understand sorption behavior of the soil. The batch sorption experiment for the clay soil can be well fitted with Freundlich isotherm (R^2 = 0.9964). As known that the sorption characteristics play an important role in controlling the migration of contaminants, especially heavy metals, therefore, the finding could be used as the representative sorpion of Cd onto clay soil. Moreover, it can be further used for evaluate the groundwater vulnerability map and selected the proper site remediation in the future.

5. Acknowledgement

Thank you International Postgraduate Program in Hazardous Substance and Environmental Management, Center of Excellence on Hazardous Substance Management (HSM), Chulalongkorn University, the International Research Integration: Chula Research Scholar program, the Ratchadaphiseksomphot Endowment Fund (GCURS-59-06-79-01), the Ratchadaphiseksomphot Endowment Fund of Chulalongkorn University (CU-59-057-CC), the Office of Higher Education Commission (OHEC) and the S&T Postgraduate Education and Research Development Office (PERDO) for the financial support of the Research Program and to the Ratchadaphiseksomphot Endowment Fund, Chulalongkorn University for funding the Research Unit. for financial support. Thank you for the Department of Geology, Faculty of science, Chulalongkorn University, also, HSM for facilities and scientific support. Moreover, thank you Mr. Narongsak Rachukan, Mr. Tewanopparit Parkchai, and Ms. Wanlapa Wisittammasri a huge assistant in field excursion. Thank you Asst. Prof. Dr. Chantra Thongcumpou, Ms. Chantana intim and Ms. Chanaya Permchati for teaching, helping, and many advices during laboratory work in HSM. Thank you, my friends and my seniors, for some advice and encouragement during laboratory work.

6. References

[1] Järup L 2003 Hazards of heavy metal contamination *British Medical Bulletin* **68**(1) 167-82
[2] Information on http://www.who.int/ipcs/features/cadmium.pdf?ua=1
[3] Chotpantarat S, Wongsasuluk P, Siriwong W, Borjan M and Robson M 2014 Non-Carcinogenic Hazard Maps of Heavy Metal Contamination in Shallow Groundwater for Adult and Aging Populations at an Agricultural Area in Northeastern Thailand *Human and Ecological Risk Assessment: An International Journal* **20**(3) 689-703
[4] Wongsasuluk P, Chotpantarat S, Siriwong W and Robson M 2017 Using urine as a biomarker in human exposure risk associated with arsenic and other heavy metals contaminating drinking groundwater in intensively agricultural areas of Thailand *Environmental Geochemistry and Health* 1-26
[5] Simmons RW, Pongsakul P, Saiyasitpanich D and Klinphoklap S 2005 Elevated levels of cadmium and zinc in paddy soils and elevated levels of cadmium in rice grain downstream of a zinc mineralized area in Thailand: implications for public health *Environmental Geochemistry and Health* **27**(5-6) 501-11
[6] Kosolsaksakul P, Farmer J G, Oliver I W and Graham M C 2014 Geochemical associations and availability of cadmium (Cd) in a paddy field system, northwestern Thailand *Environmental Pollution* **187** 153-61
[7] Akkajit P and Tongcumpou C 2010 Fractionation of metals in cadmium contaminated soil: Relation and effect on bioavailable cadmium *Geoderma* **156**(3-4) 126-32

[8] Tahervand S and Jalali M 2016 Sorption, desorption, and speciation of Cd, Ni, and Fe by four calcareous soils as affected by pH *Environmental Monitoring and Assessment* **188**(6) 322
[9] ASTM D422-63(2007)e2 Standard Test Method for Particle-Size Analysis of Soils (Withdrawn 2016) ASTM International West Conshohocken PA 2007 www.astm.org
[10] Walkley A and Black I A 1934 An examination of Degtjareff method for determining soil organic matter and a proposed modification of the chromic acid titration method *Soil Science* **37**(1) 29-38
[11] Information on http://www.aqion.de/site/69
[12] Langmuir I 1917 The constitution and fundamental properties of solids and liquids *Journal of the Franklin Institute* **183**(1) 102-5
[13] Saadi R, Saadi Z, Fazaeli R and Fard N E 2015 Monolayer and multilayer adsorption isotherm models for sorption from aqueous media *Korean Journal of Chemical Engineering* **32**(5) 787-99
[14] Wikiniyadhanee R, Chotpantarat S and Ong S K 2015 Effects of kaolinite colloids on Cd 2+ transport through saturated sand under varying ionic strength conditions: Column experiments and modeling approaches *J. Contam. Hydrol.* **182** 146-56
[15] Information on https://wwwbrr.cr.usgs.gov/projects/GW_Unsat/vs2di/hlp/solute/linearIsotherm.html
[16] Limousin G, Gaudet J P, Charlet L, Szenknect S, Barthès V and Krimissa M 2007 Sorption isotherms: A review on physical bases, modeling and measurement *Applied Geochemistry* **22**(2) 249-75
[17] Waleeittikul A 2016 Effects of Ionic Strength and Pore Water Velocity on Cadmium Mobility Through Contaminated Soil in Mae Sot District, Tak Province Bangkok Chulalongkorn University 31-66
[18] Karak T, Paul R K, Das D K, Boruah R K, and Sonar I 2014 Thermodynamics of cadmium sorption on different soils of West Bengal, India. *Scientific World Journal* **2014** 216451

IOP Conf. Series: Earth and Environmental Science **150** (2018) 012010 doi:10.1088/1755-1315/150/1/012010

Recovery of phosphorus compounds from thermally-processed wastes

A Czechowska-Kosacka[1], L Pawłowski[1], G Niedbala[2] and W Cel[3]

1 Institute of Environmental Protection Engineering, Faculty of Environmental Engineering, Lublin University of Technology, 40B Nadbystrzycka Street, Lublin 20-618, Poland
2 Institute of Biosystems Engineering, Faculty of Agronomy and Bioengineering, Poznań University of Life Sciences, 50 WojskaPolskiego 5Street, 60-627 Poznań 60-627, Poland
3 Institute of Renewable Energy Engineering, Faculty of Environmental Engineering, Lublin University of Technology, 40B Nadbystrzycka Street, Lublin 20-618, Poland
E-mail: a.czechowska-kosacka@pollub.pl

Abstract. Depletion of phosphorus deposits is one of the most serious global problems, which may soon lead to a crisis in food production. It is estimated that if the current living standard is maintained, the available reserves will be depleted in 130 years. Considering the principle of sustainable development, searching for alternative phosphorus sources is extremely important. The work presented the results of the research on the possibility of utilizing wastes as a source of phosphorus. The studies were conducted on poultry manure. The physicochemical properties of phosporus-rich wastes were determined as well. The fertilizing properties of ashes from poultry manure combustion – obtained from different systems, i.e. caged and barn production. The assimilability of phosphorus from the obtained ashes was determined. Potential applications of phosphorus-rich ashes were proposed as well.

1. Introduction

Constantly increasing world population, industrial development, and the amount of migrating chemical substances in the environment contribute to both local and global changes [1]. Therefore, the world may soon face the problems of overpopulation, drinking water shortage, as well as deficit of energy and natural resources. Further development of civilization will depend on the possibility of restoring the balance between the economic development and the protection of natural resources [2], [3].

For many decades, the production processes were carried out without consideration for their impact on the natural environment. This resulted in the destruction of the environment, which enforced a change in the approach towards the issues of environmental protection. The actions involving the control of pollutants discharged from the industry, including wastewater treatment plant, disposal of waste and treatment of emitted gases, enabled to mitigate the degradation of environment. However, this strategy proved to be insufficient. Limiting the availability of non-renewable energy resources became the reason for changing the environmental protection strategy and the development of cleaner methods of production.

In accordance with the principles of cleaner production, wastes should be treated as a source of raw materials for the manufacturing of new, complete products. Examples include municipal sewage sludge and livestock production wastes [4].

Phosphorus is an element of great biological significance. Due to the minimal volatility and low solubility of phosphorus compounds, its availability for the living organisms is limited [1], [5]. Phosphorus is a structural element of plants as well as all living tissues and a nutrient for all organisms [6]-[9]. The growing population, and thus the increasing food production, results in a rising demand for phosphorus fertilizers and fodder additions.
In the developed countries, the phosphorus fertilizers constitute the greatest share as far as the use of phosphorus compounds is concerned. Combined with food and fodder additions, they account for 85% of total consumption. The remaining groups include detergents, constituting 12% and preparations for phosphate conversion coating and rust protection [10], [11].
The constantly growing demand for phosphorus compounds contributes to the exploitation of natural deposits of apatites and phosphorites. The volcanic deposits containing approximately 13% phosphorus, mainly composed of fluorapatite as well as the sedimentary deposits with about 87% phosphorus, predominantly comprising francolite, are economically important.
From the geological point of view, phosphorus is abundant; however, its usefulness is determined by the cost of extraction and processing of phosphate ore. The forecasts prepared at the end of the previous century indicate that over 50% of the currently exploited phosphorus resources will be depleted in the next 60-70 years. This tendency will surely contribute to a hike in the market price of this resource. Considering the current tendencies of phosphorus use, it may be estimated that the known reserves of this resource will be depleted in 130 years.
Taking into account the principles of sustainable development, alternative sources of phosphorus should be sought. This especially pertains to the possibility of its recovery and recycling from wastes, the most important of which include the recovery of phosphorus from municipal and industrial wastewater, sewage sludge and the use of phosphorus from poultry manure.
The aim of this work is to determine the possibility of utilizing ash from manure combustion as a potential source of phosphorus.

2. Materials and methods

2.1. Materials

Investigation of the poultry manure obtained from the caged and barn production systems was conducted for two series.
The employed production system has a significant impact on the quality, composition and properties of manure. The manure utilized in the studies originated from industrial poultry farms located in the Lubelskie voivodeship. The manure from the caged and barn production systems were investigated in the research. The composition of poultry manure is dependent on numerous factors, including bird's age, weight, fodder type and the amount of supplied water. This residue mainly comprises bird's droppings, feathers and scattered fodder. Two types of production systems were utilized in the study:

- caged production system, in which poultry is kept in cages individually supplied with fodder and water,
- barn production system, in which the birds are kept on litter in pens.

The caged production system restricts the movement of birds; hence, the consumption of fodder is reduced by 10-15% in comparison to the free range system.

2.2. Methods

In order to indicate the main and trace components in the considered materials, it was necessary to transform them into liquid form through mineralization with concentrated nitric acid.
The scope and analytic methods employed in the study:

- dry mass content in the wastes was determined with the gravimetric method. Drying of wastes was conducted by means of KC-65-M laboratory dryer at 105^0C (379K),
- heat of combustion and calorific values of sewage sludge were measured in accordance with the Polish PN-81/G-04513 standard. The studies involved the use of KL-11 colorimeter manufactured by Miakdo,

- sewage sludge was incinerated in SM-046 muffle furnace produced by Czylok-Sel (Figure 1) in 600 and 750^{0}C. The pre-dried poultry manure samples were subjected to incineration,
- mineralization with nitric acid was conducted in Multiwave 3000 manufactured by Anton Paar,
- the elemental composition of poultry manure ashes was indicated by means of ICP-EOS manufactured by JOBIN YVON,
- the content of calcium in manure was indicated by means of complexometric titration with sodium hydroxide,
- the contents and phosphorus and the phosphorus dissolved in citric acid were studied in accordance with the standard for conducting investigations on fertilizers.

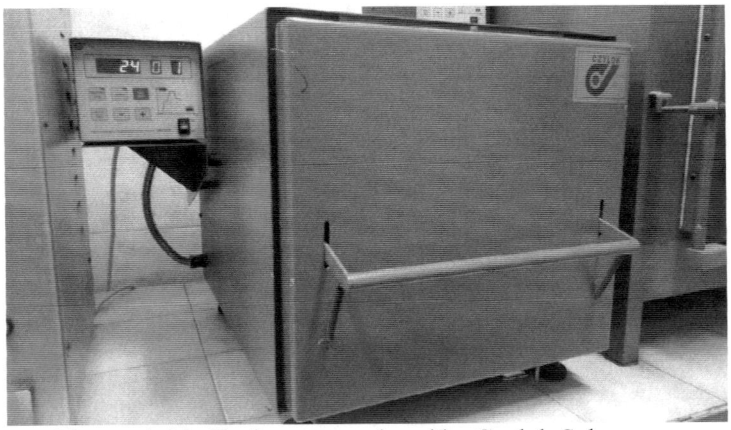

Figure 1. Muffle furnace produced by Czylok-Sel

3. Results and discussion

The collected manure was dried until dry mass was obtained. The moisture content in manure amounted to 53-55% for caged poultry and 52-61% for the poultry farmed in barn production system. The dried manure was ground and divided into two parts; the calorific value was determined for one part, whereas the other was combusted.

The mean calorific value of manure amounted to 10 000 kJ/kg for poultry farmed without litter and 12 500 kJ/kg for the poultry farmed with litter. The obtained results for poultry manure from the barn production system exhibited the calorific value which was 2500 kJ/kg higher. This difference was caused by an additional straw content in the manure. Manure was combusted in 600 and 750°C. The mass loss for manure I and II reached 58.2% and 67.4%, respectively, mainly connected with the loss of organic mass. The conducted chemical analyses on the phosphorus and calcium content in the ash from the combustion of poultry manure from caged and barn production systems were included in Table 1.

While analysing the data from Table 1, a slight increase in the percentage of total P_2O_5 and P_2O_5 dissolved in citric acid was observed for the production system with litter, as well as a slight drop in the percentage of calcium.

Table 1. Composition of poultry manure from caged and barn production systems in air-dried state.

Component	Manure I	Manure II	Manure I	Manure II
	Without litter		With litter	
Total P_2O_5 [%]	4.8	5.1	5.4	6.1
P_2O_5 dissolved in 2% citric acid [%]	1.9	2.4	3.3	4.1
Ca%	15.8	10.9	8.6	10.9

Table 2. Composition of ashes from the combustion of poultry manure from caged and barn production systems.

Ash		P_2O_5 content (%)	Ca content (%)
		Caged production system	
Manure I	600°C	12.2	32.5
	750°C	13.1	38.4
Manure II	600°C	12.5	34.2
	750°C	13.6	40.9
		Barn production system	
Manure I	600°C	8.5	29.3
	750°C	12,1	25.1
Manure II	600°C	12.5	36.3
	750°C	14.2	41.1

Table 2 contains the results of determining the composition of ash from the combustion of poultry manure depending on the combustion temperature and production system. The analyses found in the table indicate a slight but clear increasing tendency within a given type of manure, regardless of production system. The same tendency is exhibited by the calcium percentage.
The conducted analyses of ash from the calcination of the manure ash from the caged and barn production systems indicate an increase in the content of certain analyzed elements, i.e. As, Cd, and Ni. In the case of As, a significant increase in the concentration was noted from 4.2 ppm (caged production system) to 682 ppm (barn production system), Cd: 0.72 ppm (caged production system) to 382 pp, (barn production system) and Ni: in over twice as high concentration than in the case of the caged production system (Table 3). It may be purported that large differences in the concentration of As, Cd, and Ni in the case of the barn production system stem from the application of straw, which increased the concentration of the enumerated metals. The type of utilized feed is irrelevant, because all the birds were fed with the same type of fodder. Slight differences could only result from the varying batch number.

Table 3. Ash composition obtained from the calcination of manure I ash from the caged and barn production systems at 600°C.

Element	Unit	Concentration: Caged production system	Concentration: Barn production system
Al	%	0.41	0,31
As	ppm	4.2	682
B	ppm	102	95
Ba	ppm	76	74
Cd	ppm	0.72	382
Cr	ppm	25	33
Cu	ppm	231	241
Fe	%	0.48	0.53
Hg	ppb	2.21	7

K	%	4.5	4.4
Mg	%	1.3	1.4
Na	%	0.5	0.88
Ni	ppm	43	109
Pb	ppm	-	-
Sr	ppm	326	332
Zn	%	0.115	0.11

However, the chemical analysis of litter and utilized feed were not studied, due to the unavailability of data.

The analysis of data in Figure 2 indicates that along with the rising calcination temperature of poultry manure ash from caged production system, a decrease in the ash assimilability occurs. It should also be noted that it is much lower than in the case of ash subjected exclusively to drying.

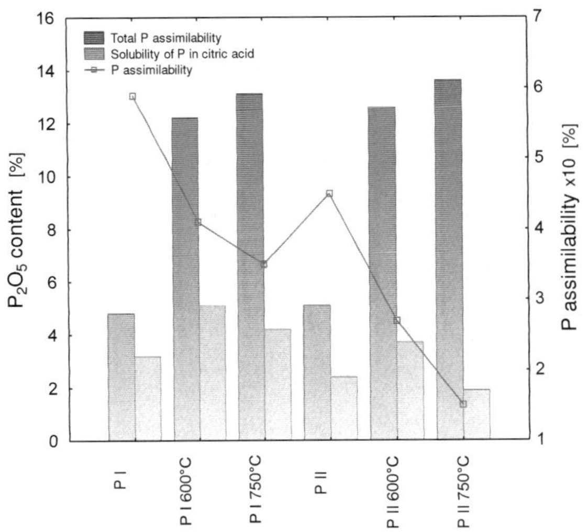

Figure 2. Impact of calcination temperature of poultry manure ash from caged production system on the total content and assimilability of phosphorus.

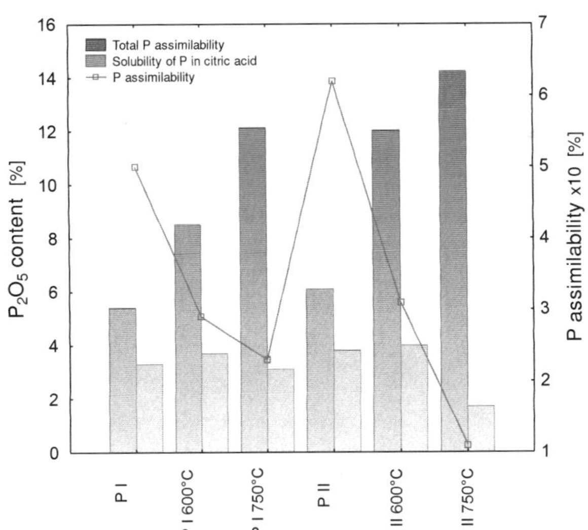

Figure 3. Impact of calcination temperature of poultry manure ash from barn production system on the total content and assimilability of phosphorus.

A similar tendency is exhibited by the poultry manure ash from barn production system. Its assimilability is much lower than in the dried manure (Figure3).

Similar analysis results were obtained by Wzorek [12], both in the case of the system with and without litter.

Therefore, direct use of non-processed poultry manure should also be considered due to the introduction of nitrogen and organic elements into soil, which are lost in the course of high-temperature processes. Therefore, the direct application of manure as a fertilizer seems preferable to their reduction to ashes.

Reduction of ash mass in relation to the original amount of manure, possibility of energy production from its combustion and removal of pathogenic organisms constitute the advantages of thermal processing.

4. Conclusion

Thus far, studies focused on the acquisition of phosphorus from the municipal and animal sewage. The conducted research showed that poultry manure may constitute a compost component, which provides valuable mineral and organic elements to soil, mainly nitrogen and calcium, as well as phosphorus, the primary sources of which are rapidly being depleted.

Management of manure is also connected with certain odour nuisance. It should be remembered that it may contain pharmaceutical drugs employed in industrial poultry farms and thus constitute a biological threat. Therefore, thermal utilization should be considered. Combustion of waste ensures a reduction in its mass by approximately 80%. High temperature eliminates all biological threats and odour nuisance.

As shown in the research, the temperature required for elimination of the organic fraction does not exceed 600°C. The obtained ashes are characterized by a low content of phosphorus, slightly exceeding 14% P_2O_5 which can subsequently be used as a fertilizer. The assimilability of phosphorus from ashes is much lower than in the case of dried manure.

Poultry manure should not be incinerated on their own due to high calcium content in relation to phosphorus. Co-incineration of poultry manure with other types of waste, e.g. sewage sludge, which also contain substantial amounts of phosphorus, appears to be advantageous. The resultant product may serve as a source of phosphorus for the chemical industry.

5. References

[1] O'Neill P 1997 Chemia środowiska Warszawa-Wrocław
[2] *Report of UN Conference on Environment and Development* 1992 Rio de Janeiro
[3] *Global 2000. Report to the President* 1980 Government Printing Office Washington DC
[4] Richards I R and Johanston A E 2001 *The effectiveness of different precipitated phosphates as sources of phosphorus for plants* Report on undertaken for CEEP, EFMA (European Fertilisier Manufacturers Association), Anglian Water UK and Berlin Wasser Betriebe, December.
[5] Toy A D F 1973 The chemisry of phosphorus vol.3 Pergamon Texts in Inorganic Chemistry, Oxford-New York-Toronto-Sydney-Paris-Braunschweig
[6] Acelas N Y, Lopez D P, Brilman D W F (W i m), Kersten S R A and Kootstra M J 2014 *Bioresource Technol.* 174 pp 167-175
[7] Blocher Ch, Niewersch C and Melin T, 2012 *Water Res.*46 pp 2009-2019
[8] Li R, Yin J, Wan g W, Li Y and Zhan g Z 2014 *Waste Manage.* 34 pp 1211-1216
[9] Xu H, He P, Gu W, Wang G and Shao L 2012 *J. Environ. Sci.* 24(8) pp 1533-1538
[10] Steen I 1998 *Phosphorus and Potassium* 27 pp 25-31
[11] Górecki H 2003 *Chemistry for Agriculture - Chemicals in Sustainable Agriculture* 4 pp1-16
[12] Wzorek Z 2008 *Inżynieria i Technologia Chemiczna* Cracov

Acknowledgments

This work was partially funded by the Polish Ministry of Science and Higher Education (research funding S-13/WIS/2017).

IOP Conf. Series: Earth and Environmental Science **150** (2018) 012011 doi:10.1088/1755-1315/150/1/012011

Mitigation of greenhouse gases emissions impact and their influence on terrestrial ecosystem.

K Wójcik Oliveira[1] and G Niedbała[2]

1 Faculty of Environmental Engineering, Lublin University of Technology, 40B Nadbystrzycka Street, 20-618 Lublin, Poland
2 Institute of Biosystems Engineering, Faculty of Agronomy and Bioengineering, Poznań University of Life Sciences, 50 Wojska Polskiego Street, 60-627 Poznań, Poland
E-mail: k.wojcik-oliveira@pollub.pl, gniewko.niedbala@gmail.com

Abstract. Nowadays, one of the most important challenges faced by the humanity in the current century is the increasing temperature on Earth, caused by a growing emission of greenhouse gases into the atmosphere. Terrestrial ecosystems, as an important component of the carbon cycle, play an important role in the sequestration of carbon, which is a chance to improve the balance of greenhouse gases. Increasing CO_2 absorption by terrestrial ecosystems is one way to reduce the atmospheric CO_2 emissions. Sequestration of CO_2 by terrestrial ecosystems is not yet fully utilized method of mitigating CO_2 emission to the atmosphere. Terrestrial ecosystems, especially forests, are essential for the regulation of CO_2 content in the atmosphere and more attention should be paid to seeking the natural processes of CO_2 sequestration.

1. Introduction

Increasing temperature on Earth, caused by a growing emission of greenhouse gases into the atmosphere, constitutes one of the most important challenges faced by the humanity in the current century [1]-[3]. Although the global warming is a natural phenomenon, the anthropogenic activity contributed to its aggravation, mainly due to an increase in the amount of combusted fossil fuels [1], as well as cement production, and changes in land use, which lead to a further increase of temperature on Earth [4]. In 2005, the total global CO_2 emission resulting from this process amounted to 26.4 billion tons [5]. Starting from the year 1750, the CO_2 emission is constantly on the rise, mainly due to the combustion processes of carbon-containing fuel, as well as due to the mining activity [6], [7].
Carbon sequestration is a growing research topic that addresses an important aspect of the overall strategy for carbon management to help mitigate the increasing emissions of CO_2 into the atmosphere. Terrestrial ecosystems, being recognized as an important component of the carbon cycle, have gained importance owing to its potential to sequester carbon [8].
Apart from the obvious emission reduction, sequestration of carbon dioxide is a chance to improve the balance of greenhouse gases. This term denotes a number of actions aimed at mitigating the emission of carbon dioxide into the atmosphere. This process, in great simplification, consists of consecutive actions: capturing and separating carbon dioxide from exhaust fumes and storing CO_2 in a particular location [5].
According to Raj and Venkata [9], carbon sequestration is the process of capture and long-term storage of atmospheric carbon dioxide (CO_2). We can distinguish between storing the carbon dioxide

in plants and soil (Terrestrial Sequestration), underground (Geological Sequestration) and deep in Ocean (Ocean Sequestration).

Sequestration by terrestrial ecosystems remains one of the under-appreciated methods of CO_2 sequestration. According to the data from IPCC reports, the annual emission of CO_2 from the anthropogenic sources approximates 9.8 billion tons, in C equivalent, or 35.9 Mg tons in CO_2 equivalent. On the other hand, the annual absorption by terrestrial plants in the course of photosynthesis roughly amounts to 123 billion tons, in C equivalent or 451 billion tons in CO_2 equivalent. This indicates that an increasing CO_2 absorption in the course of photosynthesis by 12.5% would neutralize the emission of CO_2 from the anthropogenic sources [4].

2. Flux of CO_2 in the ecosystems of Earth

According to the literature [10] (published by Le Quéré et al. in 2015) the mean emission of CO_2 into the atmosphere constantly increases by 3.1 in 1960 to 9.8 GtC/year in 2014. Two sources are mainly responsible for this phenomenon: combustion of fossil fuels (see Table 1) and cement production.

Table 1. CO_2 emissions from fossil fuel combustion. [10]

Years / Emission	1960-69	1970-79	1980-89	1990-99	2000-09	2005-14	2014
Emission [GtC/year]	3.1±0.2	4.7±0.2	5.5±0.3	6.4±0.3	7.8±0.4	9.0±0.5	9.8±0.5
Emission [$GtCO_2$ /year]	11.4±0.7	17.2±0.7	20.2±1.1	23.5±1.1	28.6±1.5	33±1.8	35.9±1.8

Table 2. CO_2 emissions caused by changes in land use, mainly including deforestation and drying of marshes [10].

Years	1960-69	1970-79	1980-89	1990-99	2000-2009	2005-2014	2014
Emission [GtC/year]	1.5±0.5	1.3±0.5	1.4±0.5	1.6±0.5	1.0±0.5	0.9±0.5	1.1±0.5
Emission [$GtCO_2$/ year]	5.5±1.8	4.8±1.8	5.1±1.8	5.9±1.8	3.7±1.8	3.3±1.8	4.0±1.8

Apart from the industrial emissions, the changes in land use are another important source of CO_2 emissions. These include drying of marshes, deforestation and volcanic eruptions (Table 2).

Simultaneously, an increase in the absorption of this compound by the terrestrial ecosystems is observed, ranging from 1.7 GtC/year in 1960 to 4.1 GtC/year in 2014.

The rising CO_2 concentration in the atmosphere, in conjunction with nitrogen oxides of anthropogenic origin precipitating onto the earth, strengthens the fertilization effect. The direct outcome of this phenomenon is an observed increase in the absorption of CO_2 by the terrestrial ecosystems. The ocean waters constitute another absorber of CO_2. Oceans absorb 92 GtC/year and release 90 GtC/year at the same time [11]. The amount of CO_2 that is absorbed by the oceans depends on various processes, including biological activity, ocean currents and water temperatures.

The comparison of CO_2 emissions into the atmosphere from the anthropogenic sources (9.8 GtTC/year in 2014) and absorption of CO_2 by the terrestrial ecosystems clearly indicates that an increase in the absorption of this compound by the terrestrial plants by 8.2% would inhibit its rising content in the atmosphere [12], [13].

Therefore, appropriate control of CO_2 by the terrestrial ecosystems is purposeful and will result in the inhibition of CO_2 concentration increase in the atmosphere.

3. Role of Earth ecosystems in the regulation of CO_2 cycle

The terrestrial ecosystems play a major role in regulating the CO_2 cycle [14]. According to the data published by Trumper et al., the greatest amount of CO_2, i.e. 547.8 GtC, was accumulated in the tropical and subtropical forests as well as peatlands [15]. Tropical rainforests constitute one of the largest ecosystems of the Earth. In total, they absorb 1.3 Gt (net) of carbon per year. This includes 0.6 Gt of carbon in the Central and South America, 0.4 Gt of carbon in Africa and 0.25 Gt of carbon in Asia [16], [17].

It is estimated that approximately 6.5 to 14.8 million ha of tropical forests is being cut down, which additionally elevates the CO_2 emission by 0.8-2.2 Gt of carbon [18], [19].

Peatlands play an extremely important role in the carbon absorption. The estimated data indicates that peatlands contain 550 Gt of carbon, with 1450 t of carbon per hectare [20]. A rapid degradation of peatlands and their dewatering carried out in order to expand the area of arable lands, leads to the release of CO_2 into the atmosphere. It is estimated, that elimination of 6 million ha of peatlands additionally releases 0.5-0.8 Gt of carbon each year [21].

It should be noted that elimination of peatlands and cultivation of coconut palms in their place – in order to obtain biodiesel fuel – does not reduce but rather increases the emission of CO_2 into the atmosphere by as much as 3-9-fold.

The tropical and subtropical meadows, bushes and savannah constitute another important ecosystem containing 463.6 Gt of carbon. These areas are vulnerable to fires, which contribute to an annual emission of CO_2 by 0.5-4.2 Gt of carbon. Nevertheless, these ecosystems still absorb 0.5 Gt of carbon per year [15].

The meadows of temperate zones also play an essential role in the regulation of CO_2 fluxes in the Earth ecosystem. They contain 183.1 Gt of carbon, which is mainly (1333 t) accumulated in soil [22].

Another important ecosystem includes the forests in temperate zones, containing 314.9 Gt of carbon. These forests are characterized by a high content of carbon, ranging from 150 to 320 tons/ha. The forests in Europe currently absorb 7-12% of CO_2 emitted from the anthropogenic sources [21]. It should be expected that a gradual increase in the area occupied by forests and appropriate silviculture will make the European forests one of the ways of conducting CO_2 sequestration. The northern forests found in the Scandinavian countries, Alaska and Russia, contain substantial amounts of carbon, reaching approximately 384.2Gt. The low temperature characterizing these areas slows down the decomposition of dead biomass, which contributes to a high carbon content in soil ranging from 116 to 343 t per hectare [23].

The tundra found in the arctic regions, i.e. northern Canada, Scandinavian countries, Russia, Greenland and Iceland forms yet another ecosystem, which contains 155.4 Gt of carbon. Similarly to the northern forests, the decomposition of dead biomass in it is very slow. It is estimated that the permafrost found under a relatively thin layer of soil contains 1600 Gt of carbon [21]. Due to a constantly increasing global temperature, there are concerns that CO_2 and CH_4 contained in the permafrost will be released. It is estimated that 100 Gt of carbon in the form of CO_2 and CH_4 will be released into the atmosphere, thus raising the carbon concentration by as much as 47 ppm [24].

The performed analyses indicate that the terrestrial ecosystems constitute an enormous carbon reservoir. It contains approximately 2000 Gt of carbon and additionally absorbs 1.5 Gt of carbon (net) each year. It should be noted that the tropical forests play an extremely important role in the carbon balance. It is estimated that until 2100, carbon sequestration by the natural ecosystems will prevent an increase in the CO_2 concentration by 40-70 ppm.

Deforestation conducted in order to create agricultural lands leads not only to a decrease in CO_2 absorption by plants, but is also conducive to the emission of this compound, which results from the oxidation of organic carbon contained in soil – the largest reservoir of carbon. The IPCC report from 2014 states that the amount of carbon in the organic compounds contained in soil approximates 1580 Gt. The above-mentioned data indicates that the terrestrial ecosystems, forests in particular, play a key role in regulating the CO_2 content in the atmosphere.

4. Conclusion

By taking various actions, e.g. accumulating carbon in terrestrial carbon sinks such as plants, plant products and soils, the emission of CO_2 into the atmosphere may be significantly reduced, and thus contribute to the mitigation of climate changes. Appropriate soil and land management practices should be applied, which will increase the carbon stocks of land ecosystems and hence reduce the losses to the atmosphere. Increasing CO_2 absorption by terrestrial ecosystems (by intensifying photosynthesis) and consequently increasing plant biomass production, is one way to reduce the atmospheric CO_2 emissions.

5. References

[1] IPCC, 2007: Climate Change 2007: The Physical Science Basis. Contribution of Working Group I to the Fourth Assessment Report of the Intergovernmental Panel on Climate Change [Solomon, S., D. Qin, M. Manning, Z. Chen, M. Marquis, K.B. Averyt, M.Tignor and H.L. Miller (eds.)]. Cambridge University Press, Cambridge, United Kingdom and New York, NY, USA.

[2] IPCC, 2013: *Climate Change 2013: The Physical Science Basis. Contribution of Working Group I to the Fifth Assessment Report of the Intergovernmental Panel on Climate Change* [Stocker, T.F., D. Qin, G.-K. Plattner, M. Tignor, S.K. Allen, J. Boschung, A. Nauels, Y. Xia, V. Bex and P.M. Midgley (eds.)]. Cambridge University Press, Cambridge, United Kingdom and New York, NY, USA.

[3] Soussana J F et al. (2007). Full Accounting of the Greenhouse Gas (CO_2, NO_2, CH_4). Budget of Nine European Grassland Sites. *Agriculture, Ecosystems, and Environment*,121, 121-134.

[4] Pawłowski A, Pawłowska M and Pawłowski L 2017, *Ecol. Chem. Eng.* S. 24 2, 213-221

[5] Mądrawski J, Ziemblińska K, Juszczak R, Zawal D and Olejnik J. 2013 *Rocz. Ochr. Sr.* 15, 2526–2545.

[6] Czechowska-Kosacka A, Lubańska Z and Dragan P, 2016, Natural Processes in Mitigation of CO_2 and CH_4 Emission, *Rocz. Ochr. Sr* 18, 1039-1048

[7] Cel W, Czechowska-Kosacka A and Zhang T, 2016, Sustainable mitigation of greenhouse gases emissions. *Problems of Sust. Develop.* 11(1), 173-176.

[8] Sheikh A Q, Skinder B M, Pandit A K and Ganai B A 2014 Terrestrial Carbon Sequestration as a Climate Change Mitigation Activity. *J Pollut Eff Cont* 2,110

[9] Raj K K V G and Venkata R M 2017 Study and Evaluation of Carbon Sequestration Using Remote Sensing and GIS: A Review on Various Techniques. *Intern. J. of Civ.Eng. and Techn.*, 8 4, 287–300

[10] Le Quéré C, Moriarty R, Andrew R. M, Peters G. P, Ciais P, Friedlingstein P, Jones S. D, Sitch S, Tans P, Arneth A, et.al, 2014, Global carbon budget 2014, *Earth Syst. Sci. Data* 7 47–85

[11] IPCC. 2014 Climate Change 2014. Impact, Adoption, and Vulnerability. Summary for Policymakers. 2014. www.ipcc.ch/report/ar5/wg2/.

[12] Beer C, Reichstein M, Tomelleri E, Ciais P, Jung M, Carvalhais N, Rödenbeck C, Arain MA, Baldocchi D, Bonan GB, Bondeau A, Cescatti A, Lasslop G, Lindroth A, Lomas M, Luyssaert S, Margolis H, Oleson KW, Roupsard O, Veenendaal E, Viovy N, Williams C, Woodward FI and Papale D 2010 *Science* 329 834-838

[13] Hilton T W, Davis K J and Keller K 2014 *Biogeosciences* 11 217-235

[14] Tans P P, Fung I Y and Takahashi T 1990 *Science* 247 1431-1438

[15] Trumper K, Bertzky M, Dickson B, van der Heijden G, Jenkins M and Manning P 2009 The Natural Fix? The Role of Ecosystems in Climate Mitigation. A UNEP rapid response assessment. United Naions Environemnt Programme, UNEP-WCMC, Cambridge, UK.

[16] Phillips O L and Lewis S L 2014 *Global Change Biology* 20 2039-2041

[17] Pan Y, Birdsey R A, Fang J, Houghton R, Kauppi P E, Kurz W A, Phillips O L, Shvidenko A, . Lewis S L, Canadell J G, Ciais P, Jackson R B, Pacala S W, McGuire A D, Piao S, Rautiainen A, Sitch S and Hayes D 2011 *Science* 333 988-993

[18] Post W M and Kwon K C 2000 *Global Change Biology* 6 317-328

[19] Yu Z. C 2011, *Eos* 92, 97-108

[20] Krüger J. P, Leifeld J, Glatzel S, Szidat S and Alewell C 2015 *Biogeosciences* 12 2861-2871

[21] Grace J, Jose S J, Meir P, Miranda H S and Montes R A 2006 *J. Biogeogr.* 33 387-400

[22] Goodale CL, Apps M J, Birdsey R A, Field C B, Heath L S, Houghton R A, Jenkins J C, Kohlmaier G H, Kurz W, Liu S, Nabuurs G J, Nilsson S and Shvidenko A Z 2002 *Ecological Applications* 12 891–899

[23] Schlesinger W H and Lichter J 2001, *Nature* 411 466-469

[24] Janssens IA, Freibauer A, Ciais P, Smith P, Nabuurs G-J, Folberth G, et al. 2003 Europe's terrestrial biosphere absorbs 7 to 12% of European anthropogenic CO2 emissions. *Science.* 300(5625),1538-1542.

IOP Conf. Series: Earth and Environmental Science **150** (2018) 012012 doi:10.1088/1755-1315/150/1/012012

Impact of saline aquifer water on surface and shallow pit corrosion of martensitic stainless steels during exposure to CO_2 environment (CCS)

Anja Pfennig[1] and Axel Kranzmann[2]

1 Dept. Mechanical Engineering, Applied Universtiy of Berlin HTW, Berlin, Germany
2 Federal Institute of Materials Research and Testing BAM, Berlin, Germany
E-mail: anja.pfennig@htw-berlin.de

Abstract. Pipe steels suitable for carbon capture and storage technology (CCS) require resistance against the corrosive environment of a potential CCS-site, e.g. heat, pressure, salinity of the aquifer, CO_2-partial pressure. Samples of different mild and high alloyed stainless injection-pipe steels partially heat treated: 42CrMo4, X20Cr13, X46Cr13, X35CrMo4 as well as X5CrNiCuNb16-4 were kept at T=60 °C and ambient pressure as well as p=100 bar for 700 h - 8000 h in a CO_2-saturated synthetic aquifer environment similar to possible geological on-shore CCS-sites in the northern German Basin. Main corrosion products are $FeCO_3$ and FeOOH. Corrosion rates obtained at 100 bar are generally much lower than those measured at ambient pressure. Highest surface corrosion rates are 0.8 mm/year for 42CrMo4 and lowest 0.01 mm/year for X5CrNiCuNb16-4 in the vapour phase at ambient pressure. At 100 bar the highest corrosion rates are 0.01 mm/year for 42CrMo4, X20Cr13 (liquid phase), X46Cr13 and less than 0.01 mm/year for X35CrMo4 and X5CrNiCuNb16-4 after 8000 h of exposure with no regard to atmosphere. Martensitic microstructure offers good corrosion resistance.

1. Introduction

During the compression of emission gasses e.g. from combustion processes of power plants into deep geological layers of a geological on-shore saline aquifer CCS-site (CCS Carbon Capture and Storage [1]), the corrosion of injection pipe steels may become an issue when CO_2 is compressed [2], [3]. CO_2-corrosion is sensitively dependent on alloy composition, contamination of alloy and media, temperature, CO_2 partial pressure, flow conditions and protective corrosion scales [4]-[10]). Suitable heat treatment [8], [11] and martensitic micro structure [12] decrease the pitting potential. Corrosion phenomena are reason for early failure of materials under cyclic load [13], [14].

Surface corrosion layers grow slowly and locally corroded samples usually show the same corrosion products as surface layers 7], acting as corrosion catalyser. Generally siderite $FeCO_3$ [2] is formed on steels exposed to CO_2-environment due to the low solubility of $FeCO_3$ in water (pKsp = 10.54 at 25 °C [14]]) leading to anodic iron dissolution. The initial formation of transient $Fe(OH)_2$ [6], [11] possibly leads to an increase of the local pH near the hydroxide film. As a consequence an internal and external ferrous carbonate film precipitates [14]. In this paper we are presenting extended and compiled data of former work [3], [7], [13], [15]-[18].

IOP Conf. Series: Earth and Environmental Science **150** (2018) 012012 doi:10.1088/1755-1315/150/1/012012

2. Experimental Procedure

2.1. Steels

Static corrosion tests at ambient pressure as well as at high pressure (100 bar) were carried out using samples of mild steels, martensitic and duplex stainless steels (Table 1-5):

1. AISI 420 (X20Cr13, 1.4021) (Table 1)
2. AISI 420C (X46Cr13, 1.4043) (Table 2)
3. No AISI (X35CrMo17, 1.4122) (Table 3)
4. AISI 630 (X5CrNiCuNb 16-4, 1.4542) (Table 4)
5. AISI A182 F51 (329LN)) SAF 2205 (X2 CrNiMoN 22-5-3 (UNS S31803) 1.4462) (Table 5)

In order to confirm the material's chemical composition, samples were analyzed via spark emission spectrometry SPEKTROLAB M and by Electron Probe Microanalyzer JXA8900-RLn (Table 1-5).

Table 1. Chemical composition of 1.4021 (X20Cr13, AISI 420, (in mass-%).

Elements	C	Si	Mn	P	S	Cr	Ni	Fe
	0.22	0.39	0.32	0.007	0.006	13.3	0.123	rest

Table 2. Chemical composition of 1.4043 (X46Cr13, AISI 420C), (in mass percent).

Elements	C	Si	Mn	P	S	Cr	Mo	Ni	Co	Fe
	0.46	0.25	0.45	0.018	0.003	13.39	0.03	0.13	0.03	85.4

Table 3. Chemical composition of 1.4122 (X35CrMo17), (in mass percent).

Elements	C	Si	Mn	P	S
acc standard [a]	0.33-0.45	<1.00	≤ 1.00	≤ 0.045	≤ 0.03

Elements	Cr	Mo	Ni	Co	Fe
acc standard [a]	15.5 - 17.5	0.8-1.3	≤ 1.00		0.20 - 0.45

a) elements as specified according to DIN EN 10088-3 in %

Table 4. Chemical composition of 1.4542 (X5CrNiCuNb16-4, AISI 630), (in mass percent).

Elements	C	Si	Mn	P	S	Cr	Mo	Ni	Cu	Nb
	0.03	0.42	0.68	0.018	0.002	15.75	0.11	4.54	3.00	0.242

Duplex steel 1.4462 (X2 CrNiMoN 22 5 3, (table 5) manufactured by means of continuous casting followed by an appropriate tempering and water quenching allowing for an equilibrium of ferrite and austenite phase. Due to its percentage of austenite 1.4462 reveals a much better corrosion resistance resulting and a PREN number (35.1) twice as high as 1.4542 [16].

Table 5. Chemical composition of 1.4462 X2 CrNiMoN 22 5 3 (UNS S31803) (in mass percent).

phases	C	Si	Mn	Cr	Mo	Ni	N
α & γ **	0.023	0.48	1.83	22.53	2.92	5.64	0.15

*PREN α =37,4, γ =32,4 ** P=0.024, S=0.008

2.2. Aquifer Water

To simulate in-situ geothermal condition the geothermal aquifer water (as known to be similar to the Stuttgart Aquifer [19], [20]: Ca^{2+}: 1760 mg/L, K^{2+}: 430 mg/L, Mg^{2+}: 1270 mg/L, Na^{2+}: 90100 mg/L, Cl^-: 143,300 mg/L, SO_4^{2-}: 3600 mg/L, HCO_3^-: 40 mg/L) was synthesized in a strictly orderly way to avoid precipitation of salts and carbonates.

IOP Conf. Series: Earth and Environmental Science **150** (2018) 012012 doi:10.1088/1755-1315/150/1/012012

2.3. Corrosion Experiments

Laboratory scale exposure tests in CO_2-saturated aquifer brine and water saturated CO_2 were carried out using coupons of the steel qualities that may be used as injection pipe. Therefore steel samples made of as received and thermally treated specimens with 8 mm thickness, 20 mm width, 50 mm length were used. To gain martensitic microstructure with sufficient hardness and toughness prior to exposure heat treatment was done following commonly used protocols [7], [15], [17].
A hole of 3.9 mm diameter was used for sample positioning. Samples of each base metal were positioned within the vapour phase and within the liquid phase. Flow control (3 NL/h) of the technical CO_2 (purity 99.995 vol.-%)) into the brine at ambient pressure was done by a capillary meter GDX600_man by QCAL Messtechnik GmbH, Munich. The exposure of the samples between 700 h to 8000 h was disposed in reaction vessels according to the conditions at the geological site at 60 °C at ambient pressure – each material in a separated reaction vessel [3] and at 100 bar [7], [13], [15] (Figure 1).

Figure 1. Reaction vessels and experimental set up [7], [17].

Before corrosion tests the surfaces of the steels were activated by grinding with SiC-Paper down to 120 μm under water. After the corrosion tests, the samples were cut partly for scale analysis with the corrosion layer and prepared partly for kinetic analysis after the scale was etched. Descaling of the samples was performed by exposure to 37% HCl. Then parts of the samples were embedded in a cold resin (Epoxicure, Buehler), cut and polished first with SiC-Paper from 180 μm to 1200 μm under water and then finished with diamond paste 6 μm and 1 μm.

2.4. Analysis

Different light optical and electron microscopy techniques were performed on specimens to investigate the layer structures and morphology of the samples. X-ray diffraction was carried out in a URD-6 (Seifert-FPM) with CoK_α-radiation with an automatic slit adjustment, step 0.03° and count 5 sec. Phase analysis was performed by matching peak positions automatically with PDF-2 (2005) powder patterns. Mainly structures that were likely to precipitate from the steels were chosen of the ICSD and refined to fit the raw-data-files using POWDERCELL 2.4 [21] and AUTOQUAN ® by Seifert FPM. Kinetics of the corrosion were determined by the corrosion rates which were calculated via mass change of the samples before and after corrosion testing according to DIN 50 905 part 1-4.

$$corrosion\ rate\left[\frac{mm}{year}\right]=\frac{8760\left[\frac{hours}{year}\right]\cdot 10\left[\frac{mm}{cm}\right]\cdot weight\ loss[g]}{area[cm^2]\cdot density\left[\frac{g}{cm^3}\right]\cdot time[hour]} \tag{1}$$

3. Results and Discussion

As expected due to the high PREN (53.1 (Pitting Resistance Equivalent (PRE = %Cr + 3.3% Mo + 16% N)) for the duplex stainless steel X2 CrNiMoN22-5-3 is reason for no corrosion phenomena at all, even after long exposure times in Stuttgart Aquifer water [19] and Northern German Basin [20] with

IOP Conf. Series: Earth and Environmental Science **150** (2018) 012012 doi:10.1088/1755-1315/150/1/012012

or without flowing technical CO_2. Neither ferrite nor austenite phase showed corrosive deterioration (Figure 2). Molybdenum specially increases the resistance to pitting and crevice corrosion.

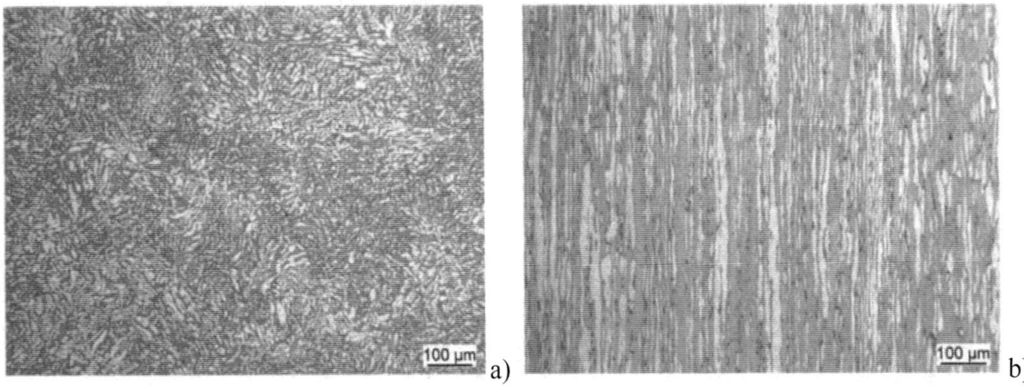

Figure 2. Microstructure of experimental alloy prior to testing; (a) transversal and (b) longitudinal sections respecting the casting direction. Etching: Beraha II [19].

In general corrosion rates are lower at 100 bar compared to ambient pressure. This applies for mild steel 42CrMo4 (Figure 3) as well as for martensitic stainless steels X46Cr13 (Figure 4), X35CrMo17 (Figure 5) and X5CrNiCuNb 16-4 (Figure 6). Corrosion rates are higher at ambient pressure than at 100 bar, which may directly be related to an open capillary system within the corrosion layer. Under pressure these are not present suppressing fast mutual diffusion of ionic species that is necessary for scale growth [18]. The corrosion rates of 42CrMo4, X20Cr13 and X46Cr13 increase with time at ambient pressure and decrease with time at 100 bar. X5CrNiCuNb16-4 and X35CrMo17 show increasing corrosion rates at ambient pressure as well as at 100 bar [18].

One reason for the decreasing corrosion rates lower gas tightness of the low pressure system with excess oxygen in the test vessels at ambient pressure leading to higher corrosion rates as a function of exposure time. Decreasing growth rates at 100 bar may be related to the fast formation of initial hydroxides [2], [6], [14], [18] then precipitating to a passivating siderite-layer with increasing thickness as a function of exposure time. This corrosion scale acts as a diffusion barrier towards mutual diffusion of ionic species to the metal surface. The corrosion rate may also decrease in the presence of iron carbonate precipitates and explains lower corrosion rates at longer exposure times.

In general corrosion rates are severely dependent on the exposure time t for specimens kept in the vapour phase. For example the corrosion rate is twice as high for 42CrMo4 after 8000 h (0.8 mm/year) in comparison to 4000 h (0.45 mm/year) when exposed at ambient pressure. This effect has less impact on specimens kept in the liquid phase or at 100 bar with no regard to atmosphere [18]. However, the atmosphere (liquid, vapour or supercritical atmosphere) does not significantly influence the general corrosive behavior.

The pressure has little influence on the corrosion rates. For 42CrMo4, X20Cr13 and X35CrMo4 the corrosion rates at ambient pressure are much higher than at 100 bar. For X20Cr13 and X5CrNiCuNb16-4 the difference is not as significant. In general the higher corrosion rates at ambient pressure may be explained by higher diffusion kinetics. The corrosion rates in the vapour phase (water-saturated CO_2) exceed those measured in the liquid phase (CO_2-saturated water) by a factor of 3–8. But, at high pressures (100 bar) corrosion rates in the liquid phases can be higher or lower than in the vapour/supercritical phase.

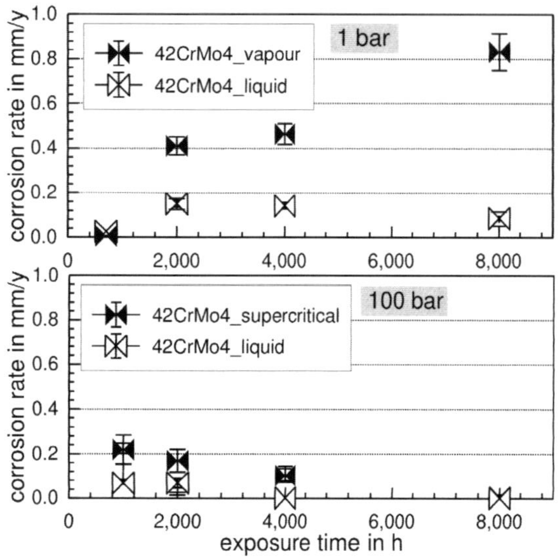

Figure 3. Comparison of corrosion rates of 42CrMo4 after 8000 h of exposure to aquifer brine water at 60 °C and ambient pressure [3] as well as at 100 bar [18].

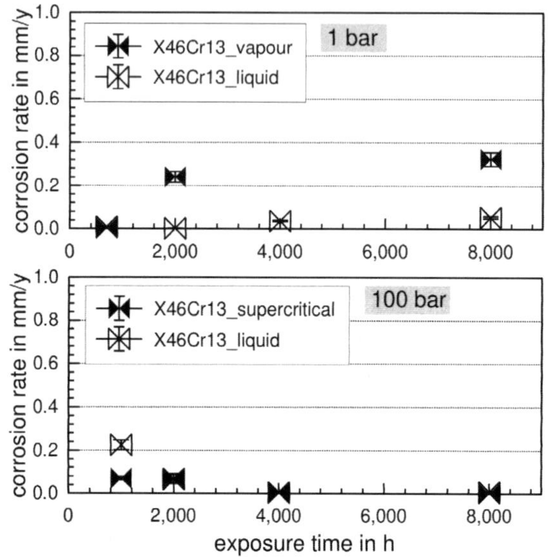

Figure 4. Comparison of corrosion rates of X20Cr13 in the liquid and vapour/supercritical phase after 8000 h of exposure to aquifer brine water at 60 °C and ambient pressure as well as at 100 bar.

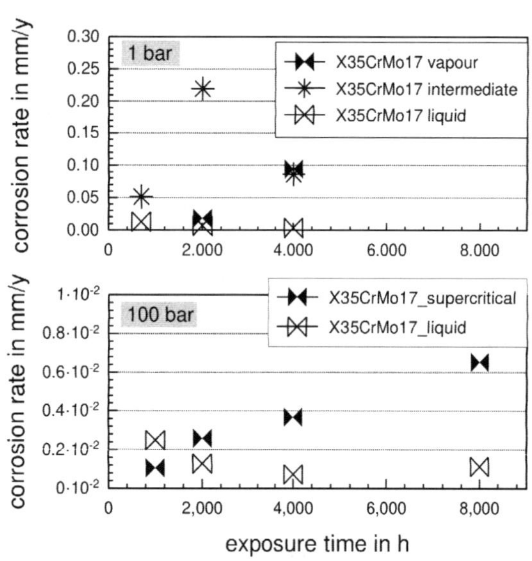

Figure 5. Comparison of corrosion rates of X46Cr13 in the liquid and vapour/supercritical phase after 8000 h of exposure to aquifer brine water at 60 °C and ambient pressure [3] as well as at 100 bar [18].

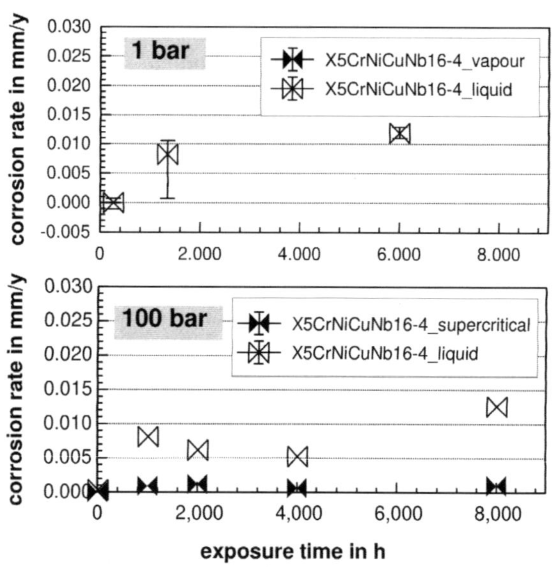

Figure 6. Comparison of corrosion rates of X35CrMo17 in the liquid and vapour/supercritical phase after 8000 h of exposure to aquifer brine water at 60 °C and ambient pressure as well as at 100 bar.

However, the corrosion rates for 42CrMo4 are higher in the liquid/vapour phase and also higher at ambient pressure compared to high pressure. This corresponds with the thickness of the precipitation layer that is greater in vapour/supercritical than in the liquid phase (Figure 3 and Figure 7).

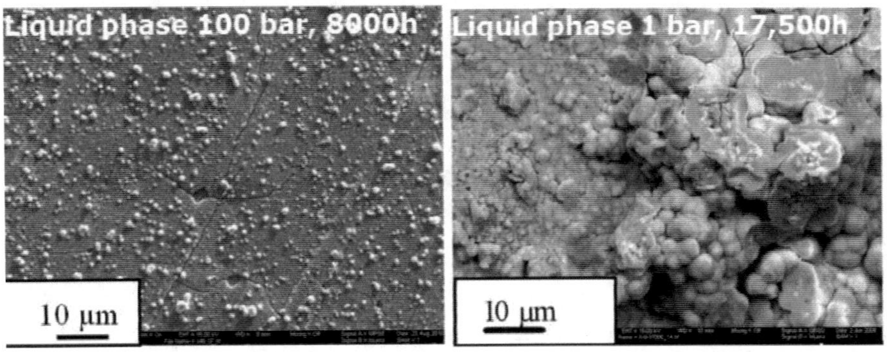

Figure 7. Surface precipitation on 42CrMo4 after exposure to CO_2 saturated saline aquifer water for 8000 h at ambient pressure and 100 bar.

The multi-layered corrosion scale precipitated on the steels at ambient pressure is complicated and has been described in detail [7], [17], [18]. This characteristic carbonate/oxide structure consists of siderite $FeCO_3$, goethite a-FeOOH, mackinawite FeS and akaganeite $Fe_8O_8(OH)_8Cl_{1.34}$ and in addition spinel-phases of various compositions. Carbides, such as Fe_3C or chromium-rich iron carbides were visible within the corrosion layer.

4. Conclusion

Immersion tests at 60°C at ambient pressure and at 100 bar were performed in saline aquifer water (Stuttgart Aquifer and Northern German Basin) at 60 °C using technical CO_2 to model the CCS-environment. According to DIN 6601 42CrMo4, X20Cr13 and X46Cr13 would be unsuitable for pressure vessel application, when being surrounded by the CO_2-saturated brine at ambient pressure.
The following conclusions for the corrosion for corrosion phenomena of different steels may be stated:

i. Main corrosion products of a continuous corrosion scale as well as pits after static corrosion tests at ambient pressure as well as 100 bar were $FeCO_3$ and FeOOH for all steel qualities. Corrosion reactions are diffusion controlled with slower diffusion in the liquid phase leading to increased siderite crystal sizes than in the vapour/supercritical phase (smaller crystals due to higher diffusion rates).
ii. Corrosion rates obtained at 100 bar are generally much lower after 8000 h of exposure than those measured at ambient pressure (depending on the atmosphere (vapour or liquid) by a factor of ca. 10–80 for 42CrMo4, a factor of ca. 1.5–30 for X46Cr13, a factor of ca. 10–100 for X20Cr13 and a factor of ca. 10 for X35CrMo17-1, exception for X5CrNiCuNb16-4 with similar surface corrosion rates in the liquid phase).
iii. Highest surface corrosion rates are 0.8 mm/year for 42CrMo4 (1% Cr), 0.3 mm/year for X46Cr13 (13% Cr), 0.3 mm/year for X20Cr13 (13% Cr), 0.1 mm/year for X35CrMo17-1 (17% Cr) and 0.01 mm/year for X5CrNiCuNb16-4 (16% Cr) in the vapour phase at ambient pressure after 8000 h of exposure. At 100 bar the highest corrosion rates are 0.01 mm/year for 42CrMo4, X20Cr13 (liquid phase) X46Cr13 and less than 0.01 mm/year for X35CrMo4 and X5CrNiCuNb16-4 after 8000 h of exposure with no regard to atmosphere.
iv. The carbon content does not show significant influence on corrosion rate. But higher chromium content results in lower surface corrosion rates and therefore better corrosion resistance. Due to high Cr, Ni and Mo content X2CrNiMoN 22-5-3 shows now surface corrosion at all under the conditions given.

5. References

[1] Van den Broek, Hoefnagels M R, Rubin E, Turkenburg, W and Faaij, E, Effects of technological learning on future cost and performance of power plants with CO_2 capture, in Projects Costs of Generating Electricity 2010 Progress in Energy and Combustion Science 177-187, ISBN 978-92-64-08430-8

[2] Nešić S 2007 Key issues related to modelling of internal corrosion of oil and gas pipelines – A review *Corrosion Science* 49 4308–4338

[3] Pfennig A and Kranzmann A 2011 Reliability of pipe steels with different amounts of C and Cr during onshore carbon dioxide injection *International Journal of Greenhouse Gas Control* 5 757–769

[4] Zhang H, Zhao Y L and Jiang Z D 2005 Effects of temperature on the corrosion behaviour of 13Cr martensitic stainless steel during exposure to CO2 and Cl- environment *Material Letters* 59 3370-3374

[5] Alhajji J N and Reda M R 1993 The effect of alloying elements on the electrochemical corrosion of low residual carbon steels instagnant CO_2-saturated brine *Corrosion Science* 34 1899-1911

[6] Choi Y S and Nešić S 2008 Corrosion behaviour of carbon steel in supercritical CO_2-water environments, No. 09256 *NACE Corrosion Conference and Expo*, New Orleans, Louisiana, USA, March 16th – 20th

[7] Pfennig A, Zastrow P and Kranzmann A 2013 Influence of heat treatment on the corrosion behaviour of stainless steels during CO_2-sequestration into saline aquifer *International Journal of Green House Gas Control* 15 213–224

[8] Bülbül Ş and Sun Y 2010 Corrosion behaviours of high Cr-Ni cast steels in the HCl solution, *Journal of Alloys and Compounds* 598 143-147

[9] Banaś J, Lelek-Borkowska U, Mazurkiewicz B and Solarski W 2007 Effect of CO_2 and H_2S on the composition and stability of passive film on iron alloy in geothermal water *Electrochimica Acta* 52 5704–5714

[10] Wei L, Pang X, Liu C and Gao K 2015 Formation mechanism and protective property of corrosion product scale on X70 steel under supercritical CO_2 environment *Corrosion Science* 100 404–420.

[11] Isfahany A N, Saghafian H and Borhani G 2011 The effect of heat treatment on mechanical properties and corrosion behaviour of AISI420 martensitic stainless steel *Journal of Alloys and Compounds* 509 3931-3936

[12] Lucio-Garciaa M A, Gonzalez-Rodrigueza J G, Casalesc M, Martinezc L, Chacon-Navaa J G, Neri-Floresa M A and Martinez-Villafañea A 2009 Effect of heat treatment on H_2S corrosion of a micro-alloyed C–Mn steel *Corrosion Science* 51 2380-2386

[13] Pfennig A, Wiegand R, Wolf M and Bork C-P 2013 Corrosion and corrosion fatigue of AISI 420C (X46Cr13) at 60 °C in CO2-saturated artificial geothermal brine *Corrosion Science* 68 134–143

[14] Han J, Yang Y, Nešić S and Brown B N 2008 Roles of passivation and galvanic effects in localized CO_2 corrosion of mild steel, Paper No. 08332, NACE Corrosion 2008, New Orleans, Louisiana, USA, March 16th – 20th

[15] Pfennig A, Wolthusen H and Kranzmann A 2017 Unusual corrosion behavior of 1.4542 exposed a laboratory saline aquifer water CCS-environment *Energy Procedia* 114 5229-5240

[16] Wolf M, Afanasiev R, Böllinghaus Th and Pfennig A 2017 Investigation of Corrosion Fatigue of Duplex Steel X2CrNiMoN22 5 3 Exposed to a Geothermal Environment under Different Electrochemical Conditions and Load Types *Energy Procedia* 114 5337-5345

[17] Pfennig A, Wolthusen H, Wolf M and Kranzmann A 2014 Effect of heat treatment of injection pipe steels on the reliability of a saline aquifer water CCS-site in the Northern German Basin *Energy Procedia* 63 5762-5772

[18] Pfennig A and Kranzmann A 2012 Effect of CO_2 and pressure on the stability of steels with different amounts of Chromium in saline water *Corrosion Science* 65 441–452

[19] Förster A, Norden B, Zinck-Jørgensen K, Frykman P, Kulenkampff,J, Spangenberg, E, Erzinger J, Zimmer M, Kopp J, Borm G, Juhlin C, Cosma C and Hurter S 2006 Baseline characterization of the CO_2SINK geological storage site at Ketzin, Germany *Environmental Geosciences* 13 145-161

[20] Förster A et al. 2010 Reservoir characterization of a CO_2 storage aquifer: The Upper Triassic Stuttgart Formation in the Northeast German Basin *Mar. Pet. Geol.* 27 2156–2172

[21] Kraus S W and Nolze G 1996 POWDER CELL – a program for the representation and manipulation of crystal structures and calculation of the resulting X-ray powder patterns *J. Appl. Cryst.* 29 301-303

Acknowledgment

This work was supported by the FNK (Fachkonferenz für wissenschaftliche Nachwuchskräfte) of the Applied University of Berlin, HTW and by IMPACT (EU-Project EFRE 20072013 2/21).

IOP Conf. Series: Earth and Environmental Science **150** (2018) 012013 doi:10.1088/1755-1315/150/1/012013

Decision making tools for selecting sustainable wastewater treatment technologies in Thailand

Praewa Wongburi[1] and Jae K. Park[2]

1 Lecturer, Faculty of Environment and Resource Studies, Mahidol University, Thailand
2 Professor, Department of Civil and Environmental Engineering, University of Wisconsin-Madison, USA
E-mail: praewa_mew@hotmail.com, praewa.won@mahidol.ac.th

Abstract. Wastewater consists of valuable resources that could be recovered or reused. Still it is under threat because of ineffective wastewater management and systems. In Thailand, less than 25% of wastewater generated may be treated while then rest is inadequately treated and sent back directly into waterbodies or the environment. Furthermore, the technologies that have been applied may be inefficient and unsustainable. Efficiency, sustainability, and simplicity are important concepts when designing an appropriate wastewater treatment system in developing countries. The objectives of this study were to review and evaluate wastewater treatment technologies and propose a method to improve or select an appropriate technology. An expert system in Excel® program was developed to determine the best solution. Sensitivity analysis was applied to compare and assess uncertainty factors. Due to the different conditions of each area, the key factor of interest was varied. Furthermore, Robust Decision Making tool was applied to determine the best way to improve existing wastewater treatment facility and to choose the most appropriate wastewater treatment technology.

1. Introduction
Wastewater is one of today's most important environmental issues that causes severe problems to humans, animals, and the environment by improper management and technologies. Wastewater is a combination of domestic, commercial, industrial, and agricultural discharge. It contains pollutants and contaminants, including nutrients, microorganisms, chemicals and other toxins. These pollutants can cause health and environmental problems when wastewater is released into body rivers improperly [1]. However, wastewater also contains reusable resources such as water, carbon and nutrients that could be recovered or reused [2]. Therefore, wastewater should be appropriately treated for removal of pollutants to meet the effluent regulatory standards. Moreover, the processes should focus on resource recovery to minimize carbon footprint, and to be self-sustainable [3].

1.1. Objectives
Since wastewater issue is critical, many engineers and scientists have been developing new technologies to investigate an efficient system to solve this problem. Basic wastewater treatment systems reduce organic compounds and suspended solids to meet the effluent permit standards [4]. The advancement of technology has developed treatment processes, which can remove dissolved organic matter and toxic substances. At present, an improvement of scientific knowledge and consciousness about the environment led to new technologies and systems, which can reduce pollutions in wastewater and recycle energy with the goal of zero discharge of pollutants [5].

However, developing countries still have problems with wastewater treatment systems because of insufficient regulations, improper management, economic situation, and selecting unsuitable technologies. Thus, the primary objective of this research paper was to determine appropriate wastewater technologies to be used in Thailand. In addition, it also suggested new methods and sustainable practices that can be used long-term to improve the overall wastewater treatment systems in the country.

1.2. Research scope

The scope of this study was to select an appropriate wastewater treatment technologies by using different methods as depicted in Figure 1. First step is to evaluate the existing wastewater facilities in Thailand by focusing on three important criteria, performance, cost, and sustainability. Performance includes two important factors, reliability and simplicity. Cost includes an initial cost and operation and maintenance costs. Resource recovery, energy management, and solid volume reduction are three main concepts of sustainability. After considering these three main factors, the study will analyse sustainable alternative wastewater technologies and develop an expert system to select the most suitable wastewater process.

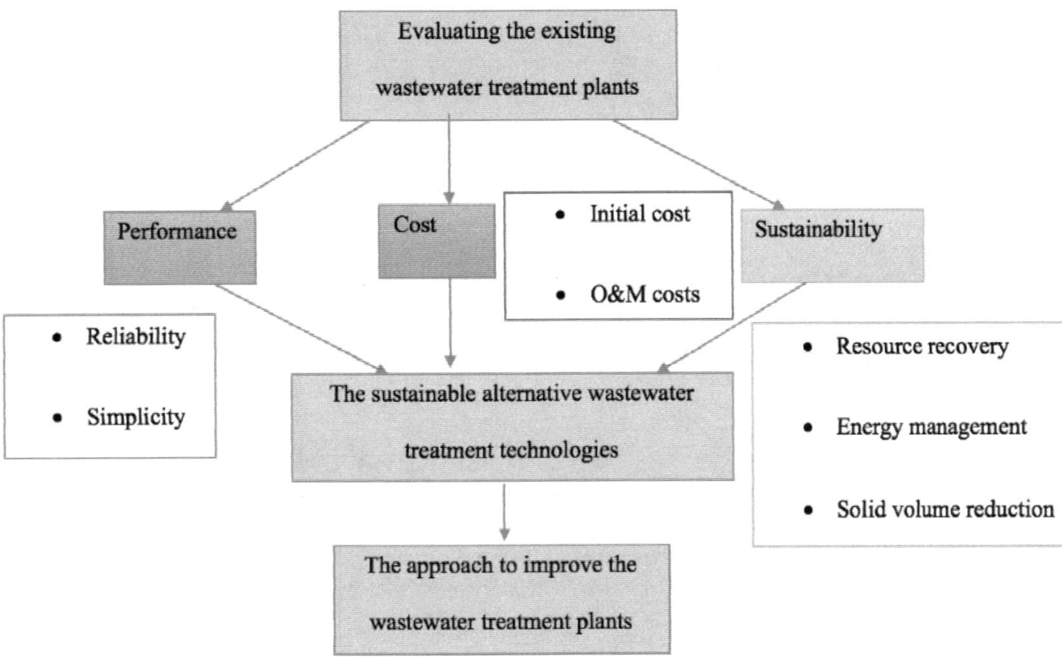

Figure 1. Research scope

2. Thailand wastewater treatment systems

There are several types of wastewater treatment systems in Thailand. The most common system is stabilization pond, which is commonly used in rural areas. Aerated lagoon and oxidation ditch, which are one type of activated sludge processes are also common processes used in Thailand's rural areas [6]. In urban areas, such as Bangkok, Pattaya, the typical system is the activated sludge process. There are other processes such as constructed wetland, rotating biological contactor (RBC), fixed film activated sludge, sequencing batch reactor (SBR) and, modified sequencing batch reactor (MSBR). Figure 2 shows the fraction of the amount of wastewater treatment systems in Thailand.

IOP Conf. Series: Earth and Environmental Science **150** (2018) 012013 doi:10.1088/1755-1315/150/1/012013

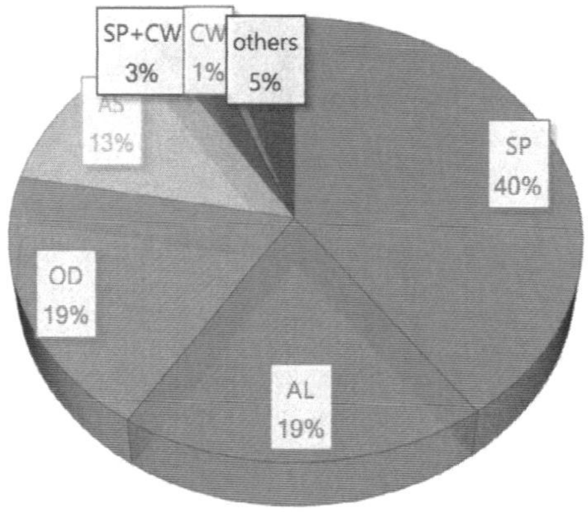

Figure 2. The percentage of the amount of wastewater treatment systems. OD: Oxidation ditch; AL: Aerated lagoon; SP: Stabilization pond; CW: Constructed wetland; AS: Activated sludge.

3. Case studies in Thailand

In this section, some examples of wastewater treatment systems in Thailand are reviewed by choosing different types of technologies and compared the efficiency and cost of the systems. Table 1 shows five different types of wastewater treatment commonly used in Thailand, which are stabilization pond, sequencing batch reactor, aerated lagoon, oxidation ditch, and activated sludge. This table provides the capacity of the plants, the actual inflow to the plant, the construction cost, the operation and maintenance costs, and BOD and SS removal ratio.

Table 1. The efficiency of examples of wastewater treatments in Thailand

Province	Type of treatment	Design Capacity (m^3/d)	Inflow (m^3/d)	Costs (million baht)		BOD (mg/l)		Removal Ratio (%)	SS (mg/l)		Removal Ratio (%)
				Construction	O&M (per year)	In	Out		In	Out	
Sukhothai	Stabilization pond	8400	11413	277	1.67	4.8	4.4	8.33	220	130	40.9
Lamphun	SBR	10000	4500	583	0.86	32.97	5.83	82.3	48.63	12.56	74.2
Angtong	Aerated lagoon	8200	3000	183	0.95	59.77	19.23	67.8	51.97	24.97	52
Laem Chabang	OD	25000	1631	194	0.45	12.28	14.99	-22.1	13.5	5.24	61.2
Si Phraya	AS	30,000	18,213	464	10.64	56	5	90.5	109	7	93.6

4. Alternative sustainable wastewater treatment systems

The study focuses on resource recovery from wastewater sector including water reuse, energy recovery, and nutrient recovery. The future goal for wastewater treatment of domestic wastewater should have a minimal carbon footprint, and be 100% self-sustainable with regards to energy, carbon, and nutrients, while achieving a discharge or reuse quality that preserves the quality of the receiving waters [7]. The goal of sustainable wastewater processes should be changed into recycling treatment to promote the conservation of water resources. This following section provides many case studies on alternative sustainable wastewater technologies, which are very useful to apply to this research.

4.1. Case studies

4.1.1 Technology roadmap for sustainable wastewater treatment plants in a carbon-constrained world.

'Resources end up in wastewater through inefficient consumption. As a result, wastewater contains potentially reusable water, carbon (energy), and nutrients (nitrogen, phosphorus, and sulfur) that could be recovered. Meanwhile, current wastewater treatment objectives are primarily to produce an acceptable quality of treated water at the lowest life cycle cost (LCC). Most current treatment processes manage carbon and nutrients as wastes to be removed, and do not attempt to capitalize on the resources inherent in wastewater' [7]. The future goal for wastewater treatment of domestic wastewater should be to have a minimal carbon footprint, and to be 100% self-sustainable with regards to energy, carbon, and nutrients, while achieving a discharge or reuse quality that preserves the quality of the receiving waters [7].

According from the article, the author presents treatment processes and pathway to improve wastewater technology.

Treatment process category are:
• Further reliance on biotechnology
• Optimization of carbon to energy within liquid and solids treatment processes
• Fixed film processes
• Nutrient recovery processes for nitrogen (N), phosphorus (P), and sulfur (S)
• Enhanced primary treatment for increased carbon diversion to solids processing
• Membrane equipment and process improvements to allow broader application
• Microconstituents/micropollutants – fate and effects research
• Develop and use new thermal sludge processes (pyrolysis and gasification)
• Control or capture air and GHG emissions

Equipment category are:
• Heat recovery from sidestreams and plant effluent
• Better aeration systems
Other category is:
• Better disinfection processes and equipment

Technology Pathways are to:
• Define whole plant treatment concepts that are energy neutral, including multi-sludge systems and sidestream treatment processes;
• Extract nutrients in forms that are amenable for agricultural reuse with minimal perception of risk to public health;
• Improve the use of biogas to produce energy;
• Identify and develop those emerging biosolids-to-energy technologies that show the greatest promise so that utilities can cost-effectively use them; and
• Explore new concepts that extract energy from wastewater, such as microbial fuel cells, or that produce or use other energy byproducts from wastewater such as hydrogen or nitrous oxide.

4.1.2. Sustainable wastewater management in developing countries

The authors presented reflections and stories about appropriate and sustainable wastewater management systems in developing countries. They also provided case studies and the implications and application. The objectives were to reflect on, discuss, and provide examples a broader use of robust, reliable, cost-effective, and efficient wastewater management systems that work in practice [8]. Basic concepts from this book were summarized below.

The authors described the sustainable ecosystem approach as follows:
1. Not about large-scale, centralized wastewater management but about appropriate, sustainable on-site systems.
2. Focus on on-site systems but improve linkages to cluster and centralized systems.
3. Not about discharge point, but about land-based wastewater management systems.

Local context lessons were as follows:
1. Local objectives, not (environmental) standards
2. Success and failure constantly interchange, producing a rather difficult context for predictions and lessons learned

3. Tales from the field as starting points for reflections
4. Cases of sense and simplicity – the ultimate way forward

5. Methodology

5.1. Robust Decision Making (RDM)

Robust Decision Making (RDM) is the adaptive or flexible decision method that can be used in the uncertain conditions [9]. Since future is uncertain and unpredictable, robust options tend to be more recommended than optimal options. On the other hand, the best option is more suitable when the future can be predicted. The purpose of RDM is to identify robust strategies, which can adapt or perform well under uncertain situations in the future. The method is helpful for decision makers in long-term consequences. The fundamental steps for Robust Decision Making are to: (1) identify the issues and set a goal; (2) select a robust strategy through investigating information, strategies, and risks; (3) take an active action towards the goals; (4) determine whether the strategy is effective; and (5) update and resolve the strategy. The last step is an essential one because if the strategy is not effective, decision makers can change strategies until they meet their goals. Figure 3 illustrates the steps of Robust Decision Making [10].

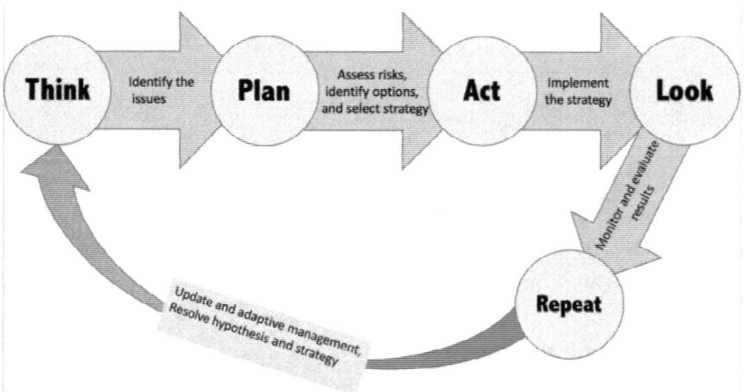

Figure 3. The Steps for Robust Decision Making.

5.2. Sensitivity Analysis (RDM)

Developing an expert system in Excel® program, setting the criteria and then weighting and ranking are important method to determine overall processes. The weights depend on an area and local conditions. Sensitivity analysis was used to evaluate the impacts of efficiency, simplicity, operation and maintenance costs, and energy efficiency on determining sustainable wastewater treatment. Common wastewater systems used in Thailand will be evaluated by selecting different factors that affect wastewater technologies selection.

Figure 4 illustrates important results. When wastewater treatment facilities concern the cost as the most important factor, stabilization pond would be the best option. However, the result is the system would be very low efficiency. This means if the wastewater treatment facilities want to produce higher wastewater treatment quality, the facilities should modify to other systems such as activated sludge or aerated lagoon. The facilities should also adapt to new alternative wastewater treatments, which are provided section 4, alternative sustainable wastewater treatment systems, in order to reuse or recycle reusable resources in wastewater. This will lead to lower their energy consumption, produce valuable resources, or even produce their own energy.

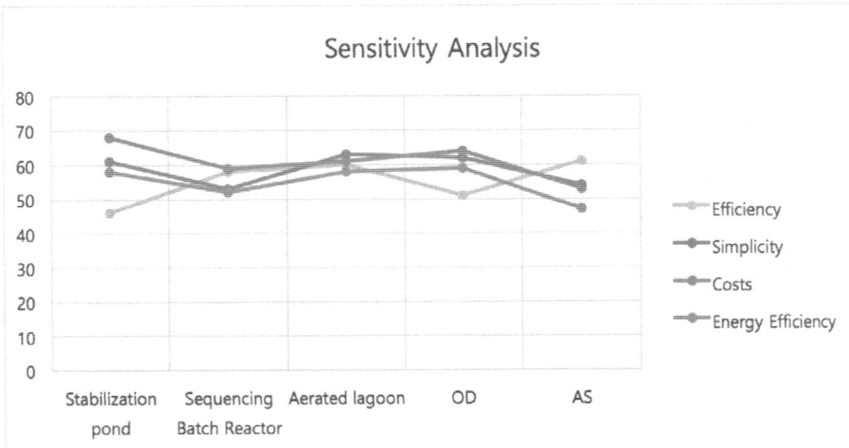

Figure 4. Sensitivity Analysis

6. Proposal of sustainable wastewater treatment

Selecting the most appropriate wastewater treatment technology is not only about implementing the best technical solution at the lowest cost, but also about sustainability including social and environmental considerations [11]. Wastewater treatment technologies in Thailand are ineffective because ease of operation and low costs were the top priorities [12]. However, one of the most important factors is effectiveness because it directly affects public health and the environment. Stabilization pond, construct wetland or other natural processes are easy to operate, but the quality of effluent and reliability of the systems tend to be poor.

6.1. An expert system in Excel® program.

An expert system was developed in Excel® program to aid in selecting the most sustainable wastewater treatment processes for a wastewater treatment plan of interest and/or comparing existing wastewater treatment technologies by a weighting and ranking system. Moreover, sensitivity analysis can be implemented to compare the technologies by changing weighting factors. Figure 5 shows the flowcharts of the developed expert system.

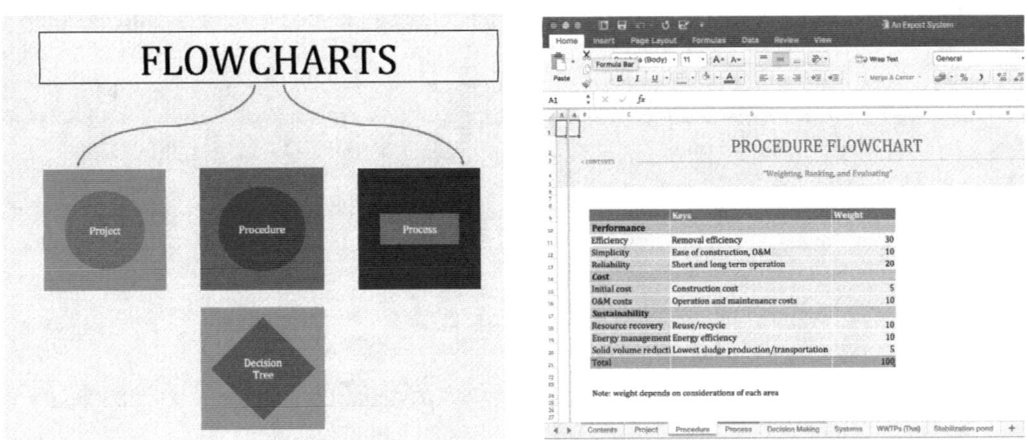

Figure 5. An expert system in Excel® program

7. Conclusions

The selection of wastewater treatment systems must be based on sustainability. The significant factors such as efficiency, simplicity, operation and maintenance costs, and energy efficiency were considered to determine the most sustainable wastewater systems. Thailand wastewater treatment systems were analyzed using the developed expert system. The key problems of wastewater in Thailand and possible solutions are described. In addition, the recommendation and future study are also stated in this chapter.

7.1. Problems and solutions

One of the major problems of wastewater treatment systems in Thailand is the lack of wastewater facilities to treat all wastewater produced from sources. Furthermore, the common systems used in Thailand are natural treatments, especially in rural area resulting in the low quality of treated water. One solution is that the government should focus on improving wastewater treatment sector in Thailand. The water quality can impact overall people's health and the environment. Thus, it is necessary to build more wastewater treatment systems.

An expert system in Excel® program was developed to provide a robust solution for any wastewater treatment plant and to aid in improving or solving a problem in the facility. Sensitivity analysis is applied in order to evaluate the uncertainties of choosing wastewater treatment technologies. The local condition affects the decision significantly. Thus, a Robust Decision Making tool was adopted to determine the way to improve existing wastewater treatment facility and to choose the most appropriate wastewater treatment system. It can aid in identifying issues, finding options, implementing strategies, monitoring results, and improving the systems efficiency.

In developing countries, local conditions, efficiency, sustainable and simplicity are important concepts when designing a wastewater treatment system [13]. Local conditions including finance, government, citizens, and the community. Efficiency of wastewater technology is also very significant. Any wastewater treatment plant should meet the regulatory standard.

Thailand wastewater treatment facilities can use anaerobic digesters to produce electricity and heat. This can reduce energy demand and lower the operational cost. Furthermore, they can recycle their sludge produced from the treatments to be used as a fertilizer. The most important production in Thailand is rice. Thus, it will be very useful and sustainable if biosolids produced from wastewater treatment plants are used as a fertilizer. This can help both the wastewater facilities and farmers. Lastly, simplicity is one of the important concepts of choosing appropriate wastewater treatment technologies. Simplicity means simplicity of the management, the technology, and the details. Simplicity is reducing the level of complexity as much as possible and cutting the number of elements that could eventually lead to the failure of the system [8], [14].

7.2. Suggestion and future study

One of the key factors in constructing a wastewater treatment plant is the community participation. Most people who live in poor rural area may not get the good water quality like the people who live in urban areas. They should call for their right to get the appropriate wastewater treatment and water quality and improve their lives. Furthermore, education on wastewater problems and management to public is an important activity. For example, a community can initiate cleaning activities by raising public awareness and strengthening community participation.

The future study should investigate about related factors that might affect wastewater activity such as social involvement, public participation, and government support. The knowledge and policy that related to wastewater management should be gathered and analyzed in order to determine the most appropriate wastewater treatment technologies.

8. References

[1] Secretariat U W 2014 *Wastewater Management- A UN-Water Analytical Brief.* Retrieved from UN Water: http://www.unwater.org/publications/publications-detail/en/c/275896/. [Accessed 12 November 2016].

[2] Jhansi S and Mishra S 2013 *Wastewater Treatment and Reuse: Sustainability Options.* Consilience: The Journal of Sustainable Development, vol. 10, no. 1, pp. 1–15.

[3] Metcalf and Eddy *1991 Wastewater Engineering: Treatment, disposal and re-use*, 3rd edition revised by George Tchobanoglous and Franklin L. Burton, McGraw Hill Inc. International.

[4] Parr J, Shaw R, and Smith M 2002 *Technical brief no. 64: wastewater treatment options*. Waterlines, an international journal of water, sanitation and waste, 15-18.

[5] Pollution Issues. (n.d.). Retrieved from http://www.pollutionissues.com/Ve-Z/Wastewater-Treatment.html. [Accessed 11 February 2017].

[6] Chokewinyoo P and Khanayai P (n.d.). *Wastewater Production, Treatment, and Use in Thailand*. 5 th Regional Workshop on Safe Use of Wastewater in Agriculture, 1-4.

[7] Crawford and George 2010 *WERF Report Series: Technology Roadmap for Sustainable Wastewater Treatment Plants in a Carbon-Constrained World*: WERF Report OWSO4R07d. IWA Publishing.

[8] Laugesen C and Fryd O 2010 *Sustainable Wastewater Management in Developing Countries*: New Paradigms and Case Studies from the Field. American Society of Civil Engineers Press, ISBN: 978-0-7844-0999-2.

[9] weADAPT 2015 Retrieved from https://www.weadapt.org: https://www.weadapt.org/knowledge-base/adaptation-decision-making/robust-decision- making. [Accessed 9 November 2016].

[10] CARPE DIEM WEST 2017 *Why Robust Decision Making Supports Risk Management and What You Can Do About It*. [ONLINE] Available at: http://www.carpediemwest.org/casestudies/robust-decision-making/. [Accessed 10 February 2017].

[11] Esparza J, Ruiz M, Utrillas M, and Perello J 2014 *Selecting a Sustainable Disinfection Technique for Wastewater Reuse Projects*. Water 2014, 2732-2747.

[12] Singhirunnusorn W and Stenstrom M 2009 *Appropriate wastewater treatment systems for developing countries: criteria and indictor assessment in Thailand.* A dissertation submitted in partial satisfaction of the requirements for the degree Doctor of Philosophy in Civil Engineering. Los Angeles.

[13] Ganda F and Ngwakwe C 2014 *Role of energy efficiency on sustainable development*. Environmental Economics, 5,1, 86-99.

[14] Rittman B and Mccarty P 2001 *Environmental Biotechnology: Principles and Applications*. New York:McGraw-Hill.

IOP Conf. Series: Earth and Environmental Science **150** (2018) 012014 doi:10.1088/1755-1315/150/1/012014

Comparison of Cu^{2+} and Zn^{2+} thermalcatalyst in treating diazo dye

Y Y Lau[1], Y S Wong[1,*], S A Ong[1], N A Lutpi[1] and L N Ho[2]

[1] School of Environmental Engineering, Universiti Malaysia Perlis, Kompleks Pusat Pengajian Jejawi 3, 02600 Arau, Perlis, Malaysia
[2] School of Materials Engineering, Universiti Malaysia Perlis, Kompleks Pusat Pengajian Jejawi 3, 02600 Arau, Perlis, Malaysia
[*] Corresponding e-mail : yswong@unimap.edu.my

Abstract. This research demonstrates the comparison between copper (II) sulphate ($CuSO_4$) and zinc oxide (ZnO) as thermalcatalysts in thermolysis process for the treatment of diazo reactive black 5 (RB 5) wastewater. $CuSO_4$ was found to be the most effective thermalcatalyst in comparison to ZnO. The color removal efficiency of RB 5 catalysed by $CuSO_4$ and ZnO were 91.55 % at pH 9.5 and 7.36 % at pH 2, respectively. From the UV-Vis wavelength scan, $CuSO_4$ catalyst is able to cleave the molecular structure bonding more efficiently compared to ZnO. ZnO which only show a slight decay on the main chemical network strands: azo bond, naphthalene and benzene rings whereas $CuSO_4$ catalyst is able to fragment azo bond and naphthalene more effectively. The degradation reactions of $CuSO_4$ and ZnO as thermalcatalysts in thermolysis process were compared.

1. Introduction

Industrial manufacturing factories, particularly those involved in colour finishing processes are major source of considerable pollution. Dominant industry contributed to the foulest contamination and pollutants is textile dye processing industries. In detail, textile industry is a group of related industries uses cotton, wool and synthetic fibers to produce fabric. Textile dyes are well known mutagens and carcinogens posing risks to various ecosystems, animals' health and agriculture. This is due to the recalcitrant nature of textile effluents which largely containing high concentrations of non-degradable dyestuffs. Presence of non-degradable dye molecule in water bodies decrease the dissemination of daylight, thus reducing the algae and other aquatic vegetation photosynthesis rate. Consequently, concentration of dissolved oxygen become lower causing wide array of toxic effects to the aquatic ecosystems and deteriorate the water quality level. Among the dye variations, the major concern of dye types are azo dyes and their derivatives which are mutagenic and carcinogenic. For instance, the complex structure of amino azo benzene dyes may lead to cancer [1]. On the other hands, reactive dyes are mostly used in dye manufacturing industry since reactive dyes are the most efficient dye to impart colour on the material. However, untreated reactive dyes are carcinogenic since it potentially boost the rate of bladder cancer [2]. In the view of these facts, treatment of detoxification and degradation of textile dyes and effluents is crucial. Development of sustainable technologies and techniques to control dye pollution such as adsorption [3], membrane filtration [4], photocatalytic [5], ozonation [6], sonophotocatalytic [7], ultrasound assisted Fenton [8] and microbial electrolysis [9] are generally employed for remediation of dye containing effluents [10]. Dye wastewater treatment by applying thermolysis process is a new insight and yet to be explored. In this research, thermolysis process catalysed by different catalysts, $CuSO_4$ and ZnO were applied in treating reactive diazo dye.

Reactive Black 5 (RB 5) was used as the dye model to study the performance of thermolysis process in treating diazo reactive typed dye. The extent of degradation of the main chemical network strands was investigated to evaluate the efficiency of thermolysis to treat the degradation intermediates.

2. Material and methods

Reactive Black 5 (RB 5) with a molecular structure $C_{26}H_{21}N_5Na_4O_{19}S_6$ was supplied by Sigma-Aldrich. 1 g of RB 5 was diluted in 1L volumetric flask with distilled water in order to obtain 1000 ppm RB 5 stock solution. After that, 100 ppm of RB 5 solution was prepared by using dilution equation. Thermolysis experiments were carried out in a rotamantle. A vertical water-cooled condenser was attached to the reactor in order to prevent any loss of water vapour. Initially, pH of 250 ml of 100 ppm RB 5 solution was measured and adjusted by adding HCl or NaOH within the range in between pH 2 to 12. The solution was stirred and heated until the desired temperature before adding catalyst to the solutions. The catalysts used are copper (II) sulphate, $CuSO_4$ and zinc oxide, ZnO. In order to determine the optimum condition of thermolysis process in treating RB 5, Hitachi U-2810 UV-Visible scanning spectrophotometry were used to identify the transmission of absorption of ultraviolet light and visible light with a sample in the wavelength between 200 to 800 nm.

3. Results and discussion

3.1. Performance analysis of catalysis $CuSO_4$ and ZnO as thermalcatalyst

In thermolysis process, apart from temperature is the critical parameter affect the treatment performance, catalyst which is used to enhance the thermolysis process is essential as an auxiliary agent for the molecular structure bonding fragmentation [11]. Therefore, selection of catalyst was conducted by comparing the colour removal efficiency between $CuSO_4$ and ZnO as thermalcatalyst under effect of pH. By keeping constant parameter for temperature of 60°C, reaction time of an hour and catalyst mass loading of 0.5 g/L for effect evaluation on pH, which varied in the range of 2 – 12. Figure 1 depicts the color removal efficiency of RB 5 catalysed by $CuSO_4$ and ZnO with respect to pH. The maximum percentage of colour removal for $CuSO_4$ and ZnO were 91.55 % at pH 9.5 and 7.36 % at pH 2, respectively. For $CuSO_4$ catalyst, the colour removal efficiency increases rapidly from pH 2 to 9.5 and the removal efficiency start to decrease after pH 9.5. In acidic condition, copper ions tend to react with chlorine ions to form tetrachlorocuprate(II) ion ($(CuCl_4)^{2-}$), which readily react with RB 5. In alkaline condition, copper ions tends to form insoluble copper hydroxide ($Cu(OH)_2$) which support the low color removal of RB 5 beyond pH 9.5 [12]. The acidic and alkaline reaction of $CuSO_4$as shown in equation (1) and (2).

$$\text{Acidic : } CuSO_4 + 4HCl \rightarrow (CuCl_4)^{2-} + 4H^+ + SO_4^{2-} \quad (1)$$

$$\text{Alkaline : } CuSO_4 + 2NaOH \rightarrow Cu(OH)_2 + Na_2SO_4 \quad (2)$$

On the other hand, ZnO only shows a removal of 7.36% at pH 2. According to Greenwood et al, (1997) ZnO will react with Cl- ions presence in the solution to form zinc chloride ($ZnCl_2$) in acidic solution [13]. Subsequently, ZnO hardly shown color removal efficiency throughout the pH ranges from 4 to 12 since zinc hydroxide ($Zn(OH)_2$) white precipitation is formed in alkaline condition.

$$\text{Acidic : } ZnO + 2HCl \rightarrow ZnCl_2 + H_2O \quad (3)$$
$$\text{Alkaline : } ZnO + NaOH \rightarrow Zn(OH)_2 + 2Na_2O \quad (4)$$

Therefore, activation of copper ions and zinc ions for the reactive dye degradation were affected by the solution pH. In comparison between both chemical catalysts, copper ions have stronger attraction forces towards reactive dye in order to destabilize the dye molecular structure compared to zinc ions.

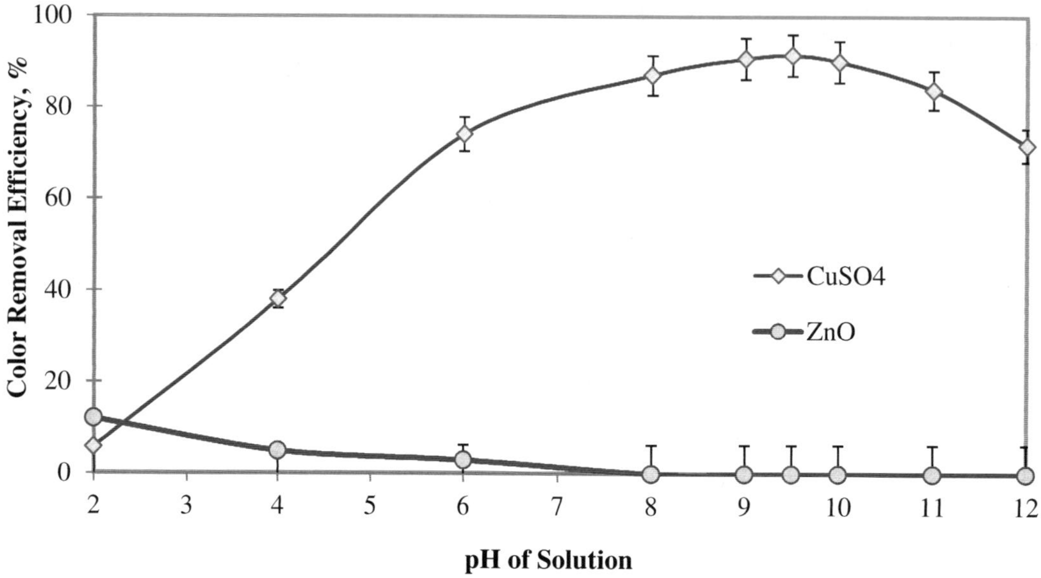

Figure 1. Color removal efficiency of RB 5 catalysed by ZnO and $CuSO_4$ with respect to pH (initial dye concentration: 100 ppm; temperature: 60°C; reaction time: 1 hour; catalyst mass loading: 0.5 g/L).

3.2. Extent of degradation analysis using characteristic light absorbance of dye

This research was further evaluated by using UV- Vis spectrum scanning from 200 – 800 nm to analyse the molecular structure changes for RB 5. Figure 2 presents the wavelength scan of RB 5 dye catalysed by ZnO at optimum pH 2 and $CuSO_4$ at optimum pH 9.5. Before thermolysis process, RB 5 wavelength scan presents three main chemical network strands: main band in the visible region located at 602 nm attributed to the azo linkage, UV region located at absorption bands of 325 and 270 nm were associated with naphthalene and benzene rings, respectively [14]. As the RB 5 was catalysed by ZnO at optimum pH 2, the three main chemical network strands show slight degradation. The visible region of the RB 5 shows slight decrement, indicating fragmentation of azo linkage which responsible for the color richness. Meanwhile, the slight decay of the absorbance at 325 nm indicating the destabilization of naphthalene ring whereas absorbance at 270 nm depicts the degradation of benzene ring. On the other hand, $CuSO_4$ catalyst is able to cleaved azo bond more efficiently compared to ZnO. The diazo bond in RB 5 has been broken down into single bond which attributed by the wavelength scan has been shifted to 530 nm. Moreover, the peak of naphthalene at 325 nm wavelength also reduced while the benzene peak at 270 nm increased sharply, attributed to the fragmentation of naphthalene to form benzene rings [15]. In thermalysis process, heat is transferred to energy in order to breakdown the RB 5 dye molecular structure. Sunsequently, with catalyst as the auxiliary agent, RB 5 dye degradation process was enhanced. From the result, $CuSO_4$ thermalcatalyst has greater catalytic characteristic and auxiliary agent for the molecular structure bonding fragmentation in treating RB 5 dye compared to ZnO.

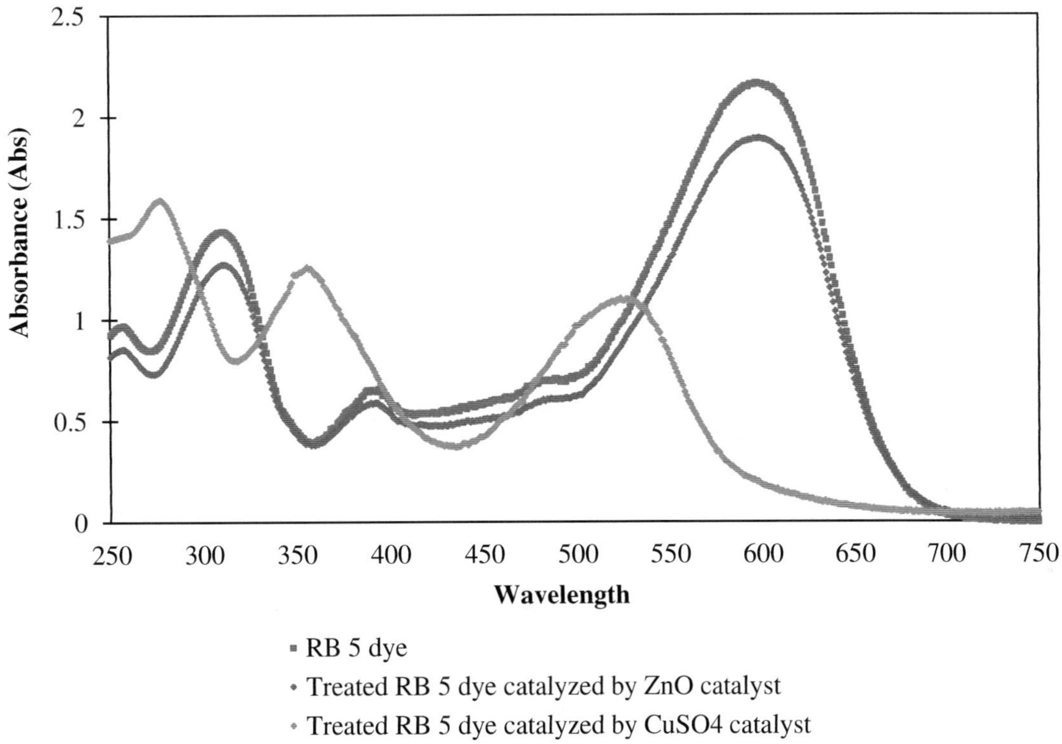

Figure 2. Wavelength scan of RB 5 dye, Treated RB 5 dye catalysed by ZnO and $CuSO_4$ catalysts (initial dye concentration: 100 ppm; temperature: 60°C; reaction time: 1 hour; catalyst mass loading: 0.5 g/L, ZnO at optimum pH 2 and $CuSO_4$ at optimum pH 9.5).

3.3. Degradation reactions comparison of CuSO4 and ZnO as thermalcatalysts in thermolysis process

Table 1 presents the comparison between performance of $CuSO_4$ and ZnO thermalcatalysts in terms of molecular structure fragmentation of RB 5 in thermolysis process. Both $CuSO_4$ and ZnO thermalcatalysts show different levels of removal percentage analysed using the dye absorbance reduction for each of the chemical strands wavelength. The bonding of azo bond was preferentially removed in the thermolysis process. $CuSO_4$ catalyst shows higher azo bond molecular structure cleavage percentage which is around 64.42% compared to ZnO catalyst. This phenomenon is owing to the vacancies of copper ions to react with the nitrogen in azo bond [16]. On the other hand, the extent of naphthalene ring molecular structure cleavage promoted by $CuSO_4$ catalyst (73.08%) shows 15.8% higher than ZnO catalyst (4.62%) due to the difference in the removal mechanisms [17]. Whereas, a different scenario was observed for benzene ring molecular structure fragmentation. ZnO has reduced 2.66% cleavage of benzene ring, however there was increment of benzene ring when $CuSO_4$ was used as catalyst. The increment of benzene ring is due to the degradation of naphthalene ring has further degraded into benzene structure. Naphthalene ring would degrade more easily and rapidly than benzene compound since naphthalene ring is not as stable as benzene ring [18]. Therefore, $CuSO_4$ catalyst promising a better degradation capability of azo bond, naphthalene and benzene ring compared to ZnO catalyst.

Table 1. Comparison molecular structure fragmentation for $CuSO_4$ and ZnO thermalcatalysts

Molecular structure fragmentation		Percentage of removal, %	
		$CuSO_4$	**ZnO**
(i) Azo bond	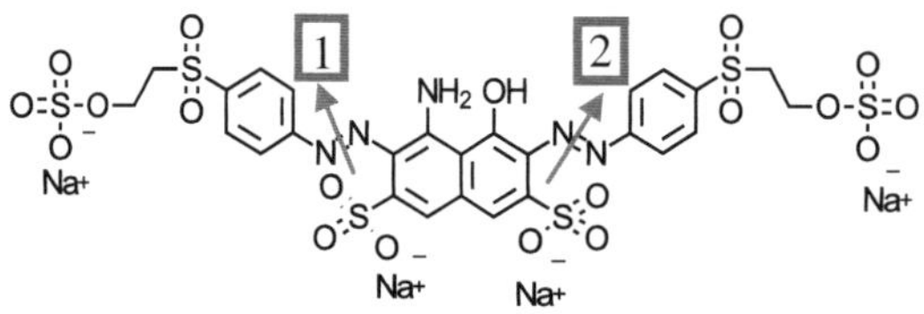	Reduce 64.42%	Reduce 8.84%
(ii) Naphthalene ring	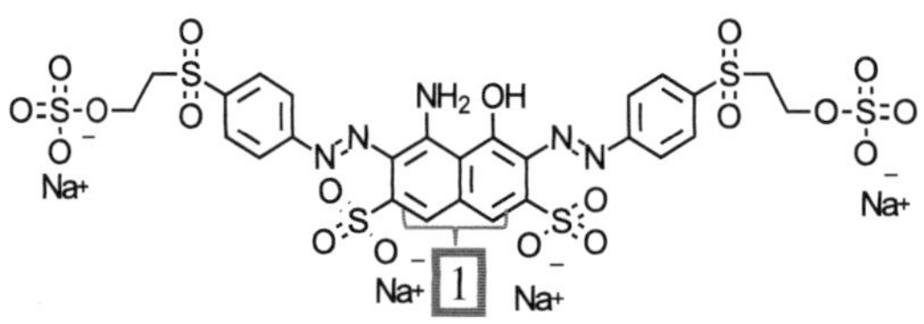	Reduce 73.08%	Reduce 4.62%
(iii) Benzene ring	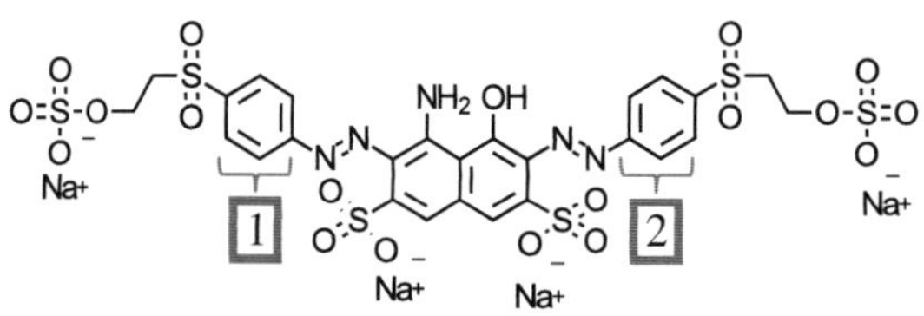	Increase 11.90%	Reduce 2.66%

4. Conclusion

In this work, the degradation of diazo reactive dye (RB 5) has been studied in a thermolysis treatment using chemical catalysts ($CuSO_4$ and ZnO). On the basis of the presented investigation it can be concluded that $CuSO_4$ catalyst exhibit effective dye degradation compared ZnO catalyst since $CuSO_4$ catalyst decolorized 91.55 % of RB 5 dye at pH 9.5 whereas ZnO catalyst decolorized 7.36 % of RB 5 dye at pH 2. Additionally, this research emphasized on the degradation of the three main chemical network strands: azo bond, naphthalene and benzene rings. $CuSO_4$ catalyst could degrade 64.42 % of azo linkage, 73.08% of naphthalene ring and 11.90 % of benzene ring. On the other hand, ZnO catalyst could only degrade 8.84 % of azo bond, 4.62 % of naphthalene ring and 2.66 % of benzene ring. The presented investigation develops a new platform to utilize $CuSO_4$ catalyst in the degradation of reactive dye.

5. References

[1] Garg A, Bhat K and Bock C 2002 *Dyes Pigments* **55** 35
[2] Khandare R V and Govindwar S P 2015 *Biotechnol. Adv.* **33** 1697
[3] Shajahan A, Shankar S, Sathiyaseelan A, Narayan K S, Narayanan V, Kaviyarasan V and Ignacimuthu S 2017 *Int. J. Biol. Macromol.* **104** 1449
[4] Naeme N and Ehsan S 2017 *Appl. Surf. Sci.* **413** 41
[5] VincenzoV, Mariantonietta M, Olga S and Diana S 2017 *Appl. Catal. B: Environ.* **209** 621
[6] Hu E, Wu X, Shang S, Tao X, jiang S and Gan L 2016 *J. Clean. Prod.* **112** 4710
[7] Reddy D R, Dinesh G K, Anandan S and Sivasankar T 2016 *Chem. Eng. Process. : Process Intensification* **99** 10
[8] Ning X, Chen H, Wu J, Wang Y, Liu J and Lin M 2014 *Chem. Eng. J.* **242** 102
[9] Han Y, Li H, Liu M, Sang Y, Liang C and Chen J 2016 *Sep. Purif. Technol.* **170** 241

[10] Serkan S. 2013 *J. Ind. Engineering Chem.* **19** 601
[11] Su C X H, Teng T T, Wong Y S, Norhashimah M and Rafatullah M 2016 *Chemosphere* **146** 503
[12] Kumar P, Prasad B, Mishra I M and Chand S 2007 *J. Hazard. Mater.* **149** 26
[13] Greenwood N N and Earnshaw A 1997 *Chemistry of the Element* 2nd ed (Elsevier Butterworth-Heinemann)
[14] Zhang Y, Gao F, Wanjala B, Li Z, Cernigliaro G and Gu Z 2016 *Appl. Catal. B: Environ.* **199** 504
[15] Zhu W, Liu P, Xiao S, Wang W, Zhang D and Li H 2015 *Appl. Catal. B: Environ.* **172** 46
[16] Gaurav H, Sandhya S and Gajanan P 2017 *Nano-Struct. Nano-Objects* **12** 1
[17] Nafaa A, Lotfi M, Nizar B and Jamel-Eddine B 2004 *J. Hazard. Mater.* **112** 207
[18] Lau Y Y, Wong Y S, Teng T T, Norhashimah M, Rafatullah M and Ong S A *Chem. Eng. J.* **246** 383

Acknowledgments
This research work was supported by the School of Environmental Engineering, Universiti Malaysia Perlis (UniMAP) and fund granted by Malaysia Ministry of Higher Education (FRGS: 9003:000516).

Electricity Generation and Community Wastewater Treatment by Microbial Fuel Cells (MFCs)

S Rakthai[1], R Potchanakunakorn[1], A Changjan[1], N Intaravicha[1], P Pramuanl[1], P Srigobue[1], S Soponsathien[1], C Kongson[1] and A Maksuwan[1]

1 Department of Environmental Science and Technology, Faculty of Science and Technology, Pathumwan Institute of Technology, 833 Rama 1 Road, Wangmai, Pathumwan, Bangkok 10330, Thailand
E-mail: nokrakthai@yahoo.com

Abstract. The attractive solution to the pressing issues of energy production and community wastewater treatment was using of Microbial Fuel Cells (MFCs). The objective of this research was to study the efficiency of electricity generation and community wastewater treatment of MFCs. This study used an experimental method completely randomized design (CRD), which consisted of two treatment factors (4x5 factorial design). The first factor was different solution containing organic matter (T) and consisting of 4 level factors including T_1 (tap water), T_2 (tap water with soil), T_3 (50 % V/V community wastewater with soil), and T_4 (100% community wastewater with soil). The second factor was the time (t), consisting of 5 level factors t_1 (day 1), t_2 (day 2), t_3 (day 3), t_4 (day 4), and t_5 (day 5). There were 4 experimental models depending on containing organic matter (T_1-T_4). The parameter measured consisted of Open Circuit Voltage (OCV), Chemical Oxygen Demand (COD), Total Dissolve Solid (TDS), acidity (pH), Electric Conductivity (EC) and number of bacteria. Data were analysed by ANOVA, followed by Duncan test. The results of this study showed that, the T_3 was the highest voltage at 0.816 V ($P<0.05$) and T_4, T_2, and T_1 were 0.800, 0.797 and 0.747 V, respectively. TheT_3 was the lowest COD at 24.120 mg/L and T_4 was 38.067 mg/L ($P<0.05$). The best model for electricity generation and community wastewater treatment by Microbial Fuel Cells was T_3. This model generated highest voltage at 0.816 V, and reduction of COD at 46.215%.

1. Introduction

Increasing scarcity of water and decreasing energy sources have been regular phenomenon in recent days [1]. Both water quantity and quality are the biggest environmental problems. Rapid population increased in urban areas and industrialization has raised the concern about appropriate wastewater management practices [2]. Surface waters are being polluted by means of wastes or effluent discharge from the industries, domestic sewage, and municipal wastes etc. Further, land application of wastewaters is now becoming one of the most economically and ecologically viable method of disposal of this water. With rapid expansion of cities and domestic water supply, quantity of grey/ wastewater is increasing in the same proportion. Overall analysis of water resource indicates that in coming years, these will be a twin edged problem to deal with deduced fresh water availability and increased wastewater generation. Like water problem, energy source scarcity is serious problem. Because the world's known oil and gas reserves are not infinite, it is necessary to evaluated alternative and renewable energy sources such as wind, solar and hydropower [3]. Energy efficiency and demand-side management—doing more with less, reducing energy consumption by substituting fuels and technologies and altering consumer behaviour is clearly the most environmentally being way to address increases in demand for energy services. Energy efficiency can include practices as diverse as switching from conventional coal power plants to combined heat and power units, lowering thermostats, better maintaining industrial boilers, and walking or cycling instead of driving. These actions not only involve very little damage to the environment [4]. Energy conservation seems to be the most effective method to mitigate both air pollution and greenhouse gases. Energy conservation measures such as demand side management have been attempted at national level. However, management development, transfers of better energy conservation technologies are still needed for further control of energy consumption and emission. Because of both serious environmental problems, microbial fuel cell (MFC) is the best optimization way for water and energy management. Microbial fuel cells provide a method of adding wastewater to the list of renew energy source. Microbial fuel cells (MFCs) provide new opportunities for the sustainable production of energy from biodegradable, reduced compounds. MFCs functions on different carbohydrates but also on complex substrates present in wastewaters [5]. As yet there is limited information available about the energy metabolism and nature of the bacteria using the anode as electron acceptor; few electron transfer mechanisms have been established unequivocally [6]. To optimize and develop energy production by MFCs fully this knowledge is essential. Depending on the operational parameters of the MFCs, different metabolic pathways are used by the bacteria. This determines the selection and performance of specific organisms. The MFCs technology is evaluated relative to current alternatives for energy generation. The objective of the research is to study the efficiency of electricity generation and community wastewater treatment of MFCs.

2. Materials and methods

2.1. Construction and designs

The sixty microbial fuel cell (MFC) was single chamber cell and were based on anode and cathode compartment. The chamber was a glass test tube (25 x 150 mm). The anode electrode was a copper plate felt on the bottom (1x7 cm), and connected with a 20 cm copper wire that used for voltage measurement. The cathode electrode was zinc plate with 1 x7 cm. The end of zinc plate was dipped in the liquid and the other side was connected with the top of the chamber.

2.2. Experimental design

The experiment was conducted using two factorials completely randomize design with three replications. The 1st factor was different solution containing organic matter (T), consisted of 4 levels, the 2nd was the time, and consisted of 5 levels sampling time, namely: Factor (t_1: day 1), (t_2: day 2), (t_3: day 3), (t_4: day 4), and (t_5: day 5). All of MFCs were tested in the same environment and were filled with different solution. The whole treatment consisted in:

2.2.1. (T_1): 55 ml of tap water

2.2.2. (T_2): 45 ml of tap water + 20 grams of soil

2.2.3. (T_3): 45 ml of 50% v/v community wastewater + 20 grams of soil

2.2.4. (T_4): 45 ml of 100% v/v community wastewater + 20 grams of soil

2.3. Parameter measurement and statistical analysis

Parameter measured consisted of Open Circuit Voltage (OCV), Chemical Oxygen Demand (COD), acidity (pH), Electrical Conductivity (EC) and Total Dissolve Solid (TDS). The experimental data obtained and analysed by using analysis of variance (ANOVA). If there was a real effect then continued with multiple range test (Duncan) with a significance level of 0.05.

3. Results and discussion

3.1. Soil properties.

Ampawa soil series from Samut Songkram province, Thailand, was used in this experiment due to properties of soil. The outstanding characteristic of this soil was silly clay loam (SiCL) that suitable for anaerobic bacteria activity. Initial properties of the soil were analyzed before the initiation of the experiment and were presented in Table 1.

Table 1. Soil properties.

Soil properties	Unit
Soil texture	silly clay loam (SiCL)
Sand (%)	16
Silt (%)	45
Clay (%)	39
pH	7.6
Organic maters (mg/L)	2.4
Phosphorus (mg/kg)	86.9
Potassium (mg/kg)	10.1
Calcium (mg/kg)	417.3
Magnesium (mg/kg)	29.5

Table 1 shown that soil particle of this experiment was silly clay loam (sand: silt: clay = 16:45:39). They were high phosphorus and organic maters, moderate calcium and low magnesium and potassium.

3.2 Community wastewater

Water from Saen Seab Canal, Thailand, was represented community wastewater in this study due to water quality analysis. The pollution control department of Thailand insisted that people should not use this water for drinking without high standard treatment process and treatment for drinking or bathing was worthless. The some water qualities were shown in Table 2.

Table 2. Community wastewater quality.

Parameters	Unit
Temperature (C)	28
pH	7.2
EC (uS/cm)	288
TDS (mg/L)	142
COD (mg/L)	36

3.3 Production of voltage

Data was collected form MFCs after one day of bacterial growth. A multi-meter was connected to each MFC and the initial voltage and current reading were taken.

Table 3. Duncan' s multiple rang test of production of voltage over time

Design Number	Voltage (V)					
	day 1	day 2	day 3	day 4	day 5	Mean
T_1	0.929	0.936	0.645	0.613	0.615	0.747^b
T_2	0.886	0.785	0.760	0.801	0.757	0.797^a

T_3	0.891	0.838	0.816	0.817	0.717	0.816 [a]
T_4	0.776	0.821	0.811	0.811	0.780	0.800 [a]
Mean	0.870 [a]	0.845 [b]	0.758 [c]	0.760 [c]	0.717 [d]	

Note: Difference letters indicate there are significant differences
T_1: MFC with 55 ml of tap water
T_2: MFC with 45 ml of tap water + 20 grams of soil
T_3: MFC with 45 ml of 50% v/v community wastewater + 20 grams of soil
T_4: MFC with 45 ml of community wastewater + 20 grams of soil

The Table 3 showed that, the average cell voltage of day 1 was significantly higher than day 2, 3, 4 and 5 ($P<0.05$) at 0.870, 0.845, 0.758, 0.760 and 0.717 V, respectively because of microbial activity. This showed that microorganism could degrade organic matter in the community wastewater into CO_2, H_2O and also electron [7]. Electrons produced by bacteria from substrates are transferred to the anode and flow to the cathode to produce electricity. Bacterial need energy source for growing and increasing then over time, the nutrient is reduced [8]. Therefore, it can be conducted that increasing time can reduce amount of voltage.

The average cell voltage of T_3, T_4 and T_2 were significantly higher than T_1 ($P<0.05$) at 0.816, 0.800, 0.797 and 0.747 V, respectively, because of soil property. This showed that soil structure include minerals and nutrients that support microbial activity. Microorganisms oxidize the organic matter in wastewater and transfer the electrons to the anode [9]. The hydrogen ions produced by the oxidation haft reaction pass through the cathode compartment where they reduced oxygen to water (the reduction half reaction). As a result of these two haft reaction, a potential difference develops between the anode and the cathode and current flows in the external circuit [10]. Therefore, it can be conducted that soil component can effected amount of voltage.

3.4 Community wastewater improvement

The chemical oxygen demand (COD) was determined by open flux method as described by Tendon. The total dissolve solid (TDS), Electricity Conductivity (EC) and acidity (pH) of water were analysed immediately after bringing the samples to the laboratory and the results as showed in Table 4.

Table 4. Duncan' s multiple rang test of water quality over time

Design Number	day 1	day 2	day 3	day 4	day 5	Mean	Percentage decline in the value
	Chemical Oxygen Demand (mg/L)						
T_1	0.000	0.000	0.000	0.000	0.000	0.000^a	0.000
T_2	0.000	0.000	0.000	0.000	0.000	0.000^a	0.000
T_3	35.267	20.800	22.933	20.800	20.800	24.120^b	41.000
T_4	42.533	38.933	36.800	36.800	35.267	38.067^c	17.000
Mean	19.450^b	14.933^a	14.933^a	14.400^a	14.017^a		
	Total Dissolve Solid (TDS (mg/L))						
T_1	52.667	48.000	10.667	10.333	12.333	26.800^d	
T_2	39.000	62.000	78.000	95.333	116.333	78.133^c	
T_3	184.000	197.667	213.667	214.667	229.000	207.800^b	
T_4	207.333	205.667	233.000	247.667	251.333	229.000^a	
Mean	120.750^e	128.333^d	133.833^c	142.000^b	152.250^a		
	Electric Conductivity (EC (mS/cm))						
T_1	106.667	96.000	21.333	21.333	24.667	54.000^d	
T_2	78.000	124.000	156.000	190.000	232.667	156.133^c	
T_3	368.000	392.000	427.333	430.000	456.667	414.800^b	
T_4	414.667	413.333	466.000	492.667	502.667	457.867^a	
Mean	241.833^e	256.333^d	267.667^c	283.500^b	304.167^a		
	pH						
T_1	7.523	7.550	8.300	8.467	9.100	8.188^a	
T_2	7.683	7.480	7.647	7.997	8.187	7.799^b	
T_3	7.380	7.350	7.707	7.833	8.033	7.661^c	
T_4	7.523	7.637	7.797	7.643	8.047	7.729^b	
Mean	7.528^d	7.504^d	7.862^c	7.985^b	8.342^a		

Note: Difference letters indicate there are significant differences
T_1: MFC with 55 ml of tap water
T_2: MFC with 45 ml of tap water + 20 grams of soil
T_3: MFC with 45 ml of 50% v/v community wastewater + 20 grams of soil
T_4: MFC with 45 ml of community wastewater + 20 grams of soil

Based on Table 4, the chemical oxygen demand of day 5, 4, 3 and day 2 were non significance but they were significantly lower than day1 ($P<0.05$). The COD of day 1 was the highest at 19.450 mg/L then reducing steadily to 14.933, 14.933, 14.400 and 14.017 mg/L, respectively, due to limiting of microbial degradation [11]. Because the microbial fuel cell generated electricity by anaerobic process, microorganism in this experiment is anaerobic or facultative aerobic bacteria. Both types of bacteria are not potential to oxidize organic compounds chemically that concentration in community wastewater [12]. Therefore, it can be conducted that increasing experimental time cannot reduce value of COD.

The COD level observation of T_4 was significantly higher than T_3 ($P<0.05$) at 38.067 and 24.120 mg/L, respectively, due to initiative substrate concentration [13]. The T_3 was able to reduce the COD level high at 41% followed by T_4 which was able to reduce at 17%.

The Total Dissolves Solid value of day 5 was significantly higher than day 4, 3, 2 and 1 ($P<0.05$) at 152.250, 142.000, 133.833, 128.333 and 120.750 mg/L, respectively, because of zinc corrosion. The TDS value increased steadily overtime because zinc ion passes through the anode compartment. The anode compartment contain high dissolve solid [14]. Therefore, it can be conducted that increased experimental time can affect the TDS value.

The TDS value of T_4 was significantly higher than T_3, T_2 and T_1 at 299.000, 207.800, 78.133 and 26.800 mg/L ($P<0.05$), respectively, due to soil property [15]. There are a lot of not only positive charges such as potassium, sodium, hydrogen, calcium and magnesium but also negative charges such as nitrate and sulfate in soil. Both ions suspend in compartment that cause all treatment with soil was high TDS value.

The Electric Conductivity value of day 5 was significantly higher than day 4, 3, 2 and 1 ($P<0.05$) at 304.167, 283.500, 267.667, 256.333 and 241.833 mg/L, respectively, due to zinc corrosion [16]. As same as TDS value, the EC value increased steadily overtime because zinc ion passes through the anode compartment. The anode compartment contain with high ion and increasing EC value [14].

The EC value of T_4 was significantly higher than T_3, T_2 and T_1 ($P<0.05$) at 457.867, 414.800, 156.133 and 54.000 mg/L, respectively, because of ion absorption [17]. The soil particle due to the presence of charges absorbs and exchange ions in solution. When zinc ion is added to water, by corrosion, it is adsorbed on the charge of soil particle. Therefore, it can be conducted that soil was important effect on EC value [18].

The pH value of day 5 was significantly higher than day 4, 3, 2 and 1 ($P<0.05$) at 8.642, 7.985, 7.863, 7.504 and 7.528, respectively, due to microbial activity [15]. Anaerobic bacteria break down organic matter in community wastewater and produce proton, electrons, alcohol and carbon dioxide. The anaerobic products are used to convert to carbonate compound such as calcium carbonate, zinc carbonate that increasing pH value [19].

The pH value of T_1 was significantly higher than T_2 T_4 and T_3, ($P<0.05$) at 8.188, 7.799, 7.729 and 7.661, respectively, due to soil property. Like TDS and EC, dissolve cations organic matter were absorbed by soil particle. Because of pH buffer of soil property, the pH value of T_2 T_4 and T_3 which contain soil in anode chamber were slightly changed [20].

4. Conclusion

The objective of this research was to study the efficiency of electricity generation and water treatment of Saen Saeb Canal by Microbial Fuel Cells (MFCs). The experimental period was 4 month and collecting data in 5 days. This study used an experimental method completely randomized design (CRD), which consisted of two treatment factors (4x5 factorial design). The first factor was different solution containing organic matter (T) and consisting of 4 level factors including T_1 (tap water), T_2 (tap water with soil), T_3 (1:1 tap water: water from Saen Saeb Canal with soil) and T_4 (water from Saen Saeb Canal with soil). The second factor was the time (t), consisting of 5 level factors t_1 (day 1), t_2 (day 2), t_3 (day 3), t_4 (day 4) and t_5 (day 5). There were 4 experimental models depending on containing organic matter (T_1–T_4). The parameter measured consisted of Open Circuit Voltage (OCV), Chemical Oxygen Demand (COD), Total Dissolve Solid (TDS), acidity (pH) and Electric Conductivity (EC). Data were analysed by ANOVA, followed by Duncan test. The results of this study showed that, the T_3 was the highest voltage at 0.81 V and T_4, T_2 and T_1 were 0.80, 0.79 and 0.74 V, respectively. The T_3 was the lowest COD at 24 mg/L and T_4 was 38 mg/L. The T_1 was the lowest TDS at 27 mg/L and T_2, T_3 and T_4 were 78, 208 and 299 mg/L, respectively. The T_3 was the lowest pH value at 7.7 and T_4, T_2 and T_1 were 7.7, 7.8 and 8.2, respectively. The T_1 was the lowest EC value at 54 μs/cm and T_2, T_3 and T_4 were 156, 415 and 458 μs/cm, respectively ($P<0.05$). The best model for electricity generated and water treatment of Saen Saeb Canal by Microbial Fuel was T_3. This model generated highest voltage at 0.81 V and reduction of COD at 46.215%.

5. Acknowledgments
The financial supports of Department of Environmental Science and Technology, Faculty of Science and Technology, Pathumwan Institute of Technology are acknowledged.

6. References
[1] Jogan H and Dasog G S 2015 *Karnataka J. Agric. Sci.* 28(4) 518
[2] Wibulswas P 2004 *Asian J.Energy Envion.* 5(4) 285
[3] Patra A 2008 *SJWP* 1 72
[4] Koyama K 2014 *EIIJ.* 1 1
[5] Bruce E L, Bert H, Rene R, Uwe S, Jurg K, Stefano F, Peter A, Willy V and Korneel R *Karnataka 2006 Environ. Sci. Technol.* 40(17) 5181
[6] Korneel R and Verstraete W 2005 *Trend Biotechnol* 23(6) 291
[7] Devid P B T S, Hemelers H V M, Jan F H S and Cees J.N. B. 2008 *Int.J.Energy Res.* 32(9) 870
[8] Zhenhua G, Wu H, Shi L, Zhang X, Sheng, Yin F and Gooneratne R 2017 *Anim. Nutri.* 3 109
[9] Logan B E and Regan J M 2006 *Environ. Sci. Technol.* 1 5172
[10] Nattawut K and Kanyarat H 2015 *KKu Engineering Journal* 42(1) 117
[11] Pongludda P Weerasak S and Chanchainarong S 2012 *RMU J.* 6(1) 129
[12] Muhammad A Abdul N and Harroon R 2016 *J. Agric. Res.* 54(1) 97
[13] Younghyun P, Cho H, Yu J, Min B, Kim H S, Kim B G and Lee T 2017 *Bioresour. Technol.* 233 176
[14] Hassana L M, Bruggen JJA V and Lens P N L 2015 *Ecol Eng* 84 588
[15] Irene M U, Kennedy A C and Muzika R M 1009 *Appl. Soil Ecol* 42 1
[16] Payel C, Uddy U S P, Mahata N, Tiwari O N, Ray R N, Bandyopdhyay T K and Bhunia B 2017 *Renew. Sustainable Energy Rev.* 79 372
[17] Abbasi H, Salehi S, Ghorbani R, Torabi F and Amidpour M 2013 *IJEE.* 4(2) 110
[18] Irene M U, Motavalli P P and Muzika R M 2009 *Agric. Ecosyst. Environ.* 131 105
[19] Sara M, Angelo A D, Scialdone O, Canizares P, Rodrigo M A and Morales F J F 2017 *Biochem. Eng. J.* 123 38
[20] Zhang P, Zhang J, Pan G, Ahang X, Li L and Tippkotter R 2007 *Colloids Surf B Biointerfaces* 58 264

The Development of Microbial Fuel Cells (MFCs) By Haplusterts Soil (Samo - Thod Series)

N Intaravicha [1] and A Changjan [1]

1Department of Environmental Science and Technology, Faculty of Science and Technology, Pathumwan Institute of Technology, Thailand
E-mail: kung32123@gmail.com

Abstract. In this paper, we investigated on simultaneous electric energy production and organic matter was removed from synthetic wastewater by Microbial Fuel Cells (MFCs). Single chamber MFCs was made up by Haplusterts great group soil (Samo - Thod soil group) in trial design 3 x 3 factorial design in Completely Randomize Design (CRD) which 3 levels synthetic wastewater; 0, 200 and 400 mg/l of glucose and 3 levels of flooding time: 1, 3 and 5 days. The results showed the interaction significant with decreasing sugar from synthesis wastewater and Open Circuit Voltage (OCV). The maximum OCV of 200 and 400 mg/l of glucose in 3 flooding days were 131 and 142 mV and decreasing to 110 and 126 mV in 5 flooding days, respectively. The highest percent of decreased glucose approached to 80% in 5 flooding days of 0.4 g/l of glucose. The findings suggested that not only MFCs were a significantly to reduce organic matter in wastewater but also generated electric energy in the same time.

1. Introduction

It is well known that municipal wastewater was a major problem for the urban community. In municipal wastewater structures, most components consisted of carbohydrate, protein, lipid, plant and microbes nutrients [1]. When it contaminated in the water resources, oxygen gases in the water reduced dramatically. Because, microorganisms were rapid growth, that was caused of oxygen deficiency in the water. In aquatic ecosystems, living organisms, plant in to the water resources, seem to be bloomed and died, fish seem to be died too. So, municipal wastewater could be treated before draining it in to the environment [2], [3]. In the present, this problem solved by wastewater treatment construction with aerobic and/or anaerobic bioreactor. Because, this systems were high efficiency to sewage treatment but they were high cost for establishment and maintenance too [4].

Microbial Fuel Cells (MFCs) could be used in wastewater treatment facilities to mineralize organic and inorganic matters and generate electricity as a by-product [5]-[8]. The advantages of MFCs application in wastewater treatment system was higher COD removal (40–80%) compared to 20–50% COD removal by anaerobic treatment at the same retention time [9]. However, a limitation of MFC development was that the use of expensive materials such as the metal platinum or gold catalyst, proton exchange membranes (PEMs), mediators, and graphite electrodes [10].

In this paper, we designed low operation cost single chamber MFCs by homemade equipment and investigated the effect of synthesis municipal wastewater concentration and periods of wastewater flooding time, concern with wastewater treatment and generated direct current electrical in laboratory scale.

2. Materials and Methods

Soil sample

Collected soil from with soil taxonomy, very-fine, smectitic, isohyperthermic Chromic Haplusterts (Samo-Thod series) in Thailand. 15 centimetres depth of soil, air dried in room temperature, grinded and sifted with 2 millimetre diameter of sieve, analyzed pH, organic matter (OM), soil particles, CEC, available phosphorus (P), exchangeable potassium (K) calcium (Ca) and magnesium (Mg) that suggested by USDA [11]. As shown in Table 1.

Table 1. Soil properties (Very-fine, smectitic, isohyperthermic Chromic Haplusterts: Samo-Thod series) [10].

	Soil sample
pH	5.50
OM (%)	1.94
Clay (%)	75.70
Silt (%)	21.50
Sand (%)	2.80
CEC (me/100g)	45.06
P (ppm)	61.88
K (ppm)	126
Ca (ppm)	8,120
Mg (ppm)	280

Electric potential experiment

10 grams of soil sample per unit in test tube 20x150 millimetre, 3x3 Factorial in CRD that was designed with 3 replications. This experiment was incubated with, 25 ml synthesis wastewater by glucose, 3 levels of synthesis wastewater such as 0, 200 and 400 mg/l of glucose, 3 levels of flooding times such as 1, 3 and 5 days, incubated 25 ^{o}C in room temperature, with 2 electrodes per unit by electric wire, copper wire 0.5 millimetre diameter, 20 centimetres length per each, as shown in Figure 1.

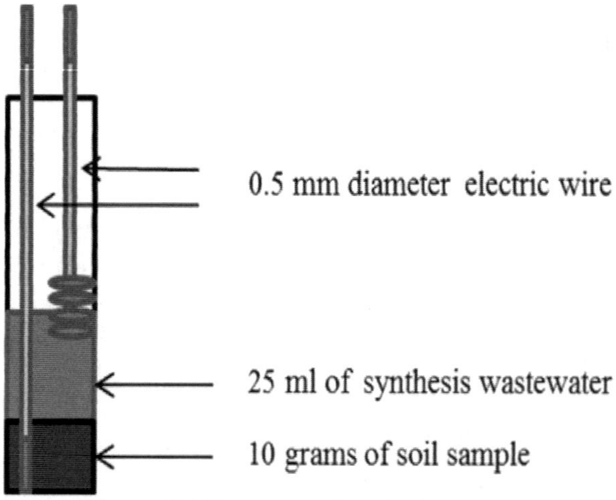

Figure 1. Wastewater incubation per unit.

The glucose remaining detected by phenol & sulfuric method [12], 1 ml of wastewater aliquot, added 1 ml of 5% w/w phenol solution, Mixed it together and left 20 minutes reaction. Put 5 ml sulfuric acid directly in the test tube, mixed it together and waited until the solution cool in room temperature. Analyzed by UV/VIS spectrophotometer (Biochrom model: Libra S12) at wavelength 410 nm. The electric potential discovered with Open Circuit Voltage (OCV) in direct current electrical (mV) from 2 electric wires. The statistics were calculated with ANOVA for 3x3 Factorial in CRD experimental design, Duncan's multiple range test, this observation was provided significant levels at 0.01.

3. Results and discussions

In Figure 2, after the first days, glucose in synthesis wastewater at 200 mg/l and 400 mg/l of glucose treatment was slightly decreased, until remaining 84.2 and 86.3 percentage respectively. While after the third days, glucose was dramatic decreased to 41.5 and 43.1 percentage respectively. After the fifth days, was declined slightly, glucose remaining were 15.5 and 11.6 percentage respectively. The ANOVA analyzed this experiment discovered highly significant different ($P<0.01$) with 3 levels of synthesis wastewater and 3 levels of the days incubation. In addition, this trial found synthesis wastewater and the days incubation interaction too.

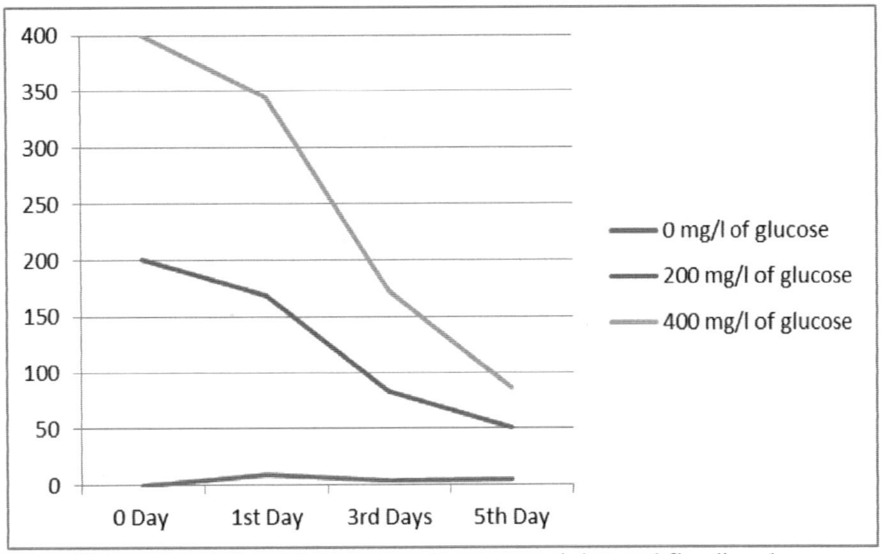

Figure 2. Relationship between glucose remaining and flooding days.

Next, focus on the effect of glucose levels and incubation time interaction with electricity product. In glucose 0 mg/l treatment, OCV was slightly increased from 119.3 to 125 mV. While, glucose 200 mg/l treatment and 400 mg/l treatment, the OCV in 3rd day were raised from 1st day, at 122.7 to 131.0 and 133.3 to 142.3 mV respectively. However, in 5th day, the OCV were dramatic decreased in both experiments at 131.0 to 109.7 and 142.3 to 126.7 mV respectively as shown as in Figure 3. In part of pure effect with incubation days, highly significant different, 3rd day were highest of electricity production at 131.4 mV, 1st day were electricity production at 125.1 mV and 5th day were electricity production at 120.4 mV respectively. In the part of pure effect with the glucose levels, glucose levels at 0 and 200 mg/l had not significant different but, they were highly significant with glucose levels at 400 mg/l, as shown the results in Table 2.

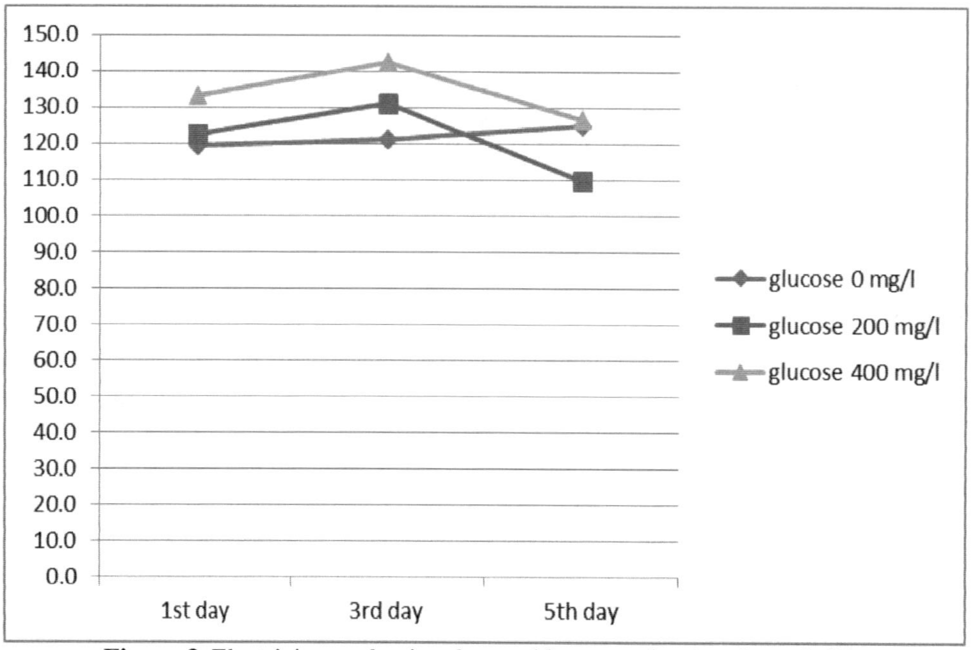

Figure 3. Electricity production detected by open circuit voltage (mV).

Table 2. Effect of glucose levels and incubation times on Open Circuit Voltage (mV).

Glucose levels (mg/l)	Incubation times 1st day	3rd day	5th day	Pure effect with glucose level (mg/l)
glucose 0 mg/l	119.3	121.0	125.0	121.8 z
glucose 200 mg/l	122.7	131.0	109.7	121.1 z
glucose 400 mg/l	133.3	142.3	126.7	134.1 y
Pure effect with incubation times	125.1 b	131.4 a	120.4 c	

R square =0.86, %CV = 3.41, Mean square error = 18.3, grand mean = 125.7.

This trial shown correlation between OCV and glucose treated that were direct variation. This phenomena could be applied to urban wastewater treatment and electricity production in the same time, which is metabolism by microorganism in wastewater and the soil. Nielssen et al. [13] were reported with microbial electricity generator in the soil, redox potential in the soil is very important factors that is efficient factor. Sorada [14] was studied concern with iron and manganese oxidizing that shown significant of electron accepters effective on the surface water, in front of Pak-Mun dam, Thailand. Tanji et al. [15] were researched in the paddy land, they discovered efficient electron accepters form in the paddy soil such as oxygen gas, nitrate, manganese dioxide, ferric hydroxide, sulfate and carbon dioxide respectively, that concern with thermodynamic sequence. The electron accepters is limiting factor for submerge microorganism [16]. While, Unger et al. [17] offered with soil microbial communities in flooding condition, they discovered with the group of microorganism, which were activities in the soil, both aerate and flooding condition that called facultative anaerobes, that can be treated sewage by natural way.

4. Conclusion

This demonstrate were summarized; glucose remaining after 5th day incubation from 0, 200 and 400 mg/l of glucose at 4.76, 50.92 and 86.42 mg/l respectively. The glucose treated was correlated with direct electricity production. The 3rd day incubation was highest OCV ineach glucose levels incubation when considered with incubation days and 400 mg/l of glucose level experiment was seems to be the highest electricity production when considered with glucose levels experiment.

5. References

[1] Hammer D A and Bastian R K 1989 Wetland Ecosystems: Natural water purifiers 5-19 D.A. *Hammer (ed.) Constructed Wetlands for Wastewater Treatment* (Lewis Publishers, Inc., Michigan)
[2] Eriksson E, Auffarth K, Henze M and Ledin A 2002 *Urban Water J* 4 85–104
[3] Taebe A and Droste R L 2004 *Sci Total Environ* 327 175–184
[4] Chan Y J, Chong M F, Law C L and Hassell D G 2009 *Chem Eng J* 155 1–18
[5] Sukkasem C, Xu S, Park S, Boonsawang P and Liu H 2008 *Water Resour* 42 4743–4750
[6] Aelterman P, Versichele M, Marzorati M, Boon N and Verstraete W 2008 *Bioresource Technol* 99 8895–8902
[7] Ahn Y and Logan B E 2010 *Bioresource. Technol* 101 469–475
[8] Kumar S S, Basu S and Bishnoi N R 2017 *Biochem Eng J* 121 17-24
[9] Logan B E 2008 *Microbial Fuel Cells* (John Wiley & Sons, Inc.,)
[10] Deng Q, Li X, Zuo J, Ling A and Logan B E 2010 *J Power Sources* 195 1130–1135
[11] United States Department Agricultural 2009 Soil Survey field and laboratory methods manual *Soil Survey Investigations Report* (51 Lincoln, Nebraska USA)
[12] Dubois M, Gilles K A, Hamilton J K, Rebers P A and Smith F 1956 *Anal Chem* 28 (3) 350–356
[13] Nielssen J, Harnische F, Rosenbaum M, Schroder U and Scholz F 2006 *Electrochem. Commun.* 8 869–873
[14] Sorada P 1999 Study on iron and manganese oxidizing bacteria at Pak-Mun hydro-power plant aera M.Sc. thesis (Kasetsart University Bangkok Thailand)
[15] Tanji K K, Gao S, Scardaci S C and Chow A T 2003 *Geoderma* 114 333– 353
[16] Ponnamperuma F N 1965 *Adv In Agron* 24 29-96
[17] Unger I M, Kennedy K C and Muzika R M 2009 *Appl Soil Ecolo* 42 1–8

IOP Conf. Series: Earth and Environmental Science **150** (2018) 012017 doi:10.1088/1755-1315/150/1/012017

The influence of reaction time on hydrogen sulphide removal from air by means of Fe(III)-EDTA/Fiban catalysts

H Wasag, W Cel, M Chomczynska and J Kujawska

Faculty of Environmental Engineering, Lublin University of Technology, Nadbystrzycka 40B, 20-618 Lublin, Poland
E-mail: h.wasag@gmail.com

Abstract. The paper deals with a new method of hydrogen sulphide removal from air by its filtration and selective catalytic oxidation with the use of fibrous carriers of Fe(III)-EDTA complex. The basis of these filtering materials includes fibrous ion exchangers with the complex immobilized on their functional groups. It has been established that the degree of catalytic hydrogen sulphide decomposition depends on the reaction time. Thus, the required degree of hydrogen sulphide removal from air could be easily controlled by applying appropriate thickness of the filtering layer under a given filtering velocity. It allows applying very thin filtering layers of the Fe(III)-EDTA/Fiban AK-22 or Fiban A-6 catalysts. The obtained results of the research confirm the applicability of these materials for deep air purification from hydrogen sulphide.

1. Introduction

Hydrogen sulphide is a highly toxic compound. It has a very typical smell and can be perceived at concentrations as low as 0.5 ppb. At low concentration the gas has an unpleasant odour, whereas at higher it could be life-threatening. Breathing hydrogen sulphide at a concentration higher than 500 ppm can be fatal after few breaths, due to its broad spectrum toxicity [1].

The means of removing hydrogen sulphide depend on its concentration. Various well-established techniques are available for the removal of hydrogen sulphide from sour gases. On the industrial scale, removal is performed by a widely used Claus reaction, whereas at lower concentrations, adsorption or oxidation to sulphur oxides is preferred [2]. The electrochemical (anode oxidation) method was recently introduced as an alternative way for sulphide control in geothermal brines, sewage, and caustic solutions. Sulphide species are electrochemically active and the oxidation decreases the sulphide species concentration by forming a wide spectrum of intermediates and final products [3]. Adsorbents, especially activated carbons, play an important role in limiting the emission of sulphur compounds (in particular H_2S) to the atmosphere. Thanks to their unique physicochemical properties, they can not only be effective adsorbents, but also catalysts or their carriers. Hence, the removal of gases containing sulphur can take place by physical or chemical adsorption or by catalytic oxidation on the carbon surface. The effectiveness of H_2S removal by activated carbons can be improved by their impregnation with caustic materials such as NaOH or KOH, copper(II) ions and by incorporation of nitrogen, sulphur and oxygen functional groups into the carbon structure [4]. $KMnO_4$ is dispersed on a porous support in order to remove hydrogen sulphide from ambient air,. Adsorbents of this type have been prepared from silica or alumina. Manganese dioxide (MnO_2) is also active in the removal of H_2S. MnO_2-impregnated sawdust has been successfully applied for the purification of ambient air [5]. A separate problem involves the removal of H_2S present in the air in trace amounts (10-100 $\mu g\ m^{-3}$) for clean rooms of fine mechanics, electronics, pharmaceutical productions as well as the purification of air from hydrogen sulphide in the concentrations exceeding 10 mg m^{-3} (maximum permissible concentration for working zones and surroundings of many productions) [6]. It is widely known that chelate complexes of trivalent iron convert hydrogen sulphide to elemental sulphur [7]. Bivalent iron formed in the reaction is oxidized by the atmospheric oxygen, so complex of trivalent iron is continuously regenerated and the overall process can be accounted as pseudo-catalytic [8]. Ferric chelates are widely used in purification of air from hydrogen sulphide i [9]. The catalyst is active only in strong alkali environment which causes serious corrosion problems [10]. In order to avoid these problems, fibrous ion exchangers were used as carriers of the catalyst. It was shown that the filtering layers with anion exchange package are much more active in the catalytic processes of hydrogen sulphide removal than cation exchanger and inert materials [11]. The latter worked only at RH (relative humidity) 80-90 % and higher. A higher catalytic activity of anion exchangers than that of cation exchanger and inert fibres proves that the Fe complex fixed on the cationic group of the fibre has a higher catalytic activity than that of the free complex. In the addition to the nature of the fibre (carrier of the catalyst), relative air humidity is a critical factor determining the efficiency of the material in purification of air from H_2S [12]. It was proven that the role of fibrous package involves improving the mass-transfer between the gas and liquid, buffering the pH of absorbing solution and catalysing the reactions of oxidation of sulphur and iron ions. The main parameters affecting the process of catalytic oxidation are: concentration of hydrogen sulphide in the air, relative humidity of the purified air, the process time and the content of Fe(III)-EDTA complex in the fibre [13]. It seems understandable that for a given set of conditions affecting kinetics of the process, the residence time in the mass-exchange unit controls the quality of the purified air. As that time simply depends on filtering layer thickness, conducting the presented research seemed reasonable.

2. Experimental

Fibrous ion exchangers Fiban used in the experiments were synthesized at the experimental production plant of the Institute of Physical Organic Chemistry National Academy of Sciences of Belarus (Minsk, Rep. of Belarus). The ion exchangers have been prepared by chemical modification of industrial polyacrylonitrile (PAN) fibre Nitron, produced in Polotsk petrochemical plant (Rep. of Belarus). The modification was performed with the staple fibres characterized by the fibre length 65 mm and effective diameter 20-22 μm. Chemically modified staple was processed to non-woven needle punctured canvas with surface density 500 g m^{-2} and 5 mm thickness [14].

Fiban A-6 is a product of aminolysis of polyacrylonitrile fiber by dimethylaminopropylamine, followed by quaternization of the tertiary amino group with epichlorohydrine. The ion exchanger contains strong and weak basic groups with the capacities 1.80 mmol g^{-1} and 0.65 mmol g^{-1}, respectively [6].

Fiban AK-22 is a polyampholyte containing primary and secondary amino groups and carboxylic acid groups. It is obtained by chemical modification of (PAN) polyacrylonitrile fibre. The predominant functional groups are R-COOH and $R\text{-}CO\text{-}NH_2\text{-}CH_2CH_2NH\text{-}CH_2CH_2NH_2$. The cationic exchange capacity is 1.5 mmol g^{-1}, the anionic exchange capacity is 4.0 mmol g^{-1} [15].

Preparation of the H_2S fibrous sorbent-catalysts was performed by soaking the canvasses or by casting the impregnation solution onto the surface of the canvas. The impregnation solution was prepared by mixing 0.3 mol L^{-1} solution of sodium salt of EDTA (ethylenediaminetetraacetic acid) and $FeCl_3 \cdot 6H_2O$ in molar ratio 1:2. Analytical grade reagents were purchased from POCH (Gliwice, Poland).

Removal of hydrogen sulphide from air was studied on the continuous flow experimental laboratory set presented in Figure 1. Hydrogen sulphide (1000 ppm $H_2S + N_2$, pure, Linde AG, Germany) was supplied from a gas cylinder (1) and it was diluted to the required concentrations by means of the air from a compressor (2). The air, before mixing with hydrogen sulphide, was humidified to obtain the relative humidity of the prepared mixture reaching approximately 90%. The relative humidity was measured by means of a hygrometer (AZ8721, Merazet, Poland). The concentrations of hydrogen sulphide in the gas stream before and after purification were measured by GFM101 (Gas Data Ltd., U.K.) gas analyser (11). The filter used in all experiments was of a frame type. This filter contains layers of the ion exchange non-woven material (impregnated with Fe(III)-EDTA complex) fixed on rigid frames arranged, as shown in Figure 2 (view from the top). The surface area of the filter was 0.04

m^2 (4 layers, 10 cm x 10 cm each). The feed air passes through the filter medium where hydrogen sulphide is oxidized to elemental sulphur. The ion exchange fibre was rinsed within 5 minutes with the impregnation solution of Fe(III)-EDTA each half an hour of the filtration process. The samples of the purified air were taken to analysis 10 minutes after rinsing three times, each in 5 minute intervals. The hydrogen sulphide concentration was calculated as an average value from three measurements.

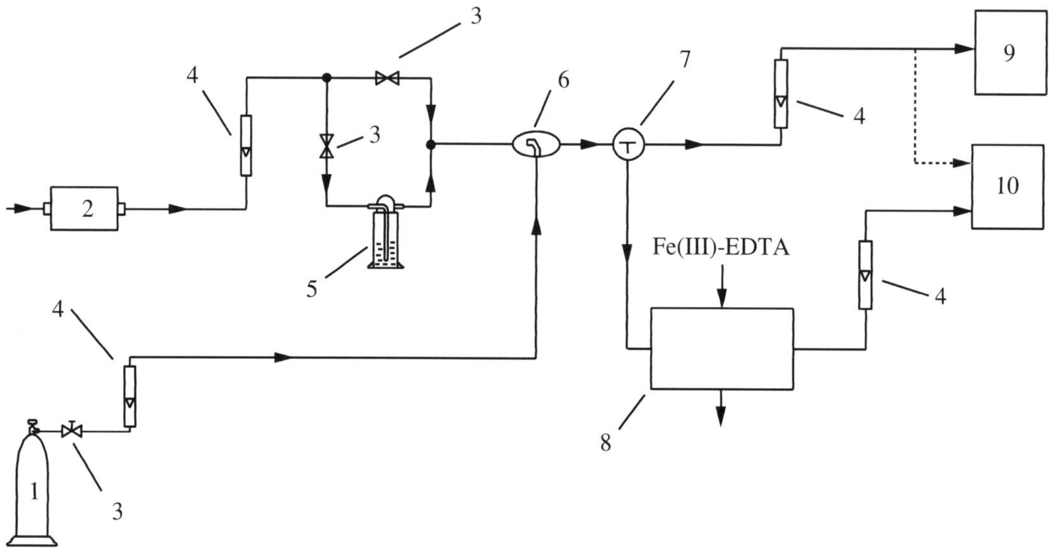

Figure 1. Scheme of experimental laboratory set-up (1 – source of hydrogen sulphide, 2 – air compressor, 3 – valve, 4 – flow meter, 5 – humidifier, 6 – mixing chamber, 7 – T-valve, 8 – filter with catalytic system Fe(III)-EDTA/Fiban, 9 – hygrometer, 10 – H_2S analyser).

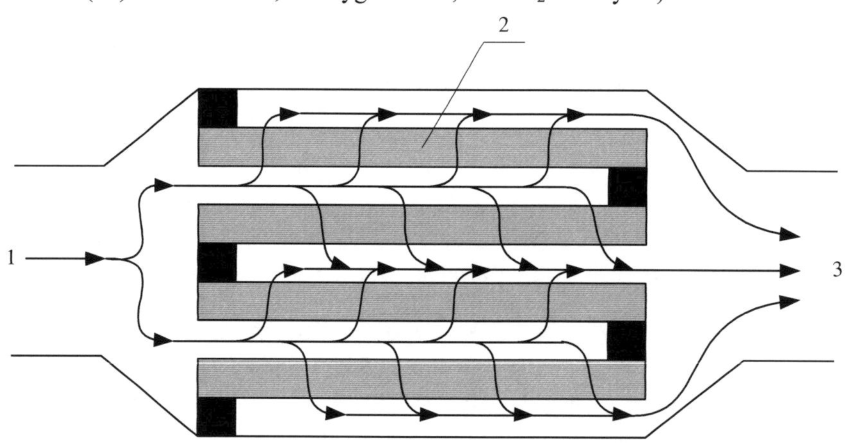

Figure 2. Scheme of the air flow through the frame type filter (1 – inlet of the air, 2 – fibrous ion exchanger layer impregnated with Fe(III)-EDTA, 3 – outlet of the purified air).

In the conducted laboratory research, the influence of filtering layer thickness on removal of hydrogen sulphide from the air was investigated for Fiban A-6 and Fiban AK-22 fibrous carriers of the catalyst for different filtration velocities and initial concentrations of hydrogen sulphide in the purified air.

3. Results and discussion

The obtained results of laboratory experiments are shown in Tables 1 - 2.

IOP Conf. Series: Earth and Environmental Science **150** (2018) 012017 doi:10.1088/1755-1315/150/1/012017

Table 1. Removal efficiency of H_2S from air by means of catalyst Fe(III)-EDTA/Fiban AK-22 depending on filtration velocity and filtering layer thickness for different initial concentrations of hydrogen sulphide in the purified air.

Initial H_2S concentration (ppm)	Flow velocity ($m\ s^{-1}$)	Filtering layer thickness (mm)	Removal efficiency (%)
30	0.005	2	94.45
		6	96.25
		10	97.40
	0.010	2	93.40
		6	95.65
		10	96.35
	0.020	2	91.15
		6	94.50
		10	95.30
60	0.005	2	92.46
		6	95.71
		10	96.96
	0.010	2	91.29
		6	93.63
		10	95.54
	0.020	2	87.98
		6	91.46
		10	93.75

Table 2. Removal efficiency of H_2S from air by means of catalyst Fe(III)-EDTA/Fiban A-6 depending on filtration velocity and filtering layer thickness for different initial concentrations of hydrogen sulphide in the purified air.

Initial H_2S concentration (ppm)	Flow velocity ($m\ s^{-1}$)	Filtering layer thickness (mm)	Removal efficiency (%)
30	0.005	2	98.97
		6	99.42
		10	99.58
	0.010	2	98.55
		6	99.00
		10	99.25
	0.020	2	97.83
		6	98.53
		10	98.95
60	0.005	2	98.67
		6	99.25
		10	99.46
	0.010	2	98.23
		6	98.88
		10	99.08
	0.020	2	97.37
		6	98.48
		10	98.83

Table 1 presents the influence of filtering layer thickness of Fe-EDTA/Fiban AK-22 catalyst on the concentration of hydrogen sulphide in the purified air under three different filtration velocity settings and for two different initial concentrations of hydrogen sulphide. In the case of Fe-EDTA/Fiban A-6 material, the results of the same experiments were summarized in Table 2.

From the data presented one could see that the concentration of hydrogen sulfide in the purified air decreases with increasing filtering layer thickness for both tested catalytic systems. If relative humidity of the purified air is not lower than 90%, then 2 mm filtering layer of Fe-EDTA/Fiban A-6 material allows removing more than 97% of H_2S. In the case of Fe-EDTA/Fiban AK-22 material, assuming the same process parameters, it is possible to obtain the degrees of hydrogen sulphide removal exceeding 90% and the filtering layer thickness of at least 80 mm. While analyzing the results, it could be seen that residence time of the purified air in the filtering layer plays a decisive role in the process. This time, it is a simple function of filtering layer thickness (s) and flow velocity (v):

$$t = s/v \qquad (1)$$

The dependence of the degree of hydrogen sulphide removal by Fe(III)-EDTA/Fiban catalysts on the residence time of the purified air in the mass-exchange unit for Fiban A-6 fibrous carrier is presented in Figure 3 and for Fiban AK-22 fibre – in Figure 4.

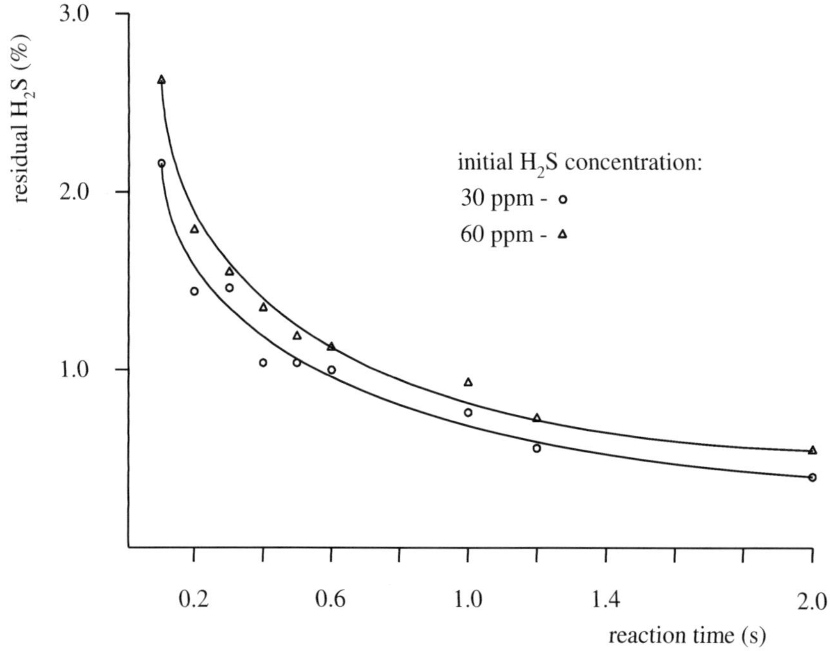

Figure 3. The influence of reaction time on residual hydrogen sulphide in the air purified by means of Fe(III)-EDTA/Fiban A-6 catalyst.

As could be seen from the results depicted in Figure 3, the degree of conversion of hydrogen sulphide to sulphur was higher than 99% if the residence time of the purified air in the mass-exchange unit was not under one second. Even for very short times of reaction, reaching 0.2 s, the residual hydrogen sulphide in the purified air was less than 2% of its initial amount. Very similar results, applying catalytic system Fe(III)-EDTA/Fiban A-6, were observed for both studied initial concentrations of hydrogen sulphide. For Fiban AK-22 fibrous carrier, the efficiency of the catalytic system in hydrogen sulphide removal is lower. Additionally, in this case the observed influence of the initial impurity concentration is stronger (see Figure 4). The obtained purifying results were noticeably worse in the presence of higher amounts of hydrogen sulphide, especially for short reaction time. Nevertheless, it was possible to remove more than 95% of the impurities with the reaction time of about 0.5 s for the lower studied concentration (30 ppm) and 1 s for the higher concentration (960 ppm).

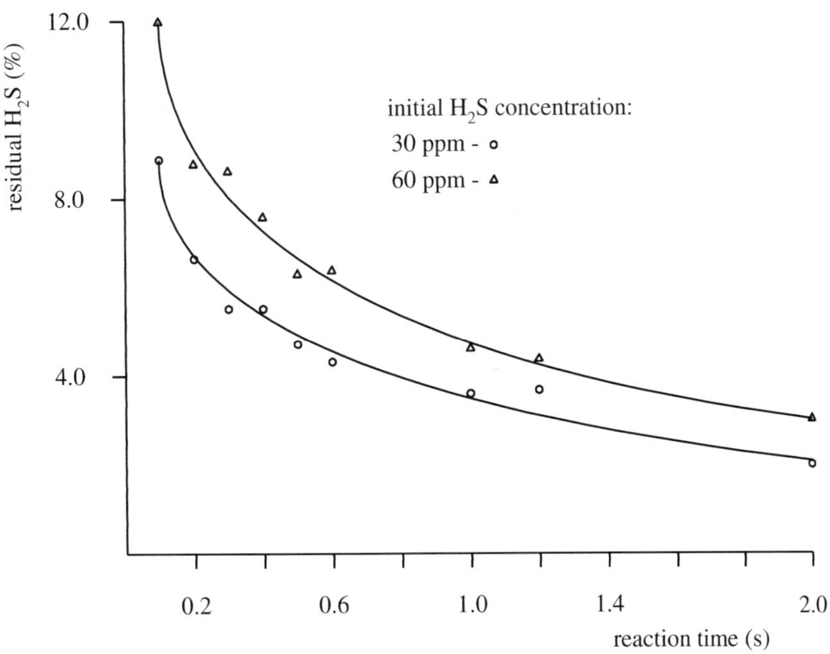

Figure 4. The influence of reaction time on residual hydrogen sulphide in the air purified by means of Fe(III)-EDTA/Fiban AK-22 catalyst.

In order to improve the performance of the catalytic system under a given set of external conditions, it is enough to increase the residence time of the purified air in the filtering material. Thus, the required degree of hydrogen sulphide removal from the purified air could be easily controlled by applying appropriate filtration velocity and thickness of the filtering layer materials, paying attention to a satisfactory level of relative humidity characterizing the purified stream.

4. Conclusions

There are many parameters affecting the removal of hydrogen sulphide from air by its catalytic conversion to elemental sulphur by means of the filter with fibrous carrier of Fe(III)-EDTA catalyst. The conducted research proved that the reaction time is one of the most important factors. The degree of hydrogen sulphide removal from air is higher than 99% if the residence time of the purified air in the mass-exchange unit is not under one second. Even for higher filtration rates, when the reaction time is very short, reaching 0.2 s, the residual hydrogen sulphide in the purified air is less than 2% of its initial amount. It has been established that the application of the Fe(III)-EDTA/Fiban catalytic system, under appropriate conditions, led to a nearly complete decomposition of H_2S. The required degree of hydrogen sulphide removal from the purified air could be easily controlled by applying appropriate filtration velocity and satisfactory thickness of the filtering layer material. For a given set of the process conditions, the layer thickness and filtration velocity could be estimated on the basis of the enclosed figures.

5. References

[1] Wu X, Kerchief A K, Schwartz V, Overbuy H S and Armstrong R T 2005 *Carbon* **43** 1084
[2] Meeyoo V, Lee J H, Trimm D L and Cant N W 1998 *Catal. Today* **44** 67
[3] Lin H, Williams N, King A and Hu B 2016 *Chem. Eng. J.* **297** 180-192
[4] Nowicki P, Skibiszewska P and Pietrzak R 2014 *Chem. Eng. J.* **248** 208-15
[5] Goldnik E and Turek T 2016 *Chem. Eng. Sci.* **151** 51-63
[6] Soldatov V S and Kosandrovich E G 2011 Ion exchangers for air purification *Ion Exchange and Solvent Extraction, A Series of Advances* **20** ed A K Sengupta (CRC Press) pp 45-117
[7] Piche S, Ribeiro N, Bacaoui A and Larachi F 2005 *Chem. Eng. Sci.* **60** 6452-61

[8] Karimi A, Tavassoli A and Nassernejad B 2010 *Chem. Eng. Res. Des.* **88** 748-56
[9] Busca G and Pistarino C 2003 *J. Loss Prev. Proc.* **16** 363-71
[10] Piche S and Larachi F 2006 *Chem. Eng. Sci.* **61** 7673-83
[11] Wasag H 2012 *Chem. Eng. Trans.* **30** 283-8
[12] Wasag H and Kosandrovich E 2013 Removal of hydrogen sulphide from air by means of Fe-EDTA/Fiban catalyst *Environmental Engineering IV*, ed M Dudzinska, A Pawlowski and L Pawlowski (London: CRC Press Taylor & Francis) pp 349-54
[13] Wasag H 2012 *World Acad. Sci. Technol.* **64** 774-8
[14] Soldatov V S, Pawlowski L, Wasag H and Schunkevich A 2004 *New materials and technologies for environmental engineering. Part I – Synthesis and structure of ion exchange fibres* (Lublin: Monographs of the Polish Academy of Sciences) **21** 127
[15] Shunkevich A A, Medjak G V, Martsinkevich R V, Grachek V I and Soldatov V S 2001 *Sorp. Chrom.* **1** 741-8

Acknowledgments
The research presented here was partially conducted within a cooperative research agreement between the Lublin University of Technology, Faculty of Environmental Engineering and the Institute of Physical Organic Chemistry, National Academy of Sciences of Belarus Republic in Minsk.
This work was partially funded by the Polish Ministry of Science and Higher Education (research funding S-13/WIS/2017).

IOP Conf. Series: Earth and Environmental Science **150** (2018) 012018 doi:10.1088/1755-1315/150/1/012018

Thermal Cracking to Improve the Qualification of the Waxes

B He[1,2], F A Agblevor[3], C G Chen[2] and J Feng[1]

1 College of Chemical and Chemistry Engineering Department, Chongqing University of Science and Technology, Chongqing 401331, China
2 College of Chemical and Chemistry Engineering Department, Chongqing University, Chongqing 400044, China
3 USTAR Bioenergy Center, Department of Biological Engineering, Utah State University, Logan, UT 84322, USA
E-mail: hbai2004@126.com

Abstract: Thermal cracking of waxes at mild conditions (430-500°C) has been reconsidered as a possible refining technology for the production of fuels and chemicals. In this study, the more moderate thermal cracking was investigated to process Uinta Basin soft waxes to achieve the required pour point so that they can be pumped to the refineries. The best thermal cracking conditions were set 420℃ and 20 minutes. The viscosity and density of the final liquid product were respectively achieved as 2.63 mP•s and 0.784 g/cm^3 at 40℃. The result of FT-IR analysis of the liquid product indicated that the unsaturated hydrocarbons were produced after thermal cracking, which was corroborated by the ^{13}C NMR spectrum. The GC analysis of the final gas product indicated that the hydrogen was produced; the dehydrogenation reaction was also proved by the elemental analysis and HHV results. The pour point of the final liquid product met the requirement.

1. Introduction

The decomposition of the waxes was normally accomplished by catalytic cracking with some cracking catalysts such as ZSM-5 at 500℃ or even higher temperature [1]-[5]. Calemma [6] processed the long chain n-alkanes with Pt/amorphous SiO_2–Al_2O_3 catalyst and investigated the temperature range of 345-380℃ with the reaction time up to 32 h and the hydrogen pressure of 8.1MPa when the catalyst concentration catalyst (g)/n-paraffin (mole) was 4.94. Not only high temperature but long reaction time was considered when Kyong-Hwan Lee [7] investigated the wax oil. The catalysts used for catalytic cracking are expensive and the catalysts should be separated from the products with some loss. The thermal cracking as a possible refining technology has many benefits, such as low gas production, without high-priced catalysts and hydrogen use, thus it is an effective and cheap way to deal with the waxes. Either to produce the fuels or chemicals, the waxes must be transported to the factories, but it's inconvenient to accomplish for the soft waxes. Arno de Klerk [8] reported a method for thermal cracking of the hard waxes, and the operating temperature ranged from 420 to 500 °C with 0-6 MPa, and 0.1-1 h residence time, and the wax properties was found without significant change at temperatures below 400 °C after 1 h residence treatment. But as for the soft waxes, more moderate reaction conditions should be considered. In this study, thermal cracking was used to process the soft waxes to modify its fluidity qualification so that it could be much easier to transport from the field to refineries.

2. Experimental

2.1. Materials

The wax provided by Uinta Basin (UT, USA) is a kind of yellow soft paste without any fluidity at room temperature. It should to be pumped to the refinery at the possible lowest temperature around 10℃ (the pour point should be lower than 10℃).

2.2. Equipment and procedure

A Parr 4848 high pressure reactor was used (Parr Instrument Company). In each experiment, wax (about 50g) was put into the reactor, then nitrogen was injected to flash the reactor; and the wax was heated up to the preset temperature and kept for various reaction times with stirring. The pressure variations with different reaction time at different preset temperatures were recorded. After the reaction, the reactor was cooled to room temperature and the residual pressures were recorded; meanwhile the final gas sample was also collected, and then the reactor was depressurized and opened to recover products for analysis.

2.3. Product analysis

Pour point was investigated by putting the cracked wax oil into a test tube(ISO 3016-1994) and cooled in ice bath until the temperature of the sample decreased to 2℃. The tube was inclined at an angle to check the liquid surface changed or not to judge if the sample passed the pour point test. The FT-IR spectrometer (Thermo Fisher Scientific Inc., Waltham, MA) was used to analyze the functional group changes before and after the cracking experiments, and a 300 MHz JEOL Carbon-13 Nuclear Magnetic Resonance (^{13}C NMR) instrument (JEOL Ltd., Tokyo) was also used to identify the major functional groups present in the cracked oil. The viscosity and density were measured at 40℃ using SVM300 Stabinger Viscometer (Anton Paar Ltd.), and the API gravity of the liquid products were calculated from their specific gravities. Gas product was collected in Supel Inert gas sampling bags (Supelco), and analyzed by the SRI 8610C gas chromatograph (SRI Instruments, CA, USA). The carbon, hydrogen, nitrogen and sulfur contents of wax and final liquid product were determined by a FLASH 2000 Series CHNS/O analyzer (Thermo Fisher Scientific Inc., Waltham, MA), and oxygen was determined by difference. The higher heating values (HHV) of the liquid samples were detected by IKA 2000 basic bomb calorimeter (IKA Works, Inc., Wilmington, NC).

3. Results and discussion

3.1. The effect of the different temperature

According to the reference researches above, the waxes were processed at mild temperature 450℃ initially, and then the temperature was changed, basing on the product properties. The cracked oil from the reaction temperature at both 450℃ and 430℃ were complete liquid and passed the pour point without doubt, the reaction temperature could be reduced even further to 410℃, but the waxes paste was found in the reactor after the reaction, so the suitable temperature should be 420℃, the pour experiment also proved that. The results of the pour point tests of the products are shown in Table 1. It is obviously that pour point test is successful when temperature is over 420℃. Hence, the suitable reaction temperature is 420℃, which not only meet the requirement of the pour point test but also meet the energy conservation.

Table 1. The pour point test of the products under different reaction temperature (30 min)

Temperature,℃	450	430	420	410
Passed the pour point or not (at 2℃)	Passed	Passed	Passed	Not passed

The viscosity and density of the cracked oil were analyzed, and the results are shown in Figure 1. From 450℃ to 410℃, the viscosity of the product increase from 0.86mP·s to 3.67 mP·s. Especially when the temperature ranged from 420℃ to 410℃, the viscosity raise greatly from 1.85 mP·s to 3.67 mP·s, the latter is almost twice as much of the former. According to the viscosity variation, we find the right reaction temperature should be 420℃ too. The densities of the products from 450℃ to 410℃ vary from 0.761 to 0.790 g/cm^3 which are all not very high.

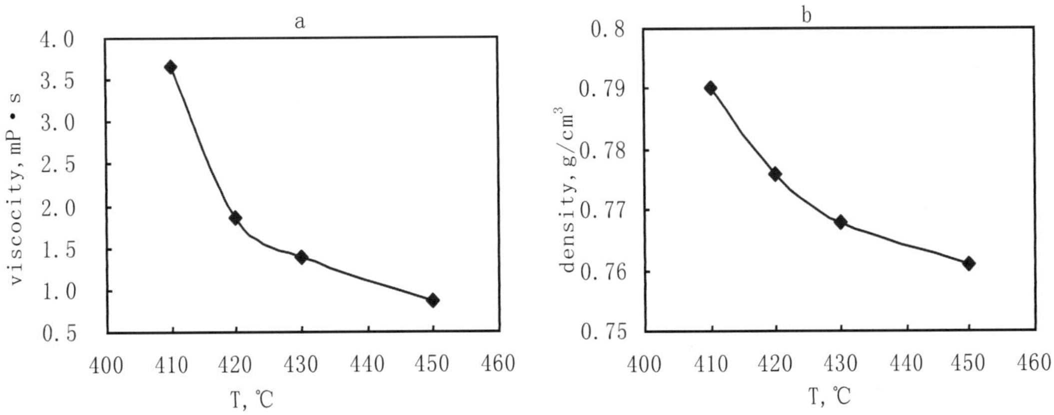

Figure 1. The viscosity and density of the cracked oil at different reaction temperature (30 min). a: the effect of temperature to the viscosity of the product; b: the effect of temperature to the density of the product.

The pressures with the reaction time are described in Figure 2 and the residual pressures are recorded in Table 2. It shows that when the reaction temperature reach to the preset 450℃, 430℃, 420℃ and 410℃, the corresponded pressures are 150psi, 110psi, 60psi and 50psi. All the pressures increase with longer reaction time at different preset temperatures, and the pressures after the reaction raise up to 250psi, 170psi, 140psi and 100pis which mean that there must be some thermal cracking reactions happened, and some more light compounds were produced, that's why the pressures keep increasing with longer reaction time at the same reaction temperature. It also shows in Figure 2 that, the pressure 140psi after the reaction at 420℃ is even lower than the pressure before the preset reaction time begins at 450℃, because there are some cracking reactions already happened even at 410℃. In fact, the actual reaction time at 450℃ is longer than the preset time due to the more heating time, and it is also because of the higher temperature lead to the higher vapor pressure. The results in Table 2 shows that the reaction with higher temperature would produce more gas, especially if T<420℃ little gas is produced, so the residual pressure could not read out by the pressure meter. Hence according to the pour point, viscosity, density, and gas production, 420℃ should be the best reaction temperature.

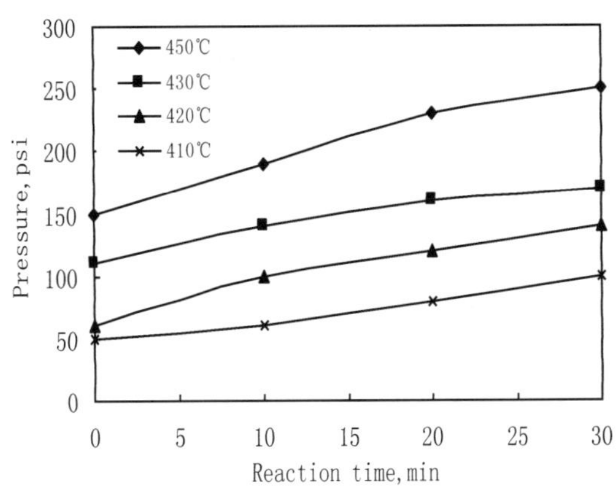

Figure 2. The pressures during different reaction times

Table 2. The residual pressure at different reaction temperature (30 minutes)

Reaction temperature, ℃	450	430	420	410
Residual pressure, psi	40	20	0	0

3.2. The effect of the different reaction time

Different reaction time from 30 minutes to 5 minutes was investigated, and the pour point test results of the cracked oil are given in Table 3. It shows that the cracked oil produced in the 30 minutes and 20 minutes could both pass the pour point tests, but the two products in 10 minutes and 5minutes couldn't pass the pour point test, which mean the reaction time is not enough. Since the cracked oil produced in 20 minutes could meet the requirement of the pour point test and also save time, 20 minutes should be the suitable reaction time.

Table 3. The pour point test of the products under different reaction time (420℃)

Reaction time, minutes	30	20	10	5
Passed the pour point or not (at 2℃)	Passed	Passed	Not passed	Not passed

The results in Figure 3 show the effect of different reaction time on the viscosity and density. With longer reaction time, the viscosities and the densities decrease especially between 20 minutes and 30 minutes, but since the pour point of the product by 20 minutes already far exceed the requirement, thus 20 minutes is enough, the viscosity is 2.63 mP•s, and the density is 0.784 g/cm^3.

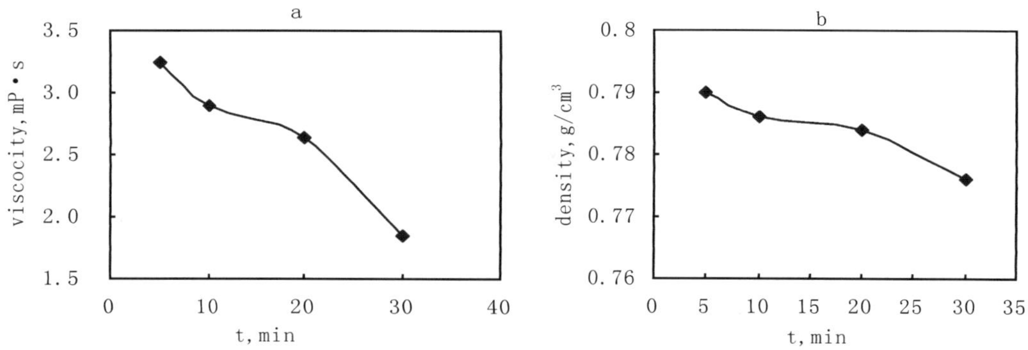

Figure 3. The viscosity and density of the cracked oil in different reaction time (420℃). a: the effect of temperature to the viscosity of the product; b: the effect of temperature to the density of the product.

The pressures before and after the reaction at 420℃ with different reaction time are described in Figure 2. It shows that the reaction the pressures read 80psi, 100psi, 120psi and 140psi after 5minutes, 10minutes, 20minutes and 30minutes, which means the more reaction time the more gas produced, so the pressure increase accordingly. The residual pressures are also recorded in Table 4. It indicate that there is no read at room temperature during the investigated reaction time from 30 minutes to 5 minutes which illustrate little gas produced, but it doesn't mean there is completely no gas produced. When opening the gas outlet valve of the reactor, there was some gas coming out.

Table 4. The residual pressure after different reaction time (420℃)

Reaction time, minutes	30	20	10	5
Residual pressure, psi	0	0	0	0

3.3. Analysis of the final product

To make sure the compound difference before and after the cracking reaction, FTIR was used to check the product(B) produced at 420℃ with 20 minutes to compare with the raw waxes(A). The results are demonstrated in Figure 4.

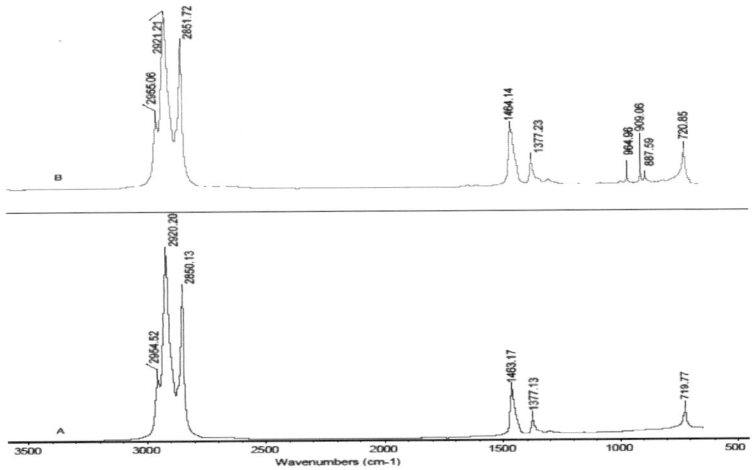

Figure 4. The FT-IR spectra of the waxes before (A)and after(B) the thermal cracking

In Figure 4, the 2955 and 1377 cm^{-1} are associated with $-CH_3$ [9], [10], and 2850-2921 cm^{-1} are corresponded to symmetric and asymmetric stretching frequency $–CH_2$ [11]-[13]; 1463cm^{-1} indicate CH_2- bending [12], [14], 720 cm^{-1} is due to the deformation vibration in ($-CH_2-$)n groups [15], both 887.59 and 964.96cm^{-1} show $-CH=CH_2$ functional groups [16], 909.06cm^{-1} indicate -CH=CH- newly produced [17]. Why the double bond hydrocarbons are produced in the cracked oil also prove the thermal cracking mechanism as follows.

$$R\text{-}CH\text{-}CH_2\text{-}R' \rightarrow R\text{-}CH{=}CH_2 + R'H \quad (1)$$

$$R\text{-}CH_2\text{-}CH_3 \rightarrow R\text{-}CH{=}CH_2 + H_2 \quad (2)$$

From the equation (1) and (2), it shows that the thermal cracking reaction typically both the carbon bond breaking and dehydrogenation lead to the unsaturated hydrocarbon newly produced, and from the residue pressure records, it comes to a conclusion that the lower temperature the lighter thermal cracking, so fewer gases are produced. It is known that the carbon bond breaking is easier than the dehydrogenation, and the longer carbon chain the easier rupture, thus to get more short carbon chain hydrocarbon liquids, less gas production and to avoid possible further more secondary cracking for more gases, lower temperature and shorter time should be better, it also support that 420℃ and 20 minutes are better parameters but not higher temperature and longer time since the acceptable product has been turned out already.

The FT-IR spectral data is corroborated by the ^{13}C NMR spectrum of the final cracked liquid product in Figure 5, which shows two major functional groups. The peaks from 10 to 40 ppm are assigned to aliphatic hydrocarbon side chains [18], while the peaks from 110-140 ppm are assigned to olefin carbon [19], and the peaks at 77 are assigned to the solvent $CDCl_3$. Thus on the evidence of the FI-IR and the ^{13}C NMR spectrum, it reaches the conclusion that there are some unsaturated hydrocarbons produced during the cracking reaction, and most of the products are saturated hydrocarbons.

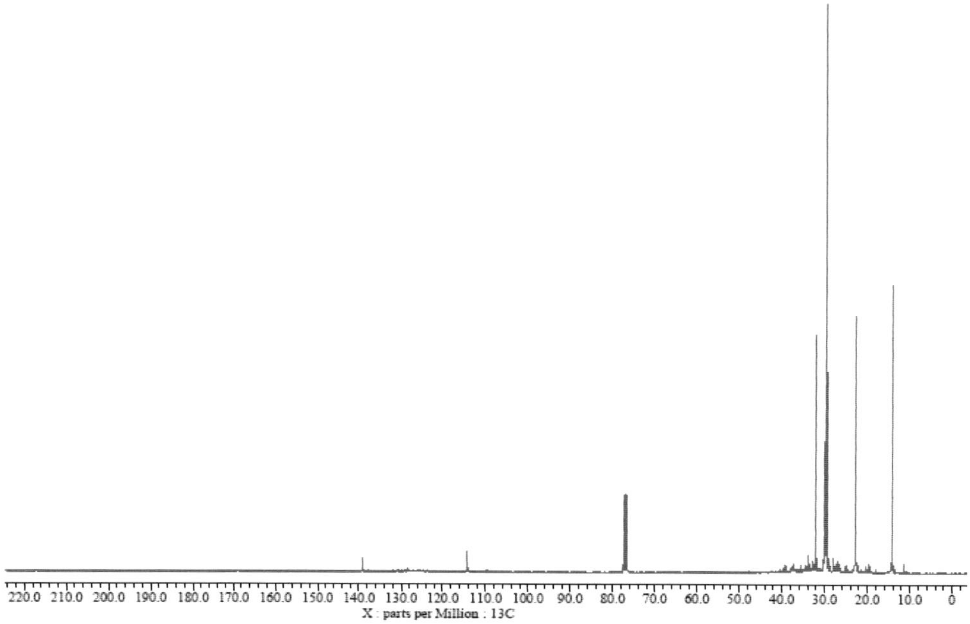

Figure 5. The ^{13}C NMR spectrum of the final liquid product

To check if the hydrogen was produced, the gas from the final experiment (420℃ and 20 minutes) was collected and analyzed, the results is recorded in Table 5. It shows that the hydrogen is produced and the volume percentage is 2.96%, which proves the dehydrogenation equation (1) above. The percentage of the hydrocarbon from methane to pentane achieves at 28.43% and far more than the volume of the hydrogen, which also indicate that the carbon bond breaking is easier than the dehydrogenation under the same condition, so it is helpful to produce more saturated hydrocarbons and less olefins.

Table 5. The gas components from the final product

Gas content	H_2	CH_4	C_2H_4	C_2H_6	C_3H_6	C_3H_8	C_4H_8	C_4H_{10}	C_5H_{10}	C_5H_{12}
Volume percentage,%	2.96	5.13	1.83	5.74	4.48	4.99	2.04	2.65	0.77	0.80

To compare the heating value difference and the elemental composition variation before and after the thermal cracking, the HHV and elemental analysis of the wax and final liquid product are recorded in Table 6. It shows that the carbon content in final liquid product is higher than the wax; the hydrogen in final product is less than the wax, which is due to the dehydrogenation of the thermal reaction. As to the nitrogen there are just 0.01% differences between the wax and final liquid product, maybe it is because of the nitrogen flash of the reactor. There is no sulfur and oxygen in wax and final liquid product. The HHV of the final liquid product is lower than the wax because the dehydrogenation reaction leads to some hydrogen loss, whose heating value is higher than carbon. Although there is about 0.56 MJ/kg higher heating value losses of the final liquid product as compared to the wax, the liquid product is more convenient to be transported or utilized. What's more, the gas product could be also used as fuels.

Table 6. The EA and HHV of the final liquid product

Elemental composition (Wt%)	Materials	
	Wax	Final liquid product
C	84.87	85.07
H	14.31	14.10
N	0.82	0.83

S	0	0
O	0	0
HHV(MJ/kg)	46.266	45.708

4. Conclusion

Moderate thermal cracking was used to investigate the soft waxes processing, and the properly reaction parameters were selected out. The FT-IR analyzing of the oil product proved that the unsaturated hydrocarbons newly appeared which was corroborated by the ^{13}C NMR spectrum. The GC analyzing showed that the hydrogen was produced, it meant the dehydrogenation reaction happened which was also proved by the elemental analysis and HHV results. The original paste waxes were converted into hydrocarbon liquids; the pour point of the final product far more achieved the required target of the factory. So non-catalytic thermal cracking to deal with the soft waxes is a convenient and cheap way to get liquid products.

5. Reference

[1] Zhou G F, Jensen P A, Le D M, Knudsen N O and Jensen A D 2016 *Green Chem.*,**18**,1965-75
[2] Rao T M, Dupain X and Makkee M 2012 *Microporous Mesoporous Mater.*, **164**,148-63
[3] Huang Y, Wei L, Julson J, Gao Y and Zhao X 2015 *J. Anal. Appl. Pyrol.*,**111**, 148–55
[4] Artetxe M, Lopez G, Amutio M, Elordi G, Bilbao J and Olazar M 2013 *Ind. Eng. Chem. Res.*, **52**, 10637-45.
[5] Kim J Y, Lee J H, Park J, Kim J K, An D, Song I K and Choi J W 2015 *J. Anal. Appl. Pyrol.*, **114**,273-80
[6] Calemma V, Peratello S and Perego C 2000 *Appl. Catal. A:Gen.*, **190**, 207–18
[7] Lee K H 2012 *J. Anal. Appl. Pyrolysis*, **94**, 209-14
[8] Klerk A D 2007 *Ind. Eng. Chem. Res.*, **46**, 5516-21
[9] Yang H and Irudayaraj J 2001 *JAOCS,* **78**,889-95;
[10] Petrakis E A, Kimbaris A C, Pappas C S, Tarantilis P A and Polissiou M G 2009 *J. Agric. Food Chem.*,**57**,10044–48;
[11] Zhu J, He H, Zhu L, Wen X and Deng F 2005 *J. Colloid Interf. Sci.*,**286**,239–44;
[12] Mahapatra S S and Karak N 2004 *Prog. Org. Coat*, **51**,103–8 ;
[13] Lieberzeit P A, Glanznig G, Leidl A, Voigt N and Dickert F L 2006 *IEEE Sens. J.*, **6**,529-35
[14] Yano J and Sato K 1999 *Food Res. Int.*,**32**, 249-59
[15] Xie R, Qu B and Hu K 2001 *Polym. Degrad. Stabil.*,**72**, 313-21
[16] Sarker M, Rashid M M 2012 *IOSR JEN,* **2**,38-49
[17] Sarker M, Rashid M M and Molla M 2011 *Int. Rev. Chem. Eng.*, **3**,340-49
[18] Agblevor F A, Mante O, Mcclung R and Oyama S T 2012 *Biomass Bioenergy*, **45**,130-37
[19] Okieimen F E, Pavithran C and Bakare I O 2005 *Eur. J. Lipid Sci. Technol.*,**107**, 330-6

Acknowledgments

Project is supported by Scientific and Technological Research Program of Chongqing Municipal Education Commission (KJ131405、KJ1601309、KJ1713342), Utah Science Technology and Research (USTAR), School Scholarship Fund of Chongqing University of Science and Technology

Biogas Production from Water Hyacinth (Eichhornia Crassipes): The Effect of F/M Ratio

W D Nugraha[1], Syafrudin[1], L L Pradita[1], H H A Matin[1] and Budiyono[2]

1 Department of Environmental Engineering, Diponegoro University, 50275 Semarang, Indonesia
2 Department of Chemical Engineering, Diponegoro University, 50275 Semarang, Indonesia
E-mail: lathifahlp@gmail.com

Abstract. Distribution of water hyacinth (Eichhornia crassipes), generally considered as a water weed, that has been a problem which can harm the environment, irrigation system, and agriculture. However water hyacinth can be used in biogas production because it has large enough amount of hemicellulose contents. The purpose of this study was to know the effect of F/M ratio to biogas production from water hyacinth waste with Liquid Anaerobic Digestion (L-AD) method. A series of laboratory experiments using biodigester were performed in batch anaerobic operation at room temperature. F/M ratio that used in each reactor was 39.76, 20.03, 13.32, and 10.01. Degradation process was done in 60 days. The result showed that F/M ratio effects to the biogas production. The best performance of biogas production from this research will be obtained if F/M ratio is in the range of 10.01-20.03 (correspond to 25%-50% of rumen fluid) with water hyacinth as the main substrate.

1. Introduction

The current global energy supply is highly dependent on fossil sources (crude oil, lignite, hard coal, natural gas). These are fossilized remains of dead plants and animals, which have been exposed to heat and pressure in the Earth's crust over hundreds of millions of years. For this reason, fossil fuels are non-renewable resources which reserves are being depleted much faster than new ones are being formed [1].

The worldwide distribution of Eichhornia crassipes (E. crassipes) generally considered as an aquatic weed, has become a persistent and expensive aquatic problem damaging the environment, irrigation system and crops [2]. Water hyacinth is type of weed which grow very fast. The growth of water hyacinth can be 1.9% a day with 0.3-0.5 m in length [3]. According to Malik (2006), water hyacinth consist of 95% water and make it consist of hollow tissue, have a high energy, consist of materials that can be fermented and have big potential to produce biogas [3]. Water hyacinth can be used in biogas production because it has high hemicellulose content compare with other single organic components. Hemicellulose is complex polysaccarides which is polymer composite which if it hydrolized produce derivative mixture product that can be treated with anaerobic digestion methode to produce two simple mixture compounds which are methane and carbon dioxide that commonly called biogas [3] .

Anaerobic digestion (AD) has been extensively used to convert organic waste streams from various sources, such as agricultural, industrial, and municipal solid waste, to biogas. The AD process can operate in both liquid and solid states in terms of total solid (TS) content. In general, the TS content of liquid AD (L-AD) systems ranges from 0.5 to 15%, while solid-state AD (SS-AD) systems usually operate at TS contents of higher than 15% [4].

Several studies of biogas production from lignocellulose biomass using water hyacinth and L-AD methods had been conducted. A research on biogas using waste of fisheries in the form of offal, filth, and gills as the main substrate had been done. Biogas production that has been resulted on F/M ratio 0.2 with total maximum methane 164.7 l / kg CODMn. Besides that, F/M ratio gave a significant effect to the retention time (HRT). Time retention that obtained from F/M ratio 0.2, 0.4, and 0.6 each were 36, 13, and 19 days [5].

Research also did on hydrogen production and methane from banana peel. The result was, the best perform from one fermentation phase was observed on F/M from 5.0. On this condition, methane yield, production level, and potential were 251.3 mL g-1 VS, 2.05 mL h-1, and 352.9 mL. The result of hydrogen and methane from 209.9 and 284.1 mL g-1 VS mL g-1 VS accomplished on F/M from 5.0 in the second step process [6].

Other research had been done about biohydrogen production and biomethane from water hyacinth. From that research resulted maximum hydrogen production was in water hyacinth on concentrate 40 g/ L resulted hydrogen 38.2 mmol H_2/L/d and maximum methane production was on concentrate 80 g/ L resulted methane 29.0 mol CH_4/L/d [4]. While other researcher did research about anaerobic digestion with water hyacinth, cabomba, and salvinia as the main substrate and water hyacinth and cabomba easily degraded, resulted 267 L biogas/ kg VS and 221 L biogas/ kg VS, with methane content 50%. While on salvinia resulted less degradation on 155 L biogas/ kg VS on methane quality 50% [7].

Research about adding rumen in biogas production also done about the effect of F/I ratio on biogas production from cow dung using rumen as the inoculum and the best result of biogas production was on F/I ratio between 17.64 to 35.27 (corresponding to 25-50% rumen fluid) [8]. That thing also supported by other researcher who said that rumen fluid that inserted into the reactor significantly effected to biogas production. The best biogas production was on rumen fluid and total solid on range 25-50% and 7.4 and 9.2%. Total solid between 7-9% and 50% rumen content will give the best result on biogas production. Therefore, this study focused on the effect of F/M ratio to biogas production from water hyacinth waste with liquid anaerobic digestion (L-AD) [9].

2. Materials and Methods

The main substrate that used in the research was water hyacinth with 15 cm in diameter. The biodigester that made from polyethylene bottle plugged with tightly rubber were 1500 ml in volume. The biodigester was equipped with valve for biogas measurement. The schematic diagram of experimental laboratory will be shown in Figure 1. Food to Microorganism ratio (F/M ratio) was varied from 39.76, 20.03, 13.32, and 10.01. Implementation of the research started with preparing water hyacinth substrate which was cut approximately 1-2 cm [10] and preparing rumen liquid, then conducting an examination of the total composition of solids and water content on each materials. After that calculating the variation of F/M ratio and preparing of anaerobic digestion batch reactor. Subsequently, the substrate was mixed with the inoculum. Samples that had been prepared could be inserted into the reactor, sealed in order to obtained anaerobic condition, and ready to operate. During the treatment process takes place, volume of biogas produced were observed in interval of two days. The observation was terminated after 60 days.

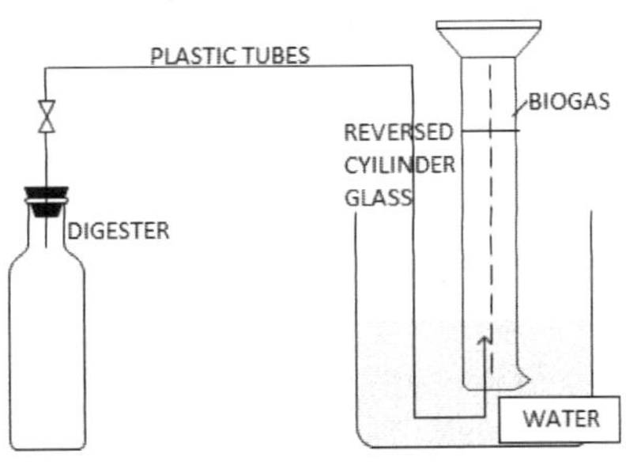

Figure 1. Schematic diagram of series laboratory batch assessment of L-AD.

3. Result and Discussion

3.1. The Effect of F/M Ratio to Biogas Production with L-AD

This research aims to know the effect of F/M ratio to biogas production from water hyacinth waste liquid anaerobic digestion (L-AD). In biogas production anaerobically, value of F/M shows comparison between the amount of substrate that contained in waste (medium) and the amount of microorganism used [5]. The variation of F/M ratio produced from the existence of rumen volume variation and total solid from each materials. The initial total solid of water hyacinth is 13.52. When it combined with different rumen volume in each reactor, total solid of water hyacinth that has been changed can be seen in Table 1.

Table 1. The Initial and Final Total Solid of Water Hyacinth.

	Initial TS (%)	Rumen Volume (ml)	Final TS (%)
F/M Ratio 39.76	13.52	50	10.82
F/M Ratio 20.03	13.52	100	9.06
F/M Ratio 13.32	13.52	150	7.73
F/M Ratio 10.01	13.52	200	6.76

If seen from total solid variation existence, from those four reactors that produced the most biogas yield was F/M ratio 10.01 with biogas yield 127.071 ml/ gr TS on total solid 6.76%. According to Brown, et al (2012), the content of higher total solid has minimal effect to the total solid efficiency and reduction of biogas production. Because on higher total solid content can cause inhibition to hydrolysis step that caused by limited mass transfer between microbe and material [11]. The product from hydrolysis step has been accumulated on substrate surface because of the limited mass transfer which finally obstruct hydrolytic enzyme absorption. The limited mass transfer make amount of hydrolysis product existence for acidogenic microbe is limited so can reduce the amount of product that has been produced in acidogenesis microbe to be converted into biogas on methanogenesis step [12]. The graphic can be seen on Figure. 2.

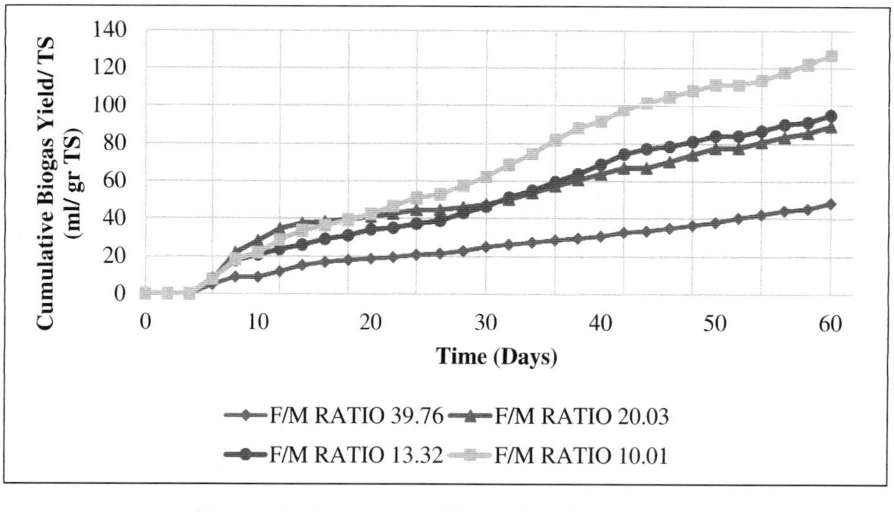

Figure 2. Cumulative Biogas Yield per gr TS.

After the research had been done, it showed that the existence of F/M ratio variation effected to biogas production. The research about the effect of F/M ratio to biogas production has been done by Sunarso, et al (2010) used manure as the main material. On that research, there were four F/M variation that has been used which were 17.64, 23.51, 35.27, and 70.54. And known that on variation 17.64-35.27 produced the most optimum biogas. Also on this research used four F/M variation which were 10.01, 13.32, 20.03, and 39.76 used water hyacinth as the main material. And it is known that on those F/M variation, the most optimum biogas production was 10.01. The optimum F/M variation on those two researches were different because there was a difference of material using, so it caused a different F/M variation.

3.2. Production Rate of Biogas from Water Hyacinth Waste with L-AD Method

The second aim from this research was to know the production rate of biogas. On Table 2 and Figure 3 show that the existence of F/M ratio variation gave effect to kinetic constants production of biogas. In a row, kinetic constants of biogas that has been formed with the treatment are biogas daily production (A), biogas production rate (U), and minimum time of biogas formation (λ). On F/M ratio 39.76 was 59.207 ml/ gr TS, 0.833 ml/ gr TS, 0.411 days. On F/M ratio 20.03 was 91.376 ml/ gr TS, 1.772 ml/ gr TS, and 0 day. Next on F/M ratio 13.32 was 121.483 ml/ gr TS, 1.878 ml/ gr TS, 3.321 days. And the last on F/M ratio 10.01 was 151.910 ml/ gr TS, 2.624 ml/ gr TS, 4.448 days.

Table 2. Kinetic Constants on The Effect of F/M Ratio to Biogas production Research

Constants	**F/M Ratio**			
	39.76	**20.03**	**13.32**	**10.01**
A, ml/(grTS)	59.207	91.376	121.483	151.910
U, ml/(grTS.day)	0.833	1.772	1.878	2.624
λ, days	0.411	0	3.321	4.448

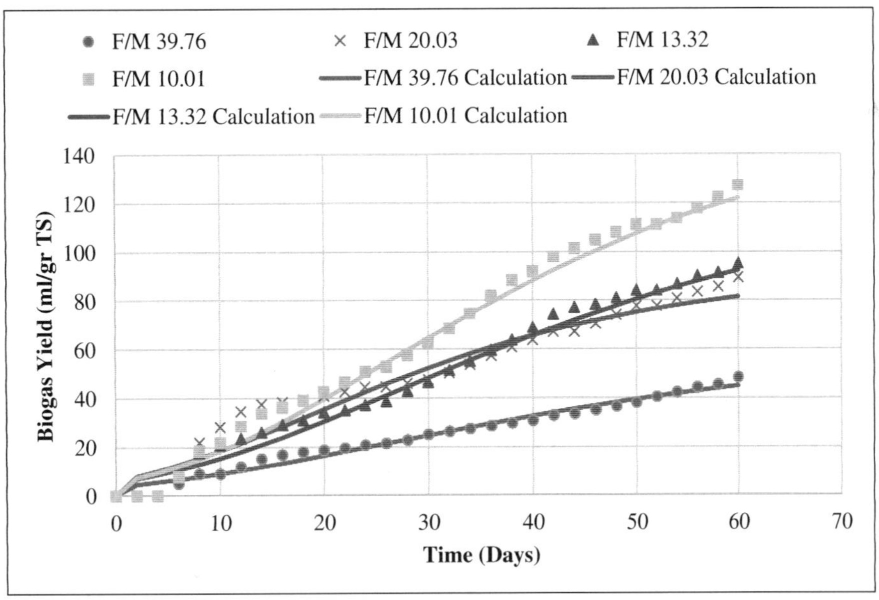

Figure 3. The Relationship between Experimental Data with The Calculation Result on The Effect of F/M Ratio to Biogas Production Research.

After doing model on F/M ratio 10.01, biogas yield on the 160th day was 151.548 ml/ gr TS. From that biogas yield, can aims for lightning, power, and also electricity from 1m^3 biogas conversion. After knowing the conversion every 1m^3 biogas, hence can be calculated for lightning, power, and electricity needs that can be produced from 151.548 ml/ gr TS of biogas yield. Knowing that if there is 1 ton water hyacinth waste, hence biogas yield that can be produced is 151,548,000 ml or 151.548 m^3. From that biogas yield can aims to produce electricity 712.276 kWh. The model biogas yield on the 160th days can be seen on Figure 4.

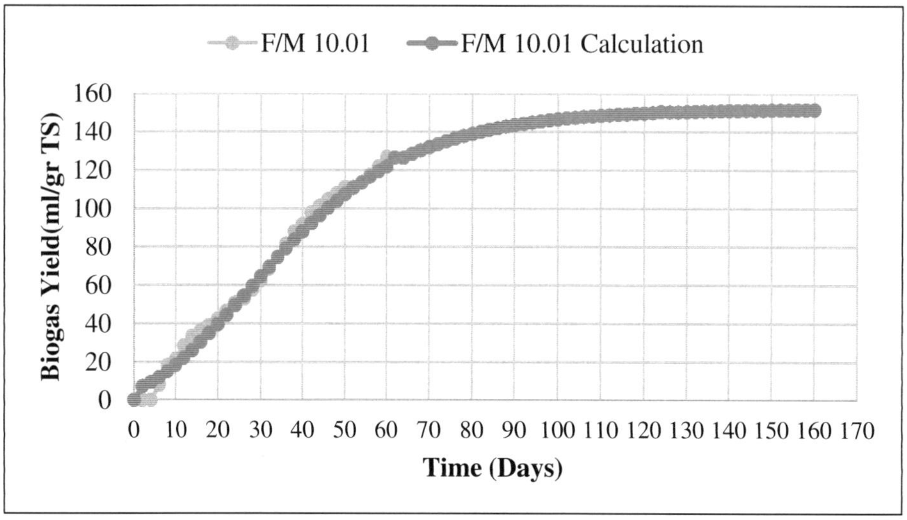

Figure 4. The Calculation Result on The Effect of F/M Ratio 10.01 on Biogas Production Research.

4. Conclusions

Water Hyacinth is one of the alternative from agricultural waste that can be used for biogas because a high hemicellulose content F/M ratio can take effect to biogas production. The smaller F/M ratio, the amount of biogas production will be higher. The highest biogas production was on F/M ratio 10.01 (content of 50% rumen). On this research, from the four F/M ratios that have been researched, the most optimum F/M ratio was on 10.01 with optimum total solid 6.76%.

The biggest biogas yield was on F/M ratio 10.01 in amount of 127.071 ml/ gr TS started on the fourth day. The daily biogas production (A) and production rate constants (U) respectively 151.910 ml/ gr TS and 2.624 ml/ gr TS.

5. References

[1] Al Seadi T, Rutz D, Prassl H, Kottner M, Finsterwalder T, Volk S and Janssen R 2008 *Biogas Handbook.* Esbjerg: University of Southern Denmark Esbjerg

[2] Chuang Y S, Lay C H, Sen B, Chen C C, K Gopalakrishnan, Wu J H, Lin C S and Lin C Y 2011 Biohydrogen and Biomethane from Water Hyacinth (Eichhornia crassipes) Fermentation : Effects of Substrate Concentration and Incubation Temperature *International Journal of Hydrogen Energy* **36** 14195-14203

[3] Yonathan A, Prasetya A R and Pramudono B 2013 Biogas Production from Water Hyacinth (Eicchornia Crassipes) : Study of Consistency and pH to Biogas Production *Journal of Chemical and Industrial Technology* **2(2)** 211-215

[4] Zhu J, Zheng Y, Xu F and Li Y 2014 Solid-state Anaerobic Co-digestion of Hay and Soybean Processing Waste for Biogas Production Bioresource Technology 154 240-247

[5] Hadiyarto A, Budiyono, Johari S, Hutama I and Hasyim W 2015 The Effect of F/M Ratio to The Anaerobic Decomposition of Biogas Production from Fish Offal Waste Waste Technology (WasTech) 3(2) 58-62

[6] Nathoa C, Sirisukpoca U and Pisutpaisal N 2014 Production of Hydrogen and Methane from Banana Peel by Two Phase Anaerobic Fermentation Energy Procedia 50 702-710

[7] O'Sullivan C, Rounsefell B, Grinham A, Clarke W and Udy J 2010 Anaerobic Digestion of Harvested Aquatic Weeds: Water Hyacinth (Eichhornis crassipes), Cabomba (Cabomba Caroliniana) and Salvinia (Salvinia molesta) Ecological Engineering 36 1459-1468

[8] Sunarso, Johari S, Widiasa I and Budiyono 2010 The Effect of Feed to Inoculums Ratio on Biogas Production rate from Cattle Manure Using Rumen Fluid as Inoculums International Journal of Science and Engineering (IJSE) 1(2) 41-45

[9] Budiyono, Widiasa I, Johari S, and Sunarso 2014 Increasing Biogas Production Rate from Cattle Manure Using Rumen Fluid as Inoculums International Journal of Science and Engineering (IJSE) 6 31-38

[10] Darnengsih, Nurjannah, and Ifa L 2016 The Effect of Raw Material Comparison to Biogas Concentration from Water Hyacinth Using Cow Dung Starter Journal of Chemical Process Engineering 01

[11] Brown D, Shil J, and Li Y 2012 Comparison of Solid-State to Liquid Anaerobic Digestion of Lignocellulosic Feedstocks for Biogas Production Bioresource Technology 124 379-386

[12] Sheets J P, Ge X and Li Y 2015 Effect of Limited Air Exposure and Comparative Performance between Thermophilic and Mesophilic Solid-State Anaerobic Digestion of Switchgrass Bioresource Technology 296-303

Acknowledgment

We would like to say thank you and appreciate to Diponegoro University through Research Professorship Program (RPP) 2017 due to the funding of this research.

Wind speed and power characteristics of Kalasin province, Thailand

Supachai Polnumtiang[1] and Kiatfa Tangchaichit[2*]

1,2 Mechanical Engineering, Faculty of Engineering, Khon Kaen University, Thailand
* E-mail: Kiatfa@kku.ac.th

Abstract. This paper presents a wind energy assessment of Kalasin province in the Upper North-Eastern region of Thailand. Four year wind data were recorded continuously from January 2012 to December 2015 at different heights of 60, 90 and 120 m above ground level (AGL). The mean wind speeds were found to be 3.14, 3.63 and 3.94 m/s at 60, 90 and 120 m AGL, respectively. The majority of wind directions for this region are distributed from the East to South directions. The highest wind power density was observed in the summer season, followed by winter and rainy seasons, in order. Four commercial wind turbines were selected to estimate energy yield output using the WAsP 10.0 software application; the results show that VESTAS with rated power of 2.0 MW was estimated to give 2,747 MWh/year with the highest capacity factor of 15.68%.

1. Introduction

Currently, the energy demand of the world is increasing rapidly [1]-[3], especially energy from fossil resources such as crude oil, natural gas and coal. The use of fossil energy causes environment damage and concern about global warming effects from carbon dioxide emission [2], [4]. Renewable energy sources, such as wind, solar and biomass energy are alternatives to replace the fossil energy demand [5]. Wind energy is the one clean and free energy which can reduce energy imports and carbon dioxide emission [6], [7]. Nowadays, wind energy is studied and developed in many countries of the world, with a total installed wind capacity increasing from 23,900 MW to 486,790 MW during the period of 2000-2016. At the end of 2016, China maintained a leadership position, with total installation of 168,690 MW, followed by USA, Germany and India which were 82,184, 50,018 and 28,700 MW [8], respectively.

Regionally, wind energy potential was studied in Hong Kong by Z.R. Shu et al. [9]. The 6- year wind data statistics were used to analyse 4 different meteorological stations; the annual Weibull scale and shape parameter varied from 2.85 to 10.19 m/s and 1.65 to 1.99, respectively. A stationary rooftop wind mast and mobile Light Detection and Ranging (LiDAR) were used to measure wind data from 3 sites in Singapore by B.R. Karthikeya et al. [10]. The offshore west coast of Taiwan was studied for wind farm installations using mesoscale and excellent wind performance was also found by Hsin-Fa Fang [11]. Wind speed and economic analysis in the central region of Thailand [12] was studied, and it was found that the annual mean wind speed was between 3 m/s and 5 m/s along the direction of tropical monsoons, which is suitable for small-scale wind turbine application. W. Huang et al. [13] used the CMO (Cell Membrane Optimization) method to place wind turbine positions in a square 2x2 km wind farm; the results show that it can reduce wake loss in the wind farm.

This paper presents micro siting of wind energy resource assessment in Kalasin province of Thailand. The 4-year mean wind speeds by season, month and day and the wind directions were analyzed.The selected wind turbines were used to estimate energy output for this region.

IOP Conf. Series: Earth and Environmental Science **150** (2018) 012020 doi:10.1088/1755-1315/150/1/012020

2. Location of wind measurement site.

Kalasin province (K site) is one wind energy resource assessment site in the upper north-east Thailand project. The distance from Bangkok to the measurement site is approximately 450 kilometers and it is located at 16.900°N, 103.329°E and 175 meter above sea level (ASL) as illustrated in Figure 1. The 4-year wind data were recorded continuously by sampling intervals of 10 min from January 2012 to December 2015. The 120 meter NRG tower mast was installed and wind measurement accessories such as the 3-cup anemometers and wind guide vanes were mounted on the mast with different heights of 60, 90 and 120 meters above ground level (AGL) as shown in Figure 2 and Table 1.

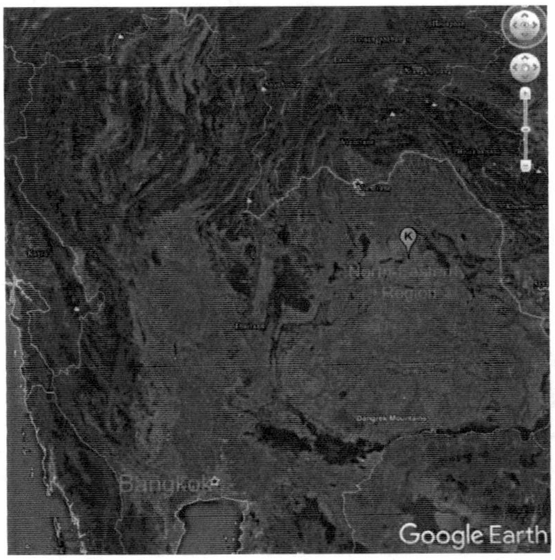

Figure 1. Wind measurement site (K)

Figure 2. Installation of wind mast and devices

Table 1. Equipment specification

Equipment	Details	Manufacturer
120 m mast tower	Guyed Lattice Mast	
NRG#40C	3-cup type, Measuring range 1-96 m/s	NRG System
Wind Vane NRG#200P	Measuring range 0-360 degree, -55 to 60 degree C	

3. Results and Discussion

3.1. Wind speed characteristics, Annual, Monthly and Daily wind speed

The annual mean wind speed in the time series chart from 2012 to 2015 of the K site at different heights of 60, 90 and 120 m AGL is illustrated in Figure 3. From the data we observe that the wind speed has not much variation over all seasons in a year. Figure 4 reveals the daily wind speed; a higher wind speed at the K site was observed at night time between 18:00 and 7:00 of the next day. The maximum wind speed was 5.01, 4.48 and 3.61 m/s at 21:00 while the minimum wind speed was 2.89, 2.74 and 2.56 m/s at 8:00 for the heights of 120, 90 and 60 m AGL, respectively. Figure 5 shows that the monthly wind speed did not fluctuate much over all seasons and all heights above ground level. The highest mean wind speed was observed in December to be 4.29 m/s and the lowest in September to be 3.32 m/s at 120 m AGL and similar trends were also observed at 90 and 60 m AGL. The annual mean wind speeds at 60, 90 and 120 m AGL were 3.14, 3.63 and 3.94 m/s respectively, as presented in Table 2. The seasons of Thailand are classified to be 3: Rainy is from June to October; Winter is from November to January and Summer is from February to May.

The mean wind speed (U_m) and standard deviation (σ) are expressed as

$$U_m = \frac{1}{N}\left[\sum_{i=1}^{N} U_i\right] \tag{1}$$

$$\sigma^2 = \frac{1}{N-1}\left[\sum_{i=1}^{N} (U_i - U_m)^2\right] \tag{2}$$

where U_m is the mean wind speed, U_i is the wind speed which is averaged over the time interval, N is the number of wind data points and σ is the standard deviation.

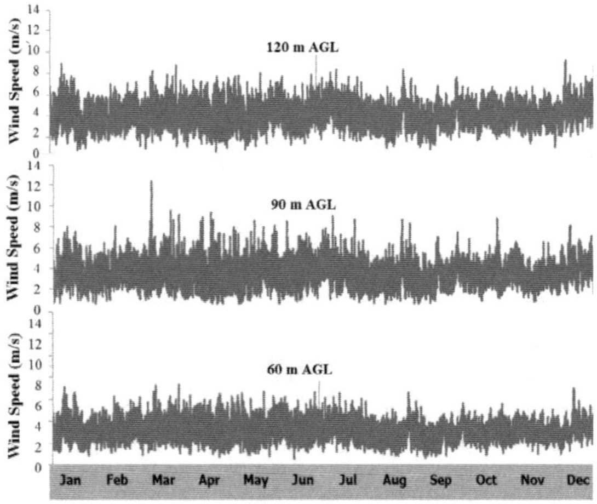

Figure 3. Wind speed in the time series

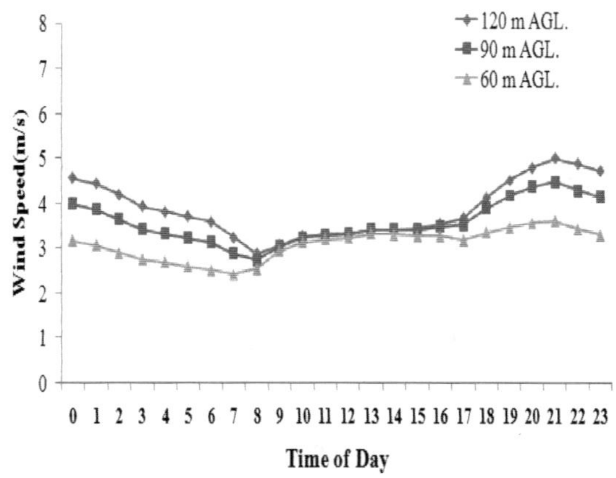

Figure 4. Daily wind speed variation

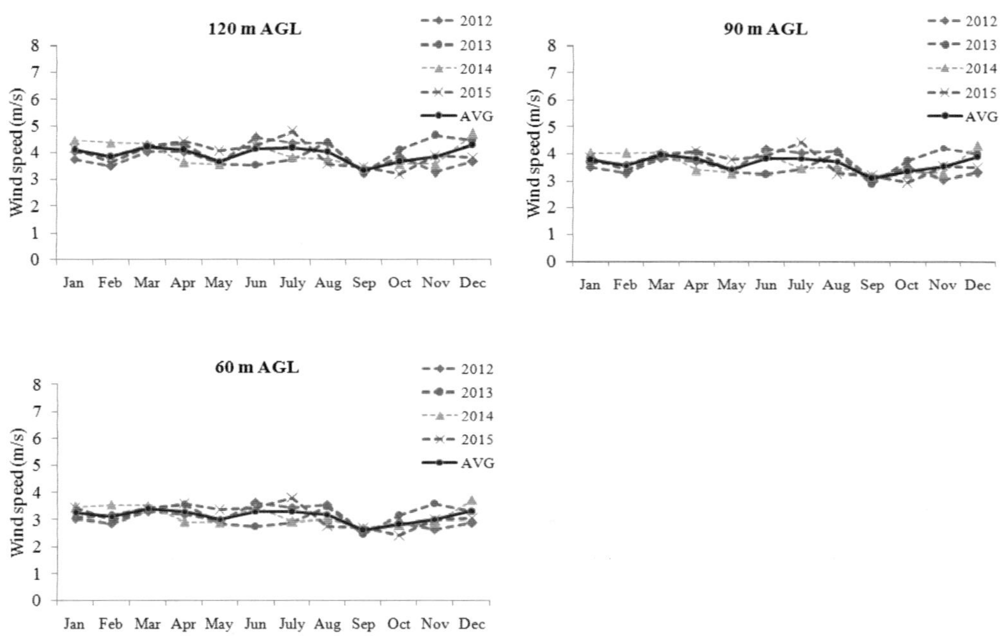

Figure 5. Monthly wind speed variation

3.2. Weilbull distribution function analysis and wind direction

The 2 parameters of the Weibull distribution function were used to characterise wind data behavior [5], [9]. This paper presents interval bins of 1 m/s and the Weibull probability function is expressed as [9]

$$f(u)=\left(\frac{k}{c}\right)\left(\frac{U}{c}\right)^{k-1}\exp\left[-\left(\frac{U}{c}\right)^{k}\right]\quad(u>0;k,c>0) \tag{3}$$

$$k=\left(\frac{\sigma}{U_m}\right)^{-1.086}\quad(0\le k\le 10) \tag{4}$$

$$c=\frac{U_m}{\Gamma(1+1/k)} \tag{5}$$

where k is the dimensionless shape parameter, c is the scale parameter (in units of m/s) and Γ is the Gamma function.

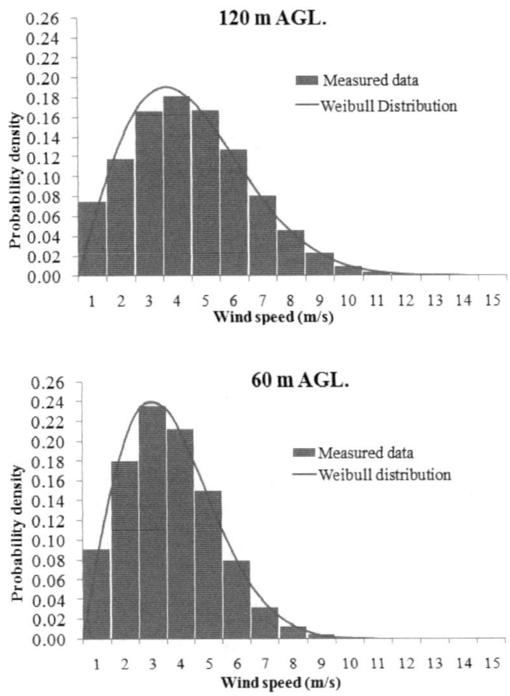

Figure 6. Actual wind speed frequency distribution and Weibull curve

Figure 6 shows the actual wind speed frequency distribution and the Weibull curve at 120, 90 and 60 m AGL. There is good correspondence between the 2 parameter Weibull curves and the actual wind speed bins. Generally, the cut in wind speed for the wind turbine is 3.0 m/s and from the Weibull curve we can estimate the percentage of wind frequency distribution above 3.0 m/s. The results show that 49%, 59% and 63% at 60, 90 and 120 m AGL respectively are above the threshold. Moreover, from the Weibull curve comparison in each season at 120 m AGL as presented in Figure 7, the shape parameters were 2.22, 2.08 and 1.93 in the winter, rainy and summer seasons, respectively. Table 2 presents a summary of the mean wind speed (U), Weibull shape parameter (k) and scale parameter (c). The maximum wind speed is shown in the winter to be 4.04 m/s at 120 m AGL. The value of the scale parameter (c) increases with height and is highest in the winter at 4.70 m/s at 120 m AGL, and the highest value of the shape parameter (k) was observed in winter. The wind rose at level

120 m AGL for this region was presented in Figure 8. The greatest frequency was observed in East to South, and similar results were also observed at 90 and 60 m AGL.

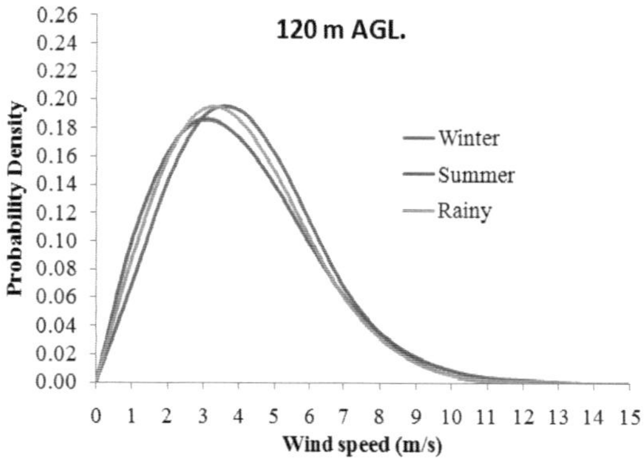

Figure 7. Weibull curve of the seasonal

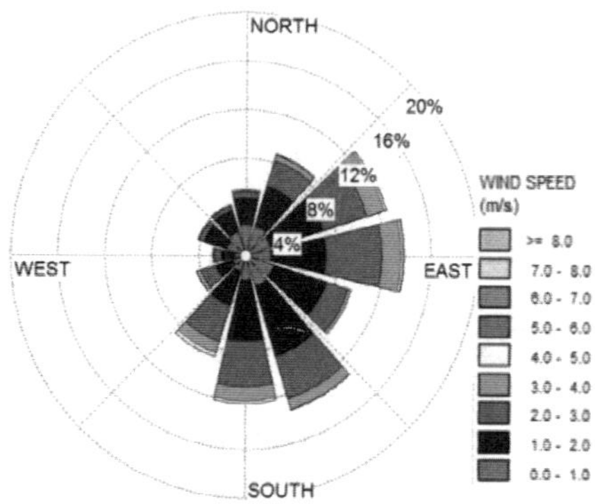

Figure 8. Wind direction at 120 m AGL

Table 2. Summary of Weibull parameters of the K site from 2012-2015

Period	Parameter	Height above ground level		
		120 m	90 m	60 m
Annual	U (m/s)	3.94	3.63	3.14
	k	1.97	2.00	1.99
	c (m/s)	4.45	4.10	3.54
Winter	U (m/s)	4.04	3.68	3.17
	k	2.22	2.24	2.18
	c (m/s)	4.70	4.20	3.60
Summer	U (m/s)	3.95	3.68	3.22
	k	1.93	1.97	1.97
	c (m/s)	4.50	4.20	3.70
Rainy	U (m/s)	3.87	3.55	3.05
	k	2.08	2.06	1.94
	c (m/s)	4.50	4.10	3.50

3.3. Wind power density and wind shear coefficient
The average of the wind power density for this area can be expressed as [12]

$$P_{av} = \int_{0}^{\infty} P(u) f(u) du \tag{6}$$

In terms of shape (k) and scale (c) parameters, the average wind power density can be determined as

$$P_{av} = \frac{\rho c^3 \Gamma(1+3/k)}{2} \tag{7}$$

Where P_{av} is the wind power density based on an air density (ρ_{air}) of $1.225\, kg/m^3$. The following equation was used to estimate the wind shear coefficient (WSC), α

$$\alpha = \frac{\ln(U_2) - \ln(U_1)}{\ln(Z_2) - \ln(Z_1)} \tag{8}$$

where U_1 and U_2 are the wind speeds at heights Z_1 and Z_2, respectively.

Figure 9 presents the wind power density for this region; the value increases with the height and the maximum value is found in the summer season to be 78 W/m^2, and to be 75 and 69 W/m^2 in the winter and rainy season at 120 m AGL, respectively. A similar trend is also observed at 90 and 60 m AGL. However the wind power density for this region can be classified in the poor condition [5].

$\overline{P}/A < 100\ W/m^2$ is poor, $\overline{P}/A \approx 400\ W/m^2$ is good and $\overline{P}/A > 700\ W/m^2$ is excellent.

Wind Shear Coefficient (α) for this region was calculated based on daily wind speed between 60 - 90 m, 90 - 120 m and also 60 - 120 m AGL. The value started increasing from 15:00 till midnight and was quite stable until 06:00, and then suddenly decreased within 3 hours from 06:00 to 9:00, and then was quite stable again from 09:00 until 15:00 as presented in Figure 10. The average value of wind shear coefficient for this region was 0.35, 0.26 and 0.31 for between 60-90, 90-120 and 60-120 m AGL, respectively. Wind shear coefficient is used to predict the wind speed profile in the vertical direction from the ground which is suitable for the hub height of the wind turbine.

3.4. Wind turbine selected and energy output

The annual energy production (AEP) and plant capacity factor (CF) can be calculated as:

$$AEP = 8760 \int_{U_{ci}}^{U_{co}} p_t(u) f(u) du \tag{9}$$

$$CF = \frac{AEP}{P_r x8760} \tag{10}$$

where U_{ci} and U_{co} are the cut in and cut out speed of the wind turbine while P_t and P_r are the power curve and rated power of the wind turbine, respectively.

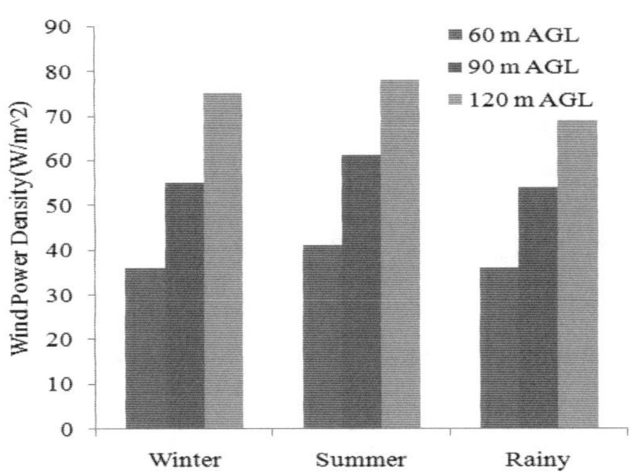

Figure 9. Wind Power Density

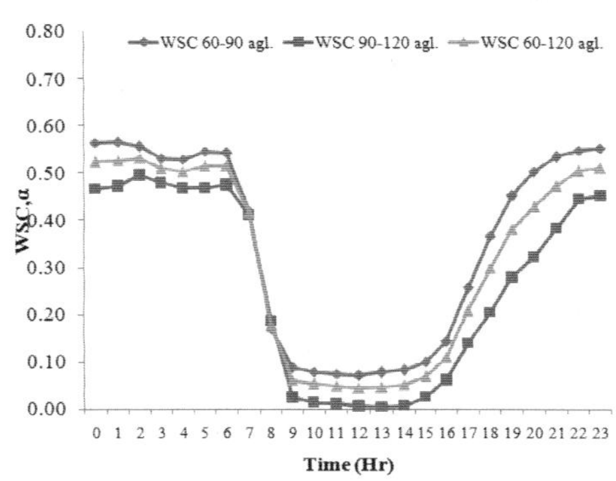

Figure 10. Wind Shear Coefficient

For this study, WAsP 10.0 software licensed by Khon Kaen University was used to estimate the energy yield output of the wind turbine plant. The selected wind turbines and technical data are presented in Figure 11 and Table 3; most wind turbines have a cut in speed of 3.0 m/s. Annual Energy Production (AEP) output and Capacity Factor (CF) of selected wind turbines are shown in Figure 12. The maximum AEP of 3,255 MWh/year can be achieved from the VESTAS 3.3 MW wind turbine, followed by VESTAS 2.0 MW, BONUS 1.3 MW and Nordex N43 wind turbines with the values of 2474, 319.1 and 248.3 MWh/year, respectively. However the capacity factor of the wind turbines have values of 15.68%, 11.16%, 4.72% and 2.80% for VESTAS 2.0 MW, VESTAS 3.3 MW, Nordex N43 and BONUS 1.3 MW, respectively. The Capacity Factor of the wind turbine was calculated based on the actual energy output in a year and the energy output at rated speed (maximum energy output) of the wind turbine. It can be used to consider the wind turbine plant installation and grid connection. However, for this region the capacity factor of selected wind turbines were quite low, so it should be recommended as a small scale wind turbine application or local application, combined with solar energy [12].

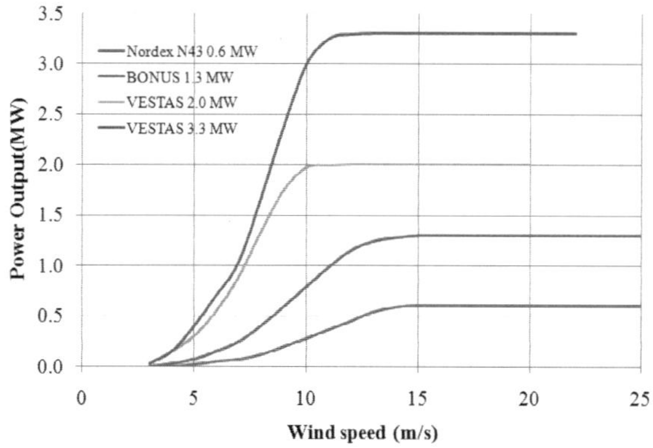

Figure 11. Power curve of selected wind turbines

Table 3. Technical data of selected wind turbines

Selected WTG	Cut-in Speed	Cut-out Speed	Hub Height	Rotor Diameter
Nordex N43 0.6 MW	3	25	78	43
BONUS 1.3 MW	3	25	60	62
VESTAS 2.0 MW	3	20	125	110
VESTAS 3.3 MW	3	22.5	117	126
(Unit)	m/s	m/s	m	m

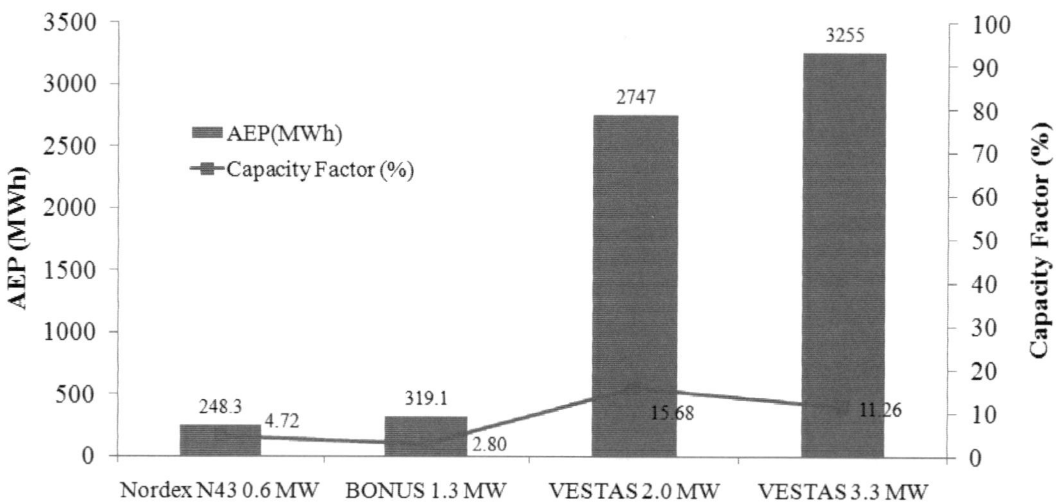

Figure 12. AEP and Capacity Factor of selected wind turbines

4. Conclusion

A wind energy assessment in Kalasin province of Thailand was made with wind data recorded for 4-years from January 2012 to December 2015. The conclusions are as follows.

- The annual mean wind speed is 3.14, 3.63 and 3.94 m/s at 60, 90 and 120 m AGL, respectively.
- Daily wind speed was higher at night time with maximum speed of 5.01 m/s at 21:00.
- Wind frequency distribution above 3.0 m/s from the Weibull curve was 49%, 59% and 60% at 60, 90 and 120 AGL, respectively.
- Wind power density for this region showed a maximum value at 78 W/m^2 at 120 m AGL in the summer season. For this region, it can be classified as a poor condition.
- The average value of the wind shear coefficient for this region was 0.35, 0.26 and 0.31, calculated based on 60-90, 90-120 and 60-120 m AGL, respectively.

IOP Conf. Series: Earth and Environmental Science **150** (2018) 012020 doi:10.1088/1755-1315/150/1/012020

- The annual energy production (AEP) from a commercial VESTAS wind turbine with rated power of 2.0 MW was estimated to be 2,747 MWh/year, with the highest capacity factor of 15.68%.
- However, the wind performance for this region was quite low. It should be recommended to be small scale for local application or be combined with solar energy.

5. References

[1] BP statistical review of world energy June 2016 Available from https://www.bp.com Accessed on 15 July 2017.
[2] Narumitr S and Chanatip P 2011 Status and outlook for Thailand's low carbon electricity development Renewable and Sustainable Energy Reviews 15 564-573.
[3] Jompop W et al 2015 Offshore wind power potential of the Gulf of Thailand Renewable Energy 81 609-626.
[4] Muyiwa et al 2014 Assessment of wind power generation along the coast of Ghana Energy Conversion and Management 77 61-69.
[5] Amir D et al 2016 Evaluation of wind energy potential in province of Bushehr, Iran Renewable and Sustainable Energy Reviews 55 455-466.
[6] Belabes B 2015 Evaluation of wind energy potential and estimation of cost using wind energy turbines for electricity generation in north of Algeria Renewable and Sustainable Energy Reviews 51 1245-1255.
[7] TiZpar A et al 2014 Wind resource assessment and wind power potential of Mil-E Nader region in Sistan and Baluchestan Province, Iran-Part1: Annual energy estimation Energy Conversion and Management 79 273-280.
[8] Global wind report annual market update 2016 Available from http://www.gwec.net Accessed on 15 July 2017.
[9] Shu Z R et al 2015 Statistical analysis of wind characteristics and wind energy potential in Hong Kong Energy Conversion and Management 101 644-657.
[10] Karthikeya B R et al 2016 Wind resource for urban renewable energy application in Singapore Renewable Energy87 403-414.
[11] Fang H F 2014 Wind energy potential assessment for the offshore area of Taiwan west coast and Penghu Archipelago Renewable Energy 67 237-241.
[12] Pham Q and Thanachai L 2015 Assessment of wind energy potential for selecting wind turbines: An application to Thailand Sustainable Energy Technologies and Assessments 11 17-26.
[13] Huang W et al 2016 Wind farm micro-siting optimization using novel cell membrane approach IOP conf. series: Earth and Environmental Science 40.

Acknowledgement
The authors are very grateful to National Research Council of Thailand (NRCT) and Center for Alternative Energy Research and Development, Khon Kaen University (AERD-KKU) for research funding.

Performance and Simulation of a Stand-alone Parabolic Trough Solar Thermal Power Plant

S T Mohammad[1], H H Al-Kayiem[1], M K Assadi[1], S I U H Gilani[1] and A K Khlief[1]

1 Mechanical Engineering Department, Universiti Teknologi PETRONAS, 32610 Bandar Seri Iskandar, Perak, Malaysia
E-mail: hussain_kayiem@utp.edu.my

Abstract. In this paper, a Simulink® Thermolib Model has been established for simulation performance evaluation of Stand-alone Parabolic Trough Solar Thermal Power Plant in Universiti Teknologi PETRONAS, Malaysia. This paper proposes a design of 1.2 kW parabolic trough power plant. The model is capable to predict temperatures at any system outlet in the plant, as well as the power output produced. The conditions that are taken into account as input to the model are: local solar radiation and ambient temperatures, which have been measured during the year. Other parameters that have been input to the model are the collector's sizes, location in terms of latitude and altitude. Lastly, the results are presented in graphical manner to describe the analysed variations of various outputs of the solar fields obtained, and help to predict the performance of the plant. The developed model allows an initial evaluation of the viability and technical feasibility of any similar solar thermal power plant.

1. Introduction

As observed in the 21st century, there are many problems that raise in perspective of energy and environment; thus, leading to a rising problems in politics and economy. A study stated that for a sustainable supply of energy and a reduction in emissions of the pollutants, there is an intense focus on renewable energy in the world [1]. Another study thoroughly discusses the trough solar thermal power plants in terms of their potentials and developments in Algeria. The authors emphasized that the process of working principles, descriptive parabolic trough power plants, and an assessment of concentrating solar power potential has provide a competitive outcome in terms of viability of CSP plants in Algeria. In addition to that, Algeria needs to consider the pre-requisites such as high-quality insolation, land and water availability, extensive transmission, and the power grid; to provide an economical CSP power generation [2]. However, Boukelia [3] studied carried out in Algeria where the authors tested on Optimization, Selection, and feasibility study of solar parabolic trough power plants. Optimization of the two parabolic trough was integrated and performed with: 1) thermal energy storage TES, and 2) fuel backup system FBS, respectively. The heat fluid used by 1st plant was Therminol VP-1 in field and 2nd plant used molten salt. In relation to that, García et al. [4] developed a simulation model that allows a recalculation of performance of 50 MWe parabolic trough power plant; which is accompanied with thermal energy storage whilst Therminol VP-1 as heat transfer fluid in the solar field. The results obtained from the model are compared with the experimental data which was obtained from operation power plants in Spain. Montes et al. [5] highlighted influences of solar multiple annual performance, natural gas consumption, and levelized cost of electricity of a 50 MWe;

using direct steam generation parabolic trough thermal power plant with thermal storage and auxiliary natural gas-fired boiler.

A study presents a thermodynamic model which allowed an estimation of performance of 100 MW hybrid parabolic trough solar thermal power plant; whilst fossil fuels backup friction required. As it was conducted in four location North Chile); it was based on minimal fossil fuel backup fraction Larraín et al. [6]. Moreover, another author used 50MW solar thermal power plant and requirement of the land area with parabolic trough system integration; and suggests this technology as the best to its industrial advantages Kalogirou [7]. A study analysed an energetic economic assessment of a solar parabolic trough power plant of 100 MW capaciate and discussed amongst the four sites chosen in Algeria Abbas et al. [8]. On the other hand, a study discusses on the technical developments of the parabolic trough solar power plant when combined with molten salt as HTF, and presents the potential of using this technology to reduced LCOE Ruegamer et al. [9]. Typically, this feasibility report was combined with optimization results, where 4E comparison of PTSTPPs with molten salt is primary HTFs was established in solar field. The results obtained were agreed upon by Xu et al. [10]; where presentation of a theoretical framework for energy and exergy analyses of a solar power tower plant whilst incorporating solar molten salt as HTF. Similarly, this analysis was also conducted in another research which accounts for parabolic trough solar thermal power plant using therminol VP-1 as working fluid Reddy et al. [11]. In order to optimize and obtain the maximum plant efficiency, the authors evaluated the energetic and exergetic losses and efficiencies through specified operating conditions.

To conclude the above discussions from literature, it is discovered that the technical disability of PTSTPP is yet to be established for a location that has a high DNI availability. In addition, there is also a need to identify on the CSP technology for Malaysia climatic condition. In this article, a proposed design of Stand-alone Parabolic Trough Power Plant with an average capacity of 1.2kW is discussed. Unversiti Teknologi PETRONAS, Perak, Malaysia has been selected as the location of operation. Theoretical Calculations have been carried out for the stated power output at MATLAB software. The results of the calculations are employed in this software as inputs/parameters in order to model & simulate the performance of the stand-alone parabolic trough solar power plant in Simulink® Thermolib software, effectively.

2. PTSTPP Analysis

This section presents the mathematical modeling Figure 1. Firstly, PTSTPP energy analysis is carried out using the equations proposed by authors in [12], [13]; hence, the solar system model is presented. Secondly, the performance assessment equations for the overall systems are shown; with assumptions that there is a steadiness in the system state.

In addition, it is negligible if there is a change in the pressure in pumps and turbines. However, the useful energy rate from the collector is defined as:

$$Q_{gain} = m_{CL} \cdot (h_7 - h_6\ ) = \eta \,. I . A_p \tag{1}$$

$$\eta = \eta_o - U_L \cdot \left(\frac{\Delta T}{I}\right) \tag{2}$$

$$\Delta T = T_m - T_a \tag{3}$$

Where m_{CL} is the mass flow rate of steam passing through Collector (kg/s), I is direct normal incidence radiation (W/m^2, A_P is aperture area of the collector (m^2), $T_m = (T_6 + T7)/2$, denotes mean temperature (°C), Ta is ambient temperature (°C), and U_L is loss co-efficient based on aperture area (W/m^2 .K). If the work input Feed Water Pump (F.W.P) and Recirculation Water Pump (RC.W.P) as well as heat losses through piping are neglected then:

$$Q_{gain} = m_{CL} \cdot (h_7 - h_6\ ) = m \,. (h_1\ - h_3\ ) \tag{4}$$

From Eqs. (1) and (4):

$$m \,. (h_1 - h_3\ ) = \eta \,. I . A_p \tag{5}$$

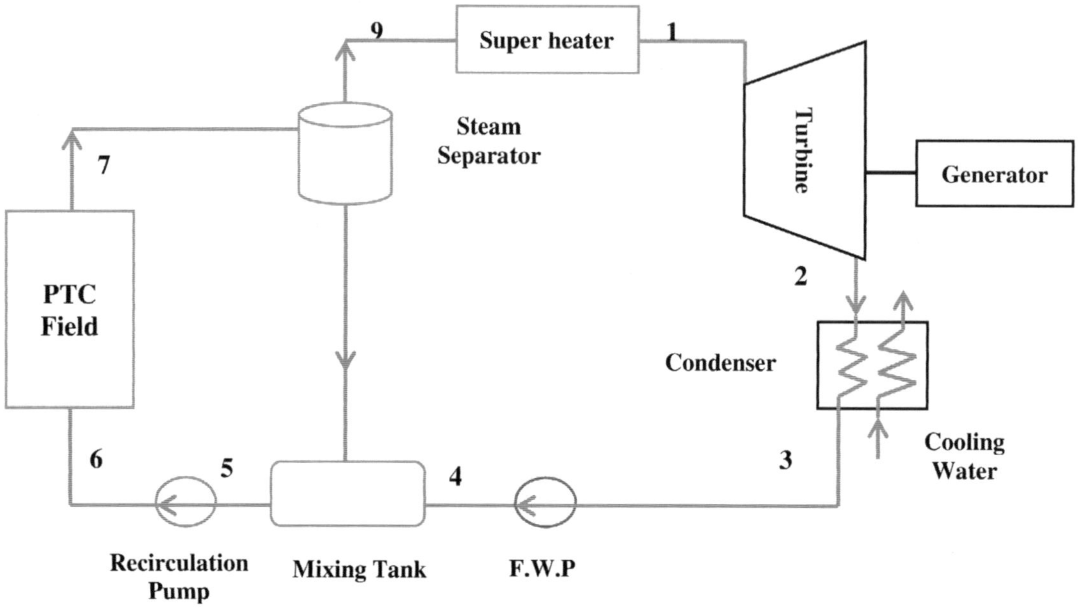

Figure 1. Simplified process flow diagram of PTC based solar thermal power plant.

In the modeling part, each component of the power plant is analyzed, defining and evaluating its efficiency by properly identifying the inputs and outputs. The stand-alone parabolic trough solar thermal power plant to be investigated is assumed to operate using the Rankine power cycle. Steam leaving the turbine and cascaded wet-vapor from the feed water heaters is condensed in the condenser. The condensate is pumped to the mixing tank pressure. The recirculating pump pressure water to the PTC solar filed. Steam separator tank is separates water from steam, steam passing through superheater and water return into mixing tank. The complete schematic diagram of the power plant is shown in Figure 2.

Table 1: Characteristics of PTC used for simulation

Parameters	Value
Length of the receiver (L)	6.4m
Width of the collector (Wa)	2.6m
Focal length (F)	0.87m
Receiver external diameter (Dao)	40mm
Receiver internal diameter (Dai)	38mm
Concentration ratio (C)	20
Rim angle (°)	72.5
Absorptance of the receiver ($\boldsymbol{\alpha}$)	0.95
Emittance of the receiver ($\boldsymbol{\epsilon_{ab}}$)	0.10
Glass envelope transmittance ($\boldsymbol{\tau}$)	0.91
Glass envelope emittance ($\boldsymbol{\epsilon_c}$)	0.91
Reflector surface reflectivity($\boldsymbol{\rho}$)	0.7
Intercept factor ($\boldsymbol{\gamma}$)	0.93
Aperture Area (m)	80

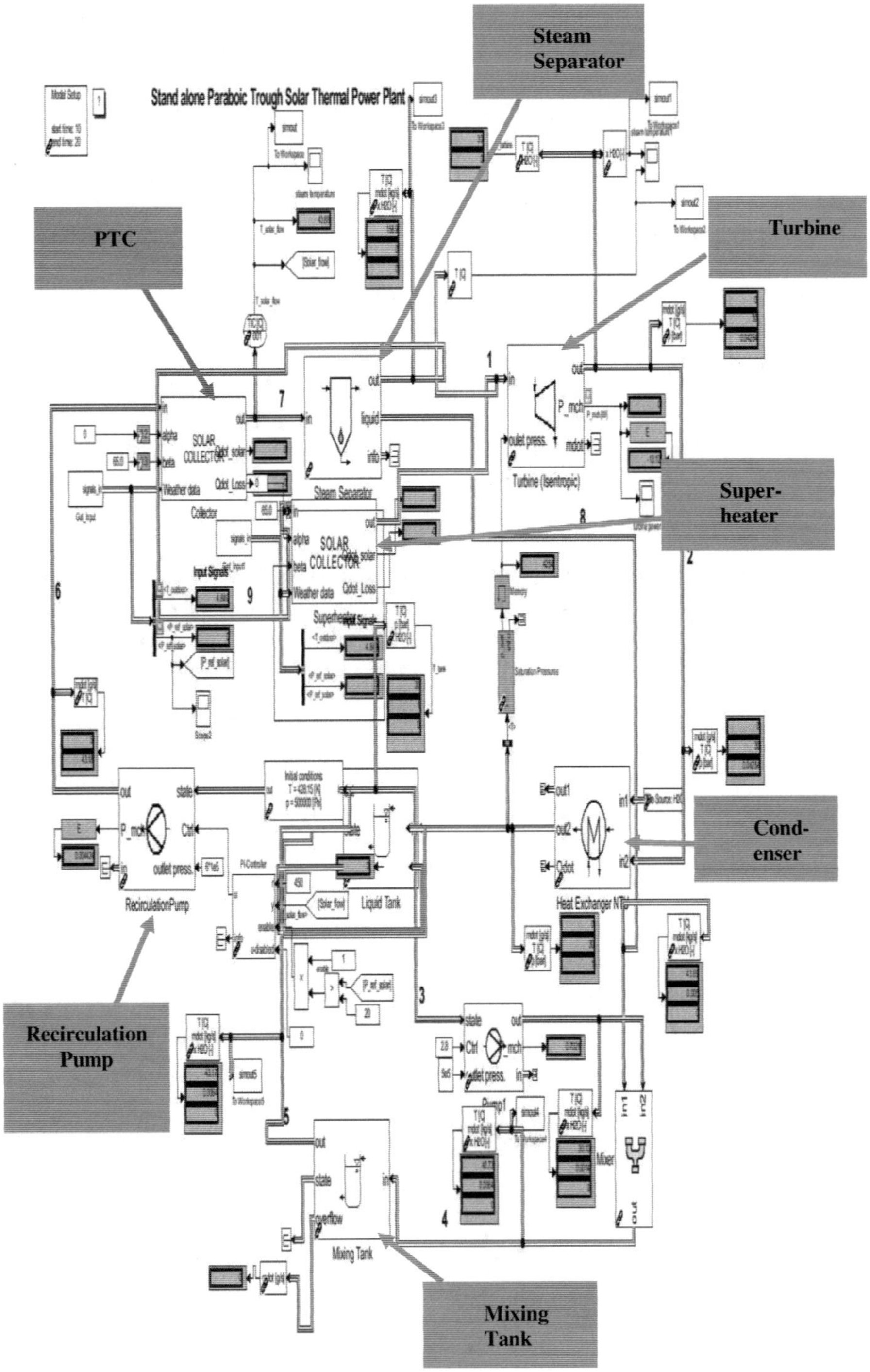

Figure 2. Block diagram of PTC based solar thermal power plant in Simulink® Thermolib.

3. Case study

In the case study, the study gathered a one whole year data to analyse system performance. However, only several days are presented in this article. The days include: February 26, June 26, September 9 and December 28, 2016. The days are chosen due to two reasons: clear and cloudy weather. Each figure below presents the solar radiation for each day, respectively. With the data collected, the performance of the system is observed through the evaluation of the results that are provided by the simulation model constructed in the study. With the assistance of the model, it is able to provide valuable data which is capable of plotting graphs, such as: 1) Solar radiation, 2) steam temperature leaving the collector field, and 3) the power output of the system. With that, there were a few system parameters which were kept constant: (1) water mass flow rate and steam's mass flows, (2) number of collectors, (3) water inlet temperature, (4) condensate water temperature and pressure and (5) the turbines' operating pressure. The SEGs was used to monitor these parameters carefully. Thus, providing a means of system control. PTC specifications and their characteristics that are considered throughout the system monitoring are shown in Table 1.

4. Results and discussion

The result of the simulation for four days of operation, two clear days (February 26, June 26) and two cloudy days (September 9, December 28) are given below. For the clear day simulation, all the mass flow rates and inlet and exit temperatures as well as the power generated are established to be equal to the design value. The cloudy day simulation, however, shows that all the results are highly dependent on DNI values for improved output from this kind of power plants.

Figure 3 shows results obtained on a clear sunny day, thus, more radiation and an optimal system performance is observed, respectively. On a clear days such as June 26, 2016, the capacity of the power plant is achieved for a period of about 9 hours (9:00 – 17:00 hours) with maximum power 1500 W. However, when solar transients due to clouds occur, it shows an overall drop in system temperature. Hence, the power output drops. If the occurrence of these transient is for a period of 5 to 10 minutes intervals, then the system is able to return to its steady power production state as the solar radiation is maintained. However, if the solar transients are prolonged (an hour or more); then the system performance is gradually reduced in a great amount and a possibility of system recovery is merely present.

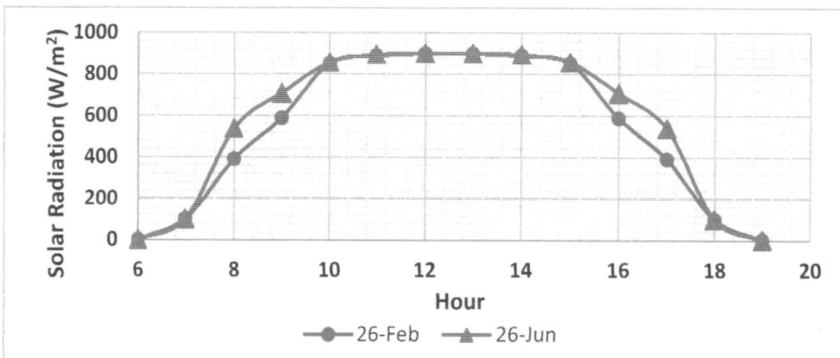

(a)

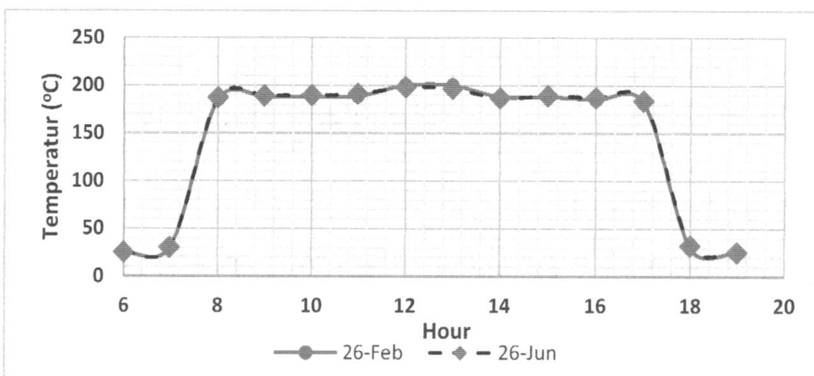

(b)

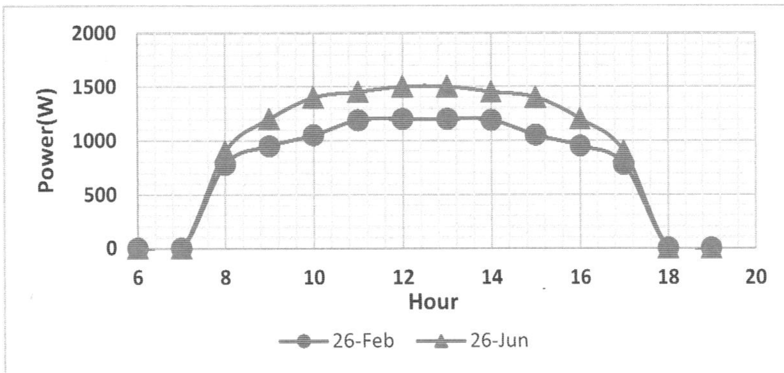

(c)

Figure 3. (a) Daily measured Solar Radiation, (b) Steam Temperature and (c) the respective Power Output for clear days: 26-Feb and 26-Jun, 2016.

Figure 4 illustrates the presence of solar transients due to long periods of rainy weather. The capacity of the power plant is achieved for a period of about 4 hours (10:00 – 15:00 hours) on a cloudy day such as September 9, 2016 with maximum power 900 W. It shows that the system has produce zero power output as the rain begins. There were various aspects taken into consideration in this study, however, they were not involved in simulation model due to maintaining the simplicity, as most cases could be negligible. This includes: (1) system's piping losses, (2) collectors' optical losses, and (3) no customization of power block.

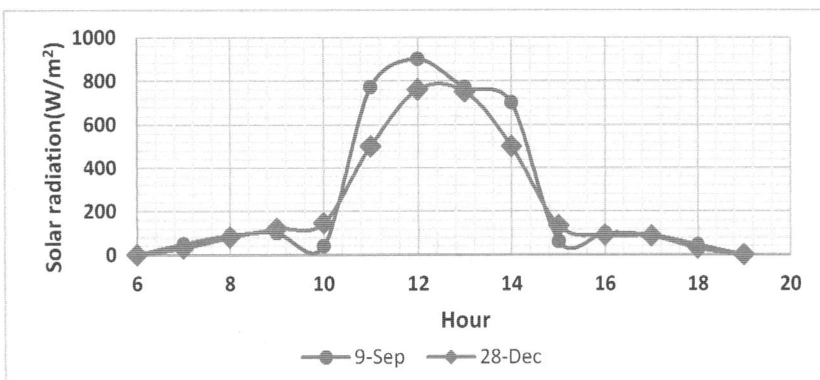

(a)

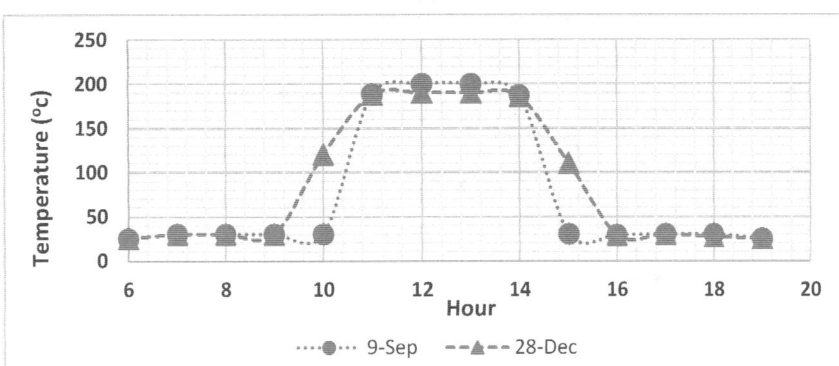

(b)

IOP Conf. Series: Earth and Environmental Science **150** (2018) 012021 doi:10.1088/1755-1315/150/1/012021

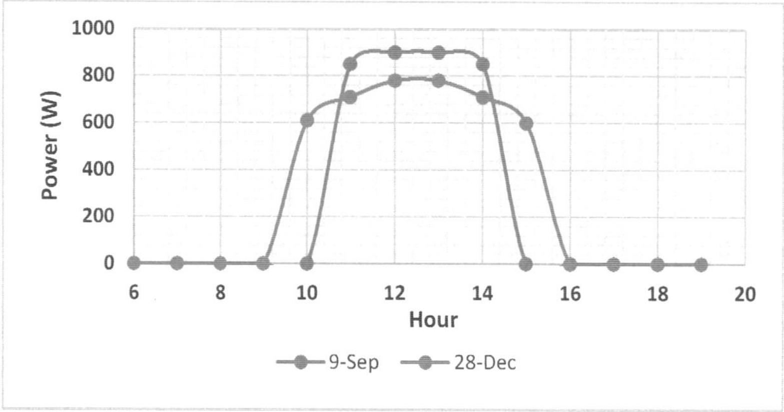

(c)

Figure 4. (a) Daily measured solar radiation, (b) steam temperature and (c) the respective power output for cloudy days: 9-Sep and 28-Dec, 2016.

5. Conclusion

To conclude this article, a 1.2 kW stand-alone parabolic trough solar power plant is designed and proposed to carry out analysis on Simulink® Thermolib. Simulink® Thermolib is used by creating a solar field model which allows prediction in the performance under various weather conditions at Perak (Malaysia) throughout a year duration period. The model is able to adapt distinctive designs by varying the parameters values that are associated with the parabolic trough; which can be easily achieved through Simulink® Thermolib. In addition to that benefits, the Thermolib model can be implemented for simulating the performance of parabolic trough field which may be located at an entirely different site with entirely distinctive weather conditions by only integrating weather data of that site. Although the data is readily provided by measurements, the comparison of theoretical and simulation is highly required to allow a justification of the results that are validate by the Thermolib Model. Lastly, Thermolib model assures to be an effective and reliable tool for confirming and analysing the average and transient performance of the solar parabolic trough efficiently.

6. References

[1] Ershu X, Dongming Z, Hui X, Shidong L, Zhiqiang Z, Zhiyong W and Zhifeng W 2015 The Badaling 1MW Parabolic Trough Solar Thermal Power pilot plant , *Energy Procedia*, 69 , 1471 – 1478

[2] Boukelia T and Mecibah M 2013 Parabolic trough solar thermal power plant: potential, and projects development in Algeria, *Renew Sustain Energy Rev*. 21, 288–297

[3] Boukelia T E, et al., 2015 Optimization, selection and feasibility study of solar parabolic trough power plants for Algerian conditions, *Energy Convers. Manag*. 101, 450–459.

[4] García L I, Álvarez J L and Blanco D 2011 Performance model for parabolic trough solar thermal power plants with thermal storage: comparison to operating plant data, *Sol. Energy* 85 (10), 2443–2460.

[5] Montes M J, Abànades A and Martınez J M-Val 2009 Performance of a direct steam generation solar thermal power plant for electricity production as a function of the solar multiple, *Sol. Energy* 83 (5) , 679–689.

[6] Larraín T, Escobar R and Vergara J 2010 Performance model to assist solar thermal power plant siting in northern Chile based on backup fuel consumption, *Renew Energy* 35 (8), 1632–1643.

[7] Kalogirou S A 2013 Solar thermoelectric power generation in Cyprus: selection of the best system, *Renew. Energy* 49, 278–281.

[8] Abbas M, Belgroun Z, Aburidah H and Merzouk N K 2013 Assessment of a solar parabolic trough power plant for electricity generation under Mediterranean and arid climate conditions in Algeria, *Energy Procedia* 42 , 93–102.

[9] Ruegamer T, Kamp H, Kuckelkorn T, Schiel W, Weinrebe G, Nava P, et al., 2014 Molten salt for parabolic trough applications: system simulation and scale effects, *Energy Procedia* 49, 1523–1532.
[10] Xu C, Wang Z, Li X and Sun F 2011 Energy and exergy analysis of solar power tower plants, *Appl. Therm. Eng.* 31 (17), 3904–3913.
[11] Reddy V S, Kaushik S C and Tyagi S K 2012 Exergetic analysis and performance evaluation of parabolic trough concentrating solar thermal power plant (PTCSTPP), *Energy* 39 (1), 258–273.
[12] Duffie J and Beckman W 2006 Solar engineering of thermal processes. *John Wiley & Sons, Inc.*
[13] Desai N B, Kedare S B and Bandyopadhyay S 2014 Optimization of design radiation for concentrating solar thermal power plants without storage, *Solar Energy* Vol. 107, pp. 98–112.

Acknowledgments
The author acknowledges the support of Universiti Teknologi PETRONAS, Malaysia, for this work.

Thermodynamics Analysis of Binary Plant Generating Power from Low-Temperature Geothermal Resource

A Maksuwan[1]

1 Department of Environmental Science and Technology, Faculty of Science and Technology, Pathumwan Institute of Technology, 833 Rama 1 Road, Wangmai, Pathumwan, Bangkok 10330, Thailand
E-mail: iiibigiii@hotmail.com

Abstract. The purpose in this research was to predict tendency of increase Carnot efficiency of the binary plant generating power from low-temperature geothermal resource. Low-temperature geothermal resources or less, are usually exploited by means of binary-type energy conversion systems. The maximum efficiency is analyzed for electricity production of the binary plant generating power from low-temperature geothermal resource becomes important. By using model of the heat exchanger equivalent to a power plant together with the calculation of the combined heat and power (CHP) generation. The CHP was solved in detail with appropriate boundary originating an idea from the effect of temperature of source fluid inlet-outlet and cooling fluid supply. The Carnot efficiency from the CHP calculation was compared between condition of increase temperature of source fluid inlet-outlet and decrease temperature of cooling fluid supply. Result in this research show that the Carnot efficiency for binary plant generating power from low-temperature geothermal resource has tendency increase by decrease temperature of cooling fluid supply.

1. Introduction

In recent years, accelerated consumption of fossil fuels has caused lots of serious environment problems. In this context, using renewable like geothermal for electricity production becomes important. Generally, the high-temperature reservoirs (> 220 °C) are the ones most suitable for commercial production of electricity with dry stream and flash stream systems. The low- and medium-temperature geothermal resources with temperatures of typically below 220 °C are by far the most commonly available resource and are highly recommended for using in local district heating [1]. Low-temperature geothermal resources, typically 150 ◦C (300 ◦F) or less, are usually exploited by means of binary-type energy conversion systems [2]-[4]. A binary cycle geothermal power plants, pumps are used to pump hot water from a geothermal well, through a heat exchanger, and the cooled water is returned to the underground reservoir [5]-[7]. A second working or binary fluid with a low boiling point, typically a butane or pentane hydrocarbon, is pumped at fairly high pressure through the heat exchanger, where it is vaporized and then directed through a turbine [8]-[10].

Production of electricity from heat requires a heat engine working between two heat reservoirs, a heat source and a heat sink. The Carnot efficiency is defined for the reversible heat engine working between two reservoirs of infinite heat capacity and constant temperature [11], [12]. Geothermal heat source is a stream of water, either in the liquid phase or as liquid-steam mixture of any quality. The streams in and out of the system have four flow properties: mass, heat capacity, enthalpy and exergy. The mass conservation is obvious, no mixing of the source and cooling streams is assumed. The heat capacity is important for the characteristics of the heat conversion, and will be treated here as a heat capacity flow, the product of fluid heat capacity and flow rate. The product of the enthalpy relative to the environmental temperature and the flow rate defines the heat flow in and out of the system. The exergy will give information on the work producing potential of the system, and is calculated in the same way as the enthalpy. Reference textbooks such as Cengel [13] give basic information on exergy and its definition, but here the analysis is as well based on Kotas [14] and Szargut [15]. Porolfsson [16] and Valdimarsson [17] apply these methods on specific geothermal applications.

The hypothesis in this research is electricity production depend on the Carnot efficiency. This efficiency is defined for the reversible heat engine working between two reservoirs based on heat and exergy flow [18], [19]. This research analyzes when the heat source and the heat sink do not have a constant temperature and focus on calculation of the combined heat and power (CHP) generation [20]-[22]. The maximum efficiency is analyzed for geothermal application of the binary plant generating power from low-temperature geothermal resource. The result in this research show that the Carnot efficiency for binary plant generating power from low-temperature geothermal resource has tendency increase by decrease temperature of cooling fluid supply.

2. Fundamental of thermodynamics

The general Second Law formulation for open systems [23], is given as

$$\dot{\varphi}_p = \frac{dS}{d\tau} - \sum_{i=1}^{n} \dot{m}\, s_i - \int_{\tau_1}^{\tau_2} \frac{1}{T}\frac{\delta Q}{\delta \tau}. \tag{1}$$

We will deal here only with steady systems therefore, all time derivatives of thermodynamic properties will vanish, leading to our working equation

$$\dot{\varphi}_p = -\sum_{i=1}^{n} \dot{m}\, s_i - \int_{\tau_1}^{\tau_2} \frac{1}{T}\frac{\delta Q}{\delta \tau}. \tag{2}$$

In analyzing any system it will always be necessary to augment equation (1) and (2) with equations expressing conservation of both energy and mass. Applying the First Law of thermodynamics for open systems operating in steady state yields the working equation

$$\dot{Q} - \dot{W}_{ss} = -\sum_{i=1}^{n} \dot{m}_i \left(h_i + \frac{v_i^2}{2} + g z_i \right). \tag{3}$$

We often refer to the state of the surroundings as the *dead state* because when fluids are in thermodynamic equilibrium with the surroundings there is no potential for doing work and the fluid may be considered dead. A consequence of the first condition is that $\dot{\varphi}_p$ in equation (2) vanishes

$$-\sum_{i=1}^{n} \dot{m}\, s_i - \int_{\tau_1}^{\tau_2} \frac{1}{T}\frac{\delta Q}{\delta \tau} = 0. \tag{4}$$

Let us consider a simplified system with only two channels one inlet (state 1) and one outlet (state 2). Also, let us for the moment ignore the effects of kinetic and potential energy. Then, Eq. (3) becomes

$$\dot{Q} - \dot{W} = \dot{m}_i \left(h_2 - h_1 \right). \tag{5}$$

where the subscript *ss* on the power in equation (3) has been dropped for clarity. Since the only heat transfer is between the system and the dead state, Eq. (5) and (4) can be rewritten, respectively, as:

$$\dot{Q}_0 - \dot{W} = \dot{m}_i \left(h_2 - h_1 \right). \tag{6}$$

and

$$-\dot{m}_i \left(s_2 - s_1 \right) - \frac{\dot{Q}_0}{T_0} = 0. \tag{7}$$

Eliminating $\dot{Q}_0$, one obtains

$$\dot{W} = \dot{m}_i \left[h_1 - h_2 - T_0 \left(s_1 - s_2 \right) \right]. \tag{8}$$

Finally, we use the exit state be identical to the dead state, and so obtain the maximum power output:

$$\dot{W}_{max} = \dot{m}_i\left[h_1 - h_0 - T_0\left(s_1 - s_0\right)\right]. \tag{9}$$

The expression in brackets is given a distinctive name, the specific exergy, e_{x1}:

$$e_{x1} \equiv h_1 - h_0 - T_0\left(s_1 - s_0\right). \tag{10}$$

The equation (10) may be used to find the specific exergy of any fluid stream at a temperature T_1 and pressure P_1, relative to a given set of ambient conditions T_o and P_o.
For a fluid flowing at a certain mass flow rate, multiplying the specific exergy by the mass flow rate results in the maximum power output theoretically obtainable from the given fluid for the given surroundings; we will call this the exergetic power. Instead of using the symbol $\dot{W}_{max}$ as in equation (9), we will henceforth use a new symbol, $\dot{E}$, for this important quantity. If kinetic or potential energy effects are important, the enthalpy h_1 should be augmented by $1/2\,v_1^2$ or gz_1, respectively. Since state 1 is really arbitrary, we can drop the subscript 1 and obtain a general expression:

$$e_x = h - h_0 - T_0\left(s - s_0\right). \tag{11}$$

where the properties at the dead state are evaluated at T_o and P_o. The power plant as a whole, the overall exergetic efficiency reduces to a very simple formula, namely, the ratio of the net power output to the exergy of the motive fluid serving as the energy source for the plant [24].

3. The combined heat and power (CHP) generation

In Figure 1, the temperature of the entering cooling fluid is taken to be the environmental temperature, the lowest temperature which can be obtained, as well as defining the thermal sink temperature for the Carnot engine efficiency.

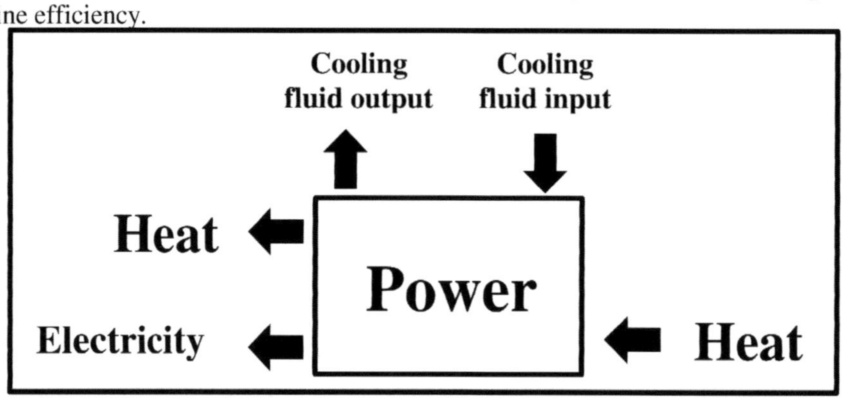

Figure 1. The generic power plant.

In the following this system will be analysed in order to gain a better understanding of the conversion of low temperature heat into electricity [25], from Figure 2.

c_h = Source fluid heat capacity, T_h = Source fluid inlet temperature

$\dot{m}_h$ = Flow rate of source fluid, T_s = Source fluid outlet temperature

c_c = Cooling fluid heat capacity, T_c = Cooling fluid outlet temperature

$\dot{m}_c$ = Cooling fluid flow rate, T_0 = Cooling fluid inlet temperature (Environmental temperature)

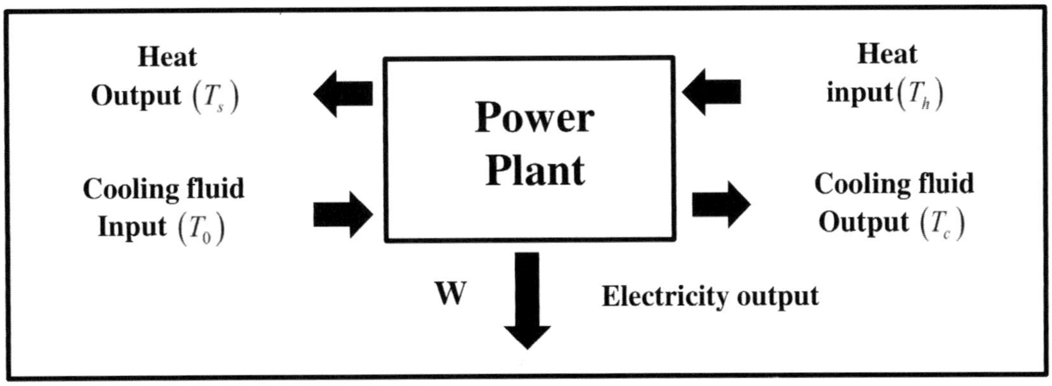

Figure 2. Model of a heat exchanger equivalent to a power plant.

The heat flow of geothermal source fluid$\left(\dot{Q}_h\right)$, geothermal outflow$\left(\dot{Q}_s\right)$ and cooling fluid supply $\left(\dot{Q}_c\right)$, respectively, are given by

$$\dot{Q}_h = \dot{m}_h c_h \left(T_h - T_0\right). \tag{12}$$

$$\dot{Q}_s = \dot{m}_h c_h \left(T_s - T_0\right). \tag{13}$$

$$\dot{Q}_c = \dot{m}_c c_c \left(T_c - T_0\right). \tag{14}$$

The exergy flow $\left(\dot{\psi} = \dot{m}e_x\right)$ of geothermal source fluid$\left(\dot{\psi}_h\right)$, geothermal outflow$\left(\dot{\psi}_s\right)$ and cooling fluid supply$\left(\dot{\psi}_c\right)$, respectively, are given by

$$\dot{\psi}_h = \dot{m}_h c_h \left[\left(T_h - T_0\right) - T_0 \ln\left(\frac{T_h}{T_0}\right)\right] = \dot{Q}_h - \dot{m}_h c_h T_0 \ln\left(\frac{T_h}{T_0}\right). \tag{15}$$

$$\dot{\psi}_s = \dot{m}_h c_h \left[\left(T_s - T_0\right) - T_0 \ln\left(\frac{T_s}{T_0}\right)\right] = \dot{Q}_s - \dot{m}_h c_h T_0 \ln\left(\frac{T_h}{T_0}\right). \tag{16}$$

$$\dot{\psi}_c = \dot{m}_c c_c \left[\left(T_c - T_0\right) - T_0 \ln\left(\frac{T_c}{T_0}\right)\right] = \dot{Q}_c - \dot{m}_c c_c T_0 \ln\left(\frac{T_h}{T_0}\right). \tag{17}$$

The energy (1. law) and exergy (2. law) balances are:

$$\dot{W}_{rev} = \dot{Q}_h - \dot{Q}_s - \dot{Q}_c - c_h \dot{m}_h T_0 \ln\left(\frac{T_h}{T_s}\right) + c_c \dot{m}_c T_0 \ln\left(\frac{T_c}{T_0}\right), \tag{18}$$

where, $\dot{W}_{rev} = \dot{\psi}_h - \dot{\psi}_s - \dot{\psi}_c$ and $\dot{W} = \dot{Q}_h - \dot{Q}_s - \dot{Q}_c$.

The energy balance is valid for all processes, ideal and real. The exergy balance gives only information on the reversible work or the largest amount of work that can be obtained from the power plant. If the power plant ideal is $\dot{W}_{rev} - \dot{W} = 0$, then the energy (1. law) and exergy (2. law) balances in the equation (18), become

$$-c_h \dot{m}_h T_0 \ln\left(\frac{T_h}{T_s}\right) + c_c \dot{m}_c T_0 \ln\left(\frac{T_c}{T_0}\right) = 0. \tag{19}$$

Then the heat capacity flow ratio for a reversible power plant is

$$C_{rev} = \left.\frac{c_c \dot{m}_c}{c_h \dot{m}_h}\right|_{rev} = \frac{\ln\left(\frac{T_h}{T_s}\right)}{\ln\left(\frac{T_c}{T_0}\right)}. \tag{20}$$

4. The efficiencies for the combined heat and power production process

The efficiencies for the combined heat and power production process are thus electrical power generation efficiencies. All the heat contained in the stream flow rate of source fluid outlet, $\dot{m}_s$ is considered a by-product, and does not enter the efficiency calculation. Product are $\dot{W}$ and $\dot{Q}_s$. Input is $\dot{Q}_h - \dot{Q}_s$. Rejected is $\dot{Q}_c$. The first law maximum efficiency as

$$\eta_{I,\max,CHP} = \frac{\dot{W}_{rev}}{\dot{Q}_h - \dot{Q}_s} = \frac{\dot{\psi}_h - \dot{\psi}_s - \dot{\psi}_c}{\dot{Q}_h - \dot{Q}_s}$$

$$= \frac{\dot{Q}_h - \dot{Q}_s - \dot{Q}_c - c_h \dot{m}_h T_0 \left(\ln\left(\frac{T_h}{T_0}\right) - \ln\left(\frac{T_s}{T_0}\right)\right) + c_c \dot{m}_c T_0 \ln\left(\frac{T_c}{T_0}\right)}{\dot{Q}_h - \dot{Q}_s}$$

$$= \frac{c_h \dot{m}_h (T_h - T_s) - \left\{c_c \dot{m}_c (T_c - T_0) + c_h \dot{m}_h T_0 \ln\left(\frac{T_h}{T_s}\right) - c_c \dot{m}_c T_0 \ln\left(\frac{T_c}{T_0}\right)\right\}}{c_h \dot{m}_h (T_h - T_s)} \tag{21}$$

By using the heat capacity flow ratio for a reversible power plant in the equation (20), $\eta_{I,\max,CHP}$ in the equation (21) can be rewrite as

$$\eta_{I,\max,CHP} = 1 - \frac{\ln\left(\frac{T_h}{T_s}\right)}{\ln\left(\frac{T_c}{T_0}\right)} \frac{(T_c - T_0)}{(T_h - T_s)}. \tag{22}$$

5. The analysis efficiencies for low-temperature geothermal power plant

From the first law maximum efficiency in the equation (22), thus

$$\eta_{I,\max,CHP} = 1 - C_{rev} \frac{(T_c - T_0)}{(T_h - T_s)}. \tag{23}$$

Let $(T_h - T_s) = a(T_c - T_0)$, $a \in \mathbb{R}$ and $a > 1$, so that $(T_h - T_s) > (T_c - T_0)$. Let $+\Delta T$ and $-\Delta T$ are the increase and decrease of temperature, respectively and $\Delta T / (T_c - T_0) \equiv \delta$. The situation are divided in 2 cases

Case1: If increase the temperature of source fluid inlet-outlet $(T_h - T_s)$ as $(T_h - T_s) + \Delta T$, let temperature of cooling fluid supply $(T_c - T_0)$ and the heat capacity flow ratio for a reversible power plant C_{rev} are constant then

$$\eta_{I,\max,CHP(1)} = 1 - C_{rev} \frac{(T_c - T_0)}{(T_h - T_s) + \Delta T},$$

$$= 1 - C_{rev} \frac{(T_c - T_0)}{a(T_c - T_0) + \Delta T},$$

$$= 1 - C_{rev} \frac{1}{a + \dfrac{\Delta T}{(T_c - T_0)}} = 1 - C_{rev} \frac{1}{(a + \delta)} = 1 - C_{rev} \frac{1}{a\left(1 + \dfrac{\delta}{a}\right)},$$

$$= 1 - C_{rev} \frac{1}{a}\left(1 + \frac{\delta}{a}\right)^{-1} \simeq 1 - C_{rev} \frac{1}{a}\left(1 - \frac{\delta}{a}\right),$$

$$= 1 - C_{rev} \left(\frac{1}{a} - \frac{\delta}{a^2}\right).$$

So that, the first law maximum efficiency in case 1, can be rewrite as

$$\eta_{I,\max,CHP(1)} = 1 - C_{rev}\left(\frac{1}{a}\right) + C_{rev}\delta\left(\frac{1}{a^2}\right). \tag{24}$$

Case2: If decrease the temperature of cooling fluid supply $(T_c - T_0)$ as $(T_c - T_0) - \Delta T$, let temperature of source fluid inlet-outlet $(T_h - T_s)$ and the heat capacity flow ratio for a reversible power plant C_{rev} are constant then

$$\eta_{I,\max,CHP(2)} = 1 - C_{rev} \frac{(T_c - T_0) - \Delta T}{(T_h - T_s)},$$

$$= 1 - C_{rev} \frac{(T_c - T_0)}{(T_h - T_s)}\left(1 - \frac{\Delta T}{(T_c - T_0)}\right),$$

$$= 1 - C_{rev} \frac{(T_c - T_0)}{a(T_c - T_0)}(1 - \delta),$$

So that, the first law maximum efficiency in case 2, can be rewrite as

$$\eta_{I,\max,CHP(2)} = 1 - C_{rev}\left(\frac{1}{a}\right) + C_{rev}\delta\left(\frac{1}{a}\right). \tag{25}$$

6. Summary

By following methodology in the section 3-5, to investigate the Carnot efficiency for binary plant generating power from low-temperature geothermal resource. The CHP calculation was compared between condition of increase temperature of source fluid inlet-outlet in case 1 and decrease temperature of cooling fluid supply in case 2. Tendency was demonstrated by the last term in the equation 24 and equation 25. The last term in both equations were demonstrated by factor a^{-2} in case 1 and by factor a^{-1} in case 2, so that $\eta_{I,\max,CHP(2)} > \eta_{I,\max,CHP(1)}$. Result in this research shown that the Carnot efficiency $\eta_{I,\max,CHP}$ has tendency increase by decrease temperature of cooling fluid supply for binary plant generating power from low-temperature geothermal resource.

7. Acknowledgments

The financial supports of Pathumwan Institute of Technology are acknowledged.

8. References

[1] Hettiarachchi H, Golubovic M, et al 2007 *Energy*. 32 p 698 – 706
[2] Liebowitz H M and Mlcak H A 1999 *Geothermal Res*. 23 p 75 – 80

[3] Mlcak H A, 2002 *Geothermal Res.* 26 707 – 713
[4] Mlcak H A, Mirolli M, Hjartarson H, Husavikur O and Ralph M. 2002. *Geothermal Res.* 26 p 715–718
[5] Abdeen M O 2008 *Renew. Sustainable Energy Rev.* 12 (2) p 344-371
[6] Madhawa Hettiarachchia H D, Mihajlo G, William M W and Yasuyuki L 2007 *Energy.* 32 (9) p 1698-1706
[7] Alessandro F and Marco V 2009 *Geothermics.* 38 (4) p 379 – 391
[8] Geothermal Technologies Program, Hydrothermal Power Systems. Geothermal Technologies Program: Technologies. U.S. DOE Energy Efficiency and Renewable Energy(EERE). 2010-07-06. Retrieved 2010-11-02.
[9] Scott W 2010 Geothermal Energy Power Plants and How They Produce Green Electricity.
[10] Cengel Y A and Michael A B 2002 Thermodynamics: An Engineering Approach, Seventh Edition. Boston: McGraw-Hill.
[11] Izumida Y and Okuda K 2014 *Phys. Rev. Lett.* 112 180603
[12] Shihong L, Ngai Y Y, Tzahi Y C, Chinedum O O and Menachem E 2014 *Environ. Sci. Technol.*48 (9) p 5306–5313
[13] Cengel B. 2002 Thermodynamics: An Engineering Approach, McGraw-Hill ISBN: 0071216898.
[14] Kotas T J 1985 The exergy method of thermal plant analysis, Butterworths, Academic Press, London.
[15] Szargut J, Morris D R and Steward F R 1988 Exergy Analysis of Thermal, Chemical and Metallurgical Processes, Springer-Verlag, Berlin.
[16] Porolfsson G 2002 Bestun a nytingu laghita jarovarma til raforkuframleioslu. (Optimization of low temperature heat utilization for production of electricity), MSc thesis, University of Iceland, Dept. of Mechanical Engineering.
[17] Valdimarsson P 2002 Cogeneration of district heating water and electricity from a low temperature geothermal source, Proc. of the 8th International Symposium on District Heating and Cooling, NEFP and the International Energy Agency, Trondheim, August 14- 16, ISBN 82-594-2341-3.
[18] Soundararajan K, Ho K. H and Su B 2014 *Appl. Energy.* 136 p 1035-1042
[19] Mohamad R B, Mortaza A, Hossein G, et al. 2017 *Power conversion. management.* 151 p 753-763
[20] Florian H and Dieter B 2010 *Appl. Ther. Eng.* 30 (11–12) p 1326 – 1332
[21] Casisia M, Pinamontia P and Reinib M 2009 *Energy.* 34 (12) p 2175 – 2183
[22] Tim S Ing, Ingo S and Ing H 2016 *Energy Procedia* 99 p 292-297
[23] Ronald D 2004 *Geothermics* 33 p 565 – 586
[24] DiPippo R and Marcille D F 1984 *Geothermal Res.* 8 p 47–52.
[25] Pall V and Scient I 2003, Production of electricity from geothermal heat – efficiency calculation and ideal cycles, Proc. Of the International Geothermal Conference, Reykjavík p 41-48

IOP Conf. Series: Earth and Environmental Science **150** (2018) 012023 doi:10.1088/1755-1315/150/1/012023

Design and Fabrication of Solar Updraft Tower and Estimation of Power Generation; Initially Focused on Bangladesh

F Ayub[1], S Akhand[1], A S Khan[1] and G Saklayen[1]

1 Faculty of Mechanical Engineering, Military Institute of Science & Technology, Bangladesh
E-mail: fayrouz.ayub@gmail.com, akhand.saarah@gmail.com

Abstract. In our studies we focused on area of sourcing, converting and delivering sustainable energy, concentrating at the potential role of solar power. Power generation through a solar updraft tower (SUT) has been a promising approach for sustainable generation of renewable energy. Developing nations are faced with many challenges. Conventional sources are insufficient to meet the increasing demand of a developing, industrious nation (e.g. Bangladesh). Our project aims in reducing electricity crisis and forming a solution for our country, Bangladesh. The electricity generated can be supplied to the national grid. This will mean reduced cost for the government in the long run and also allow the government to reduce its dependency on costly and unsustainable fossil fuel. This cost reduction benefit can be passed on to the public as reduced energy cost or preferably through nationwide energy infrastructure development. This technology will not only help with the energy concern of Bangladesh but also will help to improve the situations of other developing countries alike Bangladesh. All in all implementing this technology will pave the way towards a better world and form a part of an integrated ecosystem of sustainable energy technology.

1. Introduction

Bangladesh along with other developing nations is facing the challenge of industrialization, population increase and urbanization and is also suffering the dire consequences of these factors. These changes are causing significant amount of energy consumption which also places a great pressure on government due to scarcity of resources of conventional fossil-fuel. Alternative solutions to these problems are to depend on renewable energy resources. Solar energy is one of the renewable energy sources that represents totally non-polluting, inexhaustible, abundant energy resource and can be utilized to supply world's needs [1]. Among many solar energy technologies the solar updraft power plant is a noble promising technology which can be used to supply power to countries with large wastelands or unused desert lands [2].

A traditional solar updraft power plant system consists of a tower, collector and a turbine or several pressure based turbines [3]. During daytime, solar radiation passes through the transparent collector roof and heats the air inside the whole collector area. Heated air induces buoyancy effect and results in large pressure drop under the center of collector which causes a strong airflow through the tower, and is used to drive turbine within solar tower to generate electricity. The concept of SUTPP (solar updraft tower power plant) was first developed by Schlaich, and a prototype was put in operation during 1980s[4], [5]. Cheap building materials, negligible hazardous waste generation, low operating costs these advantages made this renewable energy technology attractive to countries with vast unused deserts and wastelands[6]. Solar energy is a great source of renewable energy in Bangladesh due to its geographical and climatic conditions. So, we tried to focus on how to utilize solar energy by modeling and fabricating a SUT prototype. Our data clearly showed that SUTPP even in a smaller size can be very effective to undermine the energy problem in rural areas of Bangladesh. It can further be useful for building up a larger SUTPP as we measured our data on various times around the year. In this paper, experimental setup and working procedure along with necessary data and calculations has been presented. The limitations that we had during this working experience and ways to overcome those limitations has also been discussed.

2. Construction and design parameter

2.1. Model-1

Before proceeding to main model of SUT(solar updraft tower),construction and simulation(Appendix-1) of a theoretical ratio base model was done, which was different from main prototype in both material and design. The tower of this model was of long hard paper roll, which was colored black externally, to absorb maximum amount of heat. It was effective in a way that it had a thermal conductivity of 0.05 W/(m K). The Collector material used in model was made of talc paper. The propeller was made of normal paper as the initial torque needed was low. There was no electrical connection in this model-1. So after the successful rotation of the propeller located inside the tower we decided to make our prototype.

2.2. Prototype

Modification of collector material was compulsory after observing model-1 collector efficiency (Appendix-2). PVC sheets of 0.9mm thickness was used for the prototype collector (Appendix-3). The tower was of PVC also and was about 1.7m in height .The pipe was black colored both internally and externally. The tower was supported by steel round frame. The whole structural layout was designed in such a way that it was strictly stable with steel as its core material. The frame was supporting the whole experimental setup. The turbine was made of polyethylene terephthalate (PET). The collector diameter Dc was selected as 1.8283m (6 ft) considering availability of space on the selected place. The ratio of collector diameter to the tower diameter is called the diameter ratio (DR) [7], [8]. The 50 kW plant in Manzanares has DR= 20 [9]. For our plant DR=12(approximately) was selected considering availability of tower material in the local market. Thus the tower diameter D_t was selected as = 0.152 m (6in).The height of the tower H_t is proportional to the efficiency of the tower. The higher the chimney is the better is the output. H_t = 2.1m was selected considering availability in the local market and ease of installation and support.

3. Experimental setup and data analysis

Firstly the open ground in front of tower-02 of MIST, Dhaka, Bangladesh was selected as the place for making the prototype considering the availability of sunlight and convenience of building the plant.

3.1. Exprimental method

The outside temperature T_o was measured by pyrometer and these temperatures were noted down day wise. The temperature T_i inside the collector area was measured by pyrometer at the same time as the outside temperature and these temperatures were also noted day wise. The velocity of air at the entrance of the tower was measured by anemometer .The densities β_o and β_i were taken according to the corresponding temperatures $T_o$and T_i (Appendix-4). The data was taken for different days of six different months and with by calculating those data, necessary results were found.

3.2. Equations and indentations

The output power of a solar updraft plant depends on several parameters (Table 1). The pressure difference between tower base and ambient is represented as,

$$\Delta P_{tot} = g \cdot (\rho_o - \rho_i) \cdot H_t$$

The output power can be calculated as,

$P_{out} = \Delta P_{tot} \cdot v_t \cdot A_{coll}$

Thus the total power available to the turbine can be calculated from this equation. As we used digital anemometer for measuring the air velocity at tower entrance it can also be calculated from this equation,

$$v_t = \sqrt{2g.H_t.\left(\frac{T_i - T_o}{T_o}\right)}$$

And the collector area is calculated according to the geometrical structure (Table 2) of the collector

$$A_{coll} = \frac{1}{3} \cdot \pi \cdot r^2 \cdot h$$

Table 1. Nomenclature

g	Gravitational acceleration (9.8 m/s^2)	D_c	Diameter of collector (m)
To	Ambient/Outside Temperature (^{o}C)	A_{coll}	Area of collector (m$^{2)}$
Ti	Temperature at tower entrance (^{o}C)	D_t	Diameter of tower (m)
ρ_0	Outside air density(kg/m^3)	A_t	Area of tower (m^2)
ρ_i	Inside air density(kg/m^3)	ΔP_{tot}	Total pressure difference(N/m^2)
H_t	Height of tower (m)		
V_t	Air velocity at tower entrance (m/s^2)		

Table 2. Technical terms

Tower height	2.1m
Tower radius	0.76m
Collector radius	1.8283m
Min roof height (at center)	0.2032m
Max roof height (at center)	0.587m
Collector slope angle	31.41^o
Typical ΔT	10K
Average output	20W
Tower material	PVC
Collector Material	PVC 0.9mm sheet
Plant frame and support material	Cast iron

Table 3. (Data table)

Month (2016)	Avg T_o (^{o}C)	Avg T_i (^{o}C)	Avg ρ (kg/m^3)	Avg ρ_i (kg/m^3)	Avg ΔP_{tot} (N/m^2)	Avg V_t (m/s^2)	Avg P_{out} (Watt)
May	32.9	50	1.1632	1.098	1.34	1.516	18.799
June	32.25	50.125	1.164	1.10225	1.3696	1.5525	19.767
August	32.8	50	1.163	1.09798	1.212	1.518	18.854
September	31.2	52	1.165	1.0796	1.6844	1.6988	26.506
October	32.4	52.78	1.1622	1.105	1.5564	1.6552	25.31
November	27.6	40.9	1.174	1.126	0.987	1.35	12.281
December	29.1	39.92	1.1672	1.1282	0.8028	1.214	8.977

4. Result and discussion

4.1. Graphs

IOP Conf. Series: Earth and Environmental Science **150** (2018) 012023 doi:10.1088/1755-1315/150/1/012023

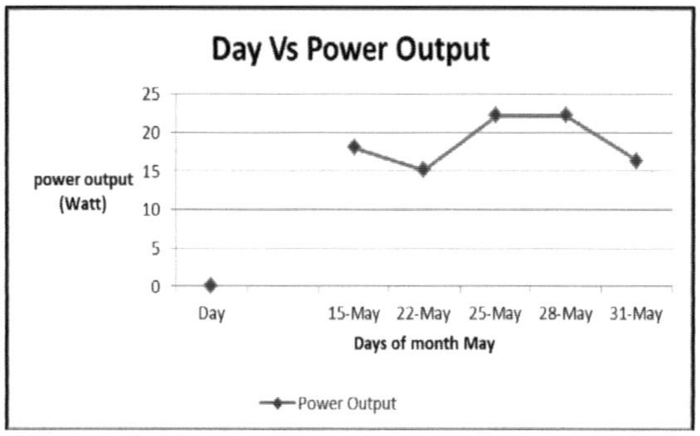

Figure 1. Composition of power outputs (May)

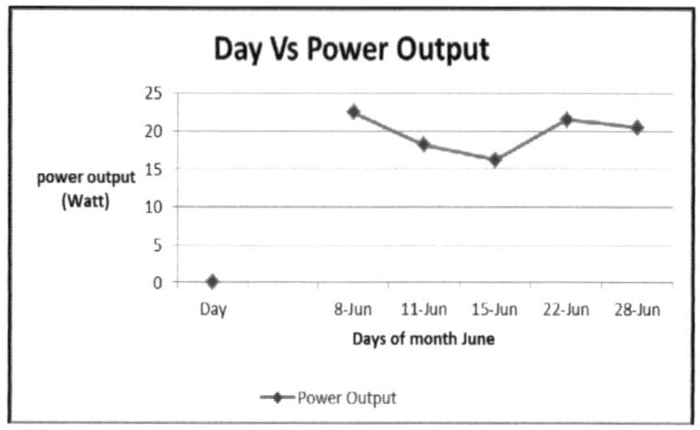

Figure 2. Composition of power outputs (June)

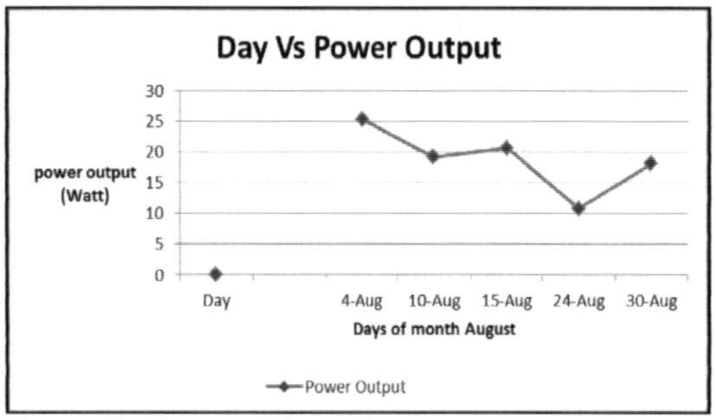

Figure 3. Composition of power outputs (August)

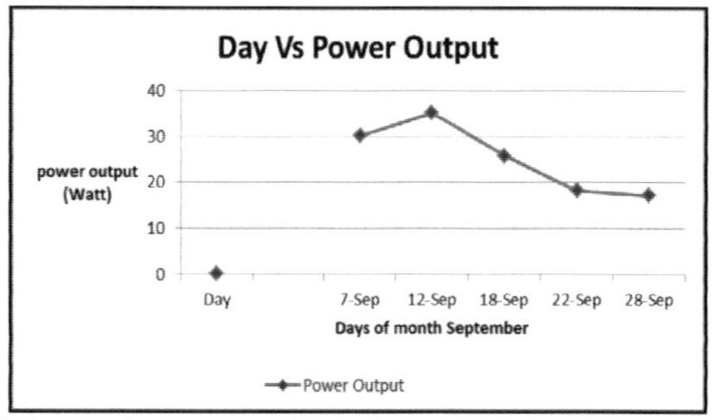

Figure 4. Composition of power outputs (Sept.)

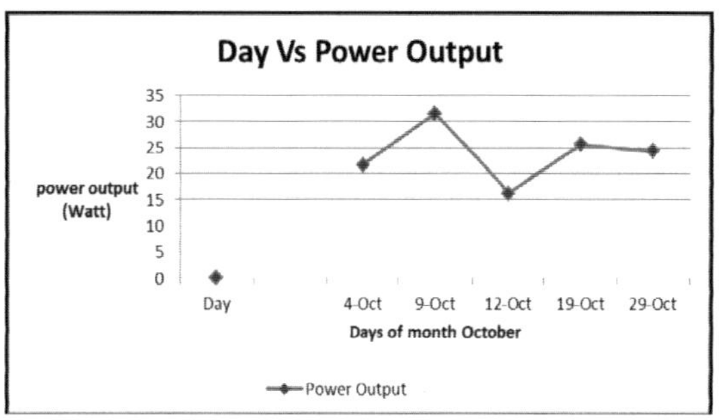

Figure 5. Composition of power outputs (Oct.)

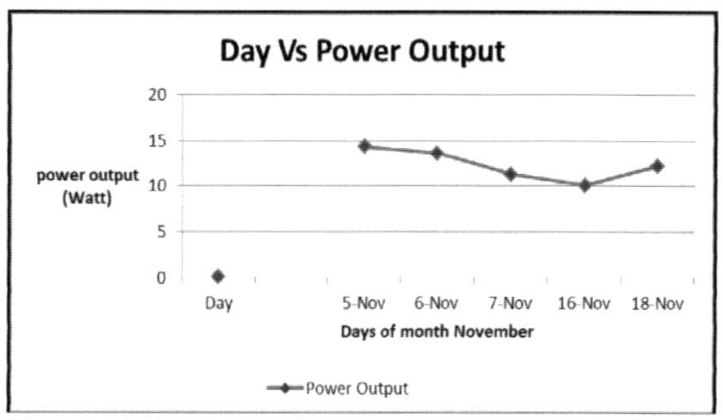

Figure 6. Composition of power outputs (Nov.)

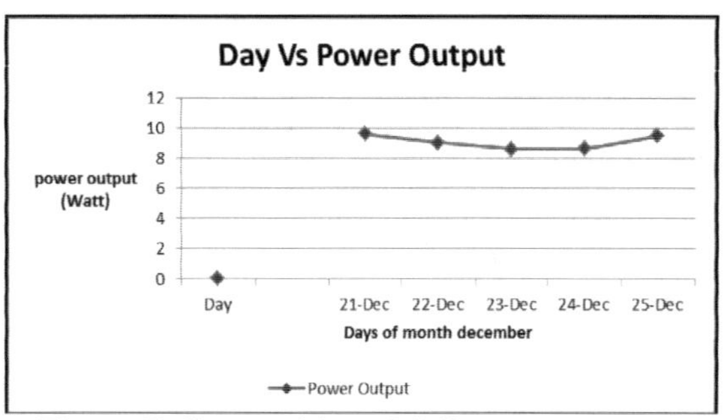

Figure 7. Composition of daily power outputs (Dec.)

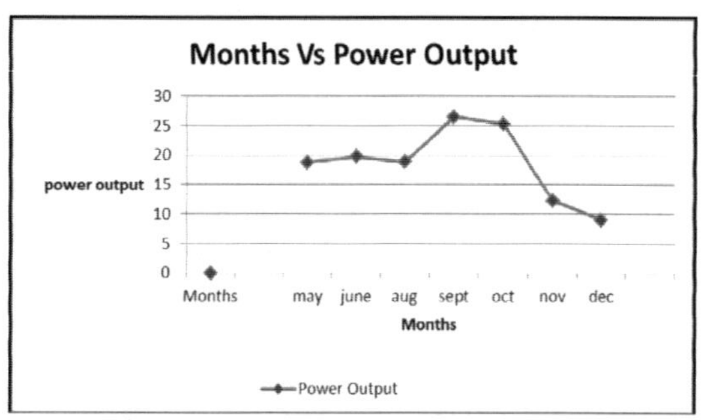

Figure 8. Months Vs Power output

Several data were collected at different times during a month (Table 3) (Figures 1-7). Average output during a month was calculated by averaging the data composed per day. From Figure 8 it is observed that highest power output is obtained between September and October and the lowest values of average power output is observed during November and December.

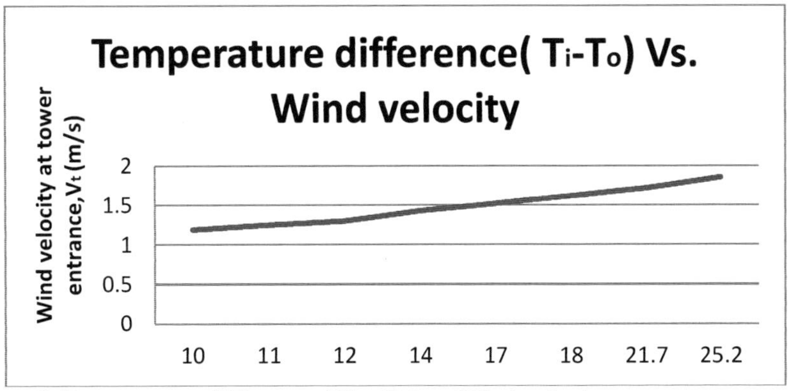

Figure 9. Temperature difference (Ti – To) vs Wind velocity at tower entrance, Vt

Temperature difference between inside and outside is directly correlated with the wind velocity at tower entrance [10]. The higher the temperature difference the higher wind velocity at tower entrance is observed (Figure 9). As the temperature of the trapped air inside the collector increases the density of air decreases and the hot air starts moving upwards with a certain velocity that depends on the temperature rise [11].

4.2. Prospect of solar updraft technology in Bangladesh

As Bangladesh is a semi-tropical country which is situated at North-eastern part of South Asia, receives abundant sunlight year round. In dry season the average bright sunshine duration is about 7.6hrs and in monsoon season the average bright sunshine is about 4.7hrs. The highest sunlight hours has been received in Khulna division which is 9.04hrs and second highest sunlight duration has been achieved in Barisal which is 8.75hrs.

Such amount of energy is greatly suitable for solar updraft technology to solve daytime energy crisis of the rural areas in Bangladesh. But to build a large scale solar updraft power plant Bangladesh will face some limitations as well. But those limitations can be eliminated if necessary and measured steps are taken by government.

5. Limitations and recommendations

Our main purpose was to observe and study the process of solar updraft principle, to do some analysis on structure and materials of components and also to estimate power generation of our pilot scale prototype. From this study and by comparing results with ideal systems we were able to find out our faults and drawbacks of the process we followed. The minimum height of the collector base from the ground level was higher than model-1 and caused cross or multidirectional air flow. When the air velocity was at a higher rate the air did not get enough time so that it could get trapped and get heated enough to move upwards rather it passed away through other sides of the collector. The ratio of collector diameter to tower diameter was small and the model would have been more effective if the ratio was in a larger range. We used PVC sheet as collector roof material which has a greater range of specific heat than that of glass. For the proper execution of "greenhouse effect" collector roof made of glass is highly recommended [12]. As this was our honors final year project we had limited resources in terms of finance and time. As we planned to introduce this model as a cost effective green technology, we had to execute this model at a very low budget so we were not actually able to modify this setup. We have started our project at the end of the summer so which hindered the potential of the project. Due to the unfavorable season we did not get expected amount of temperature difference on the other hand rainy or/and cloudy days of the winter has also affected the efficiency of the project.

6. Conclusion

Solar Updraft power plant has many features that suggest this technology to be used in rural, remote and isolated communities. SUT provides steady power output which makes this technology suitable for smaller rural communities or for a small-scale industry. The power systems of developing countries are not reliable on the other hand, small scale industries require continuous, uninterrupted power system and in this case, solar updraft power plant with heat storage system will provide the required continuous power output. In case of deserts and large wastelands solar updraft towers are most efficient at greater scale. Building solar updraft power plants could be more cost effective than building concentrated solar power plant. In that case the tower has to be built with local parts and local labor. The low maintenance requirement of SUT also ensures low cost of maintenance, which is a very important factor for constructing SUT's at rural areas. The exceptional function of solar updraft collector for drying crops can be useful for many rural communities.

Maintenance and repairing will only be an issue if there is lack of experienced technicians. In this case training programs should be held by government to train labors on maintaining the plant. During winter and monsoon seasons cloud coverage limits sunlight availability effecting overall production scheme. In this case heat storage methods and devices have to be implemented so that the continuous power generation scheme remains uninterrupted. Capacity of power generation of solar updraft tower depends on many factors like sun radiation, ambient temperature, humidity etc. but basing on these factors efficiency of SUTPP can be increased by increasing collector area and tower height proportionally.

Rapid and accelerated growth of energy demand and shortage of energy could be a limiting factor of the economic development for developing countries. There are also negative environment externalities due to energy generation by conventional fossil-fuels. In this context, it has become quintessential to develop and promote alternative renewable energy sources that can lead to sustainable sourcing of energy. Renewable energy technologies can supplement the energy needs of developing countries to a significant level. Overall, for long term sustainable development integrated energy planning approach, upgraded government policies are necessary to support continuous development of energy generation technologies from renewable resources.

7. Appendix

Appendix 1. Computerized model of SUT

Appendix 2. Final model

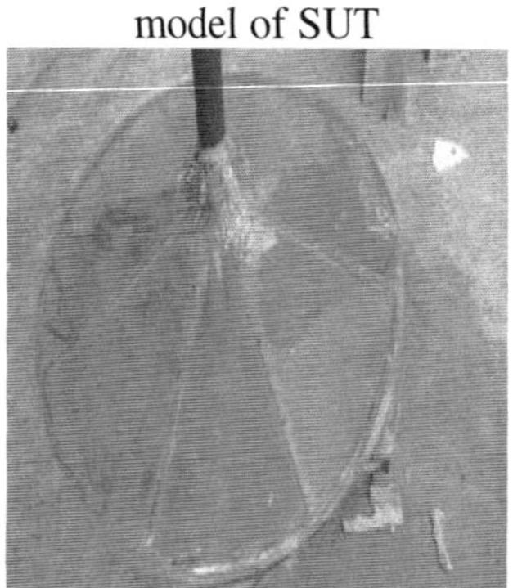

Appendix 3. PVC sheet roof

Temperature - t - (^oF)	Density - ρ - (10^{-3} slugs/ft^3)	Specific Weight - γ - (10^{-2} lb/ft^3)
-40	2.939	9.456
-20	2.805	9.026
0	2.683	8.633
10	2.626	8.449
20	2.571	8.273
30	2.519	8.104
40	2.469	7.942
50	2.420	7.786
60	2.373	7.636
70	2.329	7.492
80	2.286	7.353
90	2.244	7.219
100	2.204	7.090

Appendix 4. Density variation of air with respect to temperature.

8. References

[1] Schlaich J, Weinrebe G and Bergermann R 2013 *Solar Updraft Towers. Solar Energy,* Richter C, Lincot D, and Gueymard C A, Eds. New York, NY: Springer New York, p 658-687.
[2] Zhou X and Xu Y 2016 *Solar updraft tower power generation. Solar Energy,* 128, p 95.
[3] Zhou X, Bernardes M AD S and Ochieng R M 201. *Influence of atmospheric cross flow on solar updraft tower inflow. Energy,* 42(1), p 393-400.
[4] Ming T, Wang X, de Richter R K, Liu W, Wu T and Pan Y 2012 *Numerical analysis on the influence of ambient crosswind on the performance of solar updraft power plant system. Renewable and Sustainable Energy Reviews,* 16(8), p 5567-5583.
[5] dos Santos Bernardes M A and Zhou X 2013 *Strategies for solar updraft tower power plants control subject to adverse solar radiance conditions. Solar Energy,* 98, p 34-41.
[6] Zhou X, Yang J, Xiao B and Hou G 2007 *Experimental study of temperature field in a solar chimney power setup. Applied Thermal Engineering,* 27(11), p 2044-2050.
[7] Niroomand N and Amidpour M 2013 *New combination of solar chimney for power generation and seawater desalination. Desalination and Water Treatment,* 51(40-42), p 7401-7411.
[8] Ming T Z, Zheng Y, Liu C, Liu W and Pan Y 201. *Simple analysis on thermal performance of solar chimney power generation systems. Journal of the Energy Institute,* 83(1), p 6-11.
[9] Harte R, Höffer R, Krätzig W B, Mark P and Niemann H J, 2013. *Solar updraft power plants: Engineering structures for sustainable energy generation. Engineering Structures,* 56, p 1698-1706.
[10] Zhou X, Wang F and Ochieng R M 2010 *A review of solar chimney power technology. Renewable and Sustainable Energy Reviews,* 14(8), p 2315-2338.
[11] Kalogirou S A 2013 *Solar energy engineering: processes and systems.* Academic Press.
[12] Zhou X, Yang J, Xiao B and Hou G 2007 *Experimental study of temperature field in a solar chimney power setup. Applied Thermal Engineering,* 27(11), p 2044-2050.

IOP Conf. Series: Earth and Environmental Science **150** (2018) 012024 doi:10.1088/1755-1315/150/1/012024

Techno Economical Study of PV-Diesel Power System for a Remote Island in Indonesia : A Case Study of Miangas Island

M Rumbayan[1] and K Nagasaka[2]

1 Lecturer and Researcher, Faculty of Engineering, Sam Ratulangi University, Manado, Indonesia
2 Professor, Tokyo University of Agriculture and Technology, Tokyo, Japan
E-mail:meitarumbayan@unsrat.ac.id

Abstract. The purpose of this study is to conduct the techno economical study of PC-Diesel power system based on renewable energy available locally in a remote island. This research is a case study for Miangas island which is the border island between Indonesia and Philipines. It is located in Talaud Island regency of North Sulawesi province of Indonesia. The monthly average daily radiation in Miangas island is 5.52 kWh/m^2.The research methods used are data collection and data analysis using software HOMER. Based on the simulation result, the techno economic study of PV-Diesel power plant system based on energy demand in Miangas island can be obtained. The Cost of Energy (COE), Net Present Cost (NPC) and operating cost for proposed hybrid PV-Diesel power generation can be assessed for the design power systems uses Canadian solar Max Power C56x-325P of 150 KW PV, 18 string of Surette 6CS25P, Diesel Generator 50 kW and converter Magnum MS4448PAE 25 kW. The annual electricity production from the PV Diesel system for Miangas island is 309.589 kWh in which 80.7% electricity comes from PV, 19.3% electricity comes from diesel with the 109.063 kWh excess electricity. The cost of generating electrical energy in the term of cost of energy (COE), Net Present Cost (NPC) and operating cost are 0.318 US$/kWh, 719.673 US$ and 36.857 US$ respectively.

1. Introduction

The eastern region of Indonesia, especially the island which is the outermost island that becomes the front border requires special attention in the case of energy problems. Dependence on the supply of fossil fuels from the island to the island that have high transportation costs and produce greenhouse gases need to be minimized. For this reason the utilization of renewable energy as an alternative energy source for energy generation needs to be studied.

Miangas island was chosen as the location to be studied because of its specificity located on the border between Indonesia and the Philippines is the outermost island which became the front porch of Indonesian territory. Miangas island is situated in regency of Talaud Island. It is located at latitudes 05^0 33' 20.8'' North and longitudes 127^0 09' 6.8'' East. It takes 3 hours by boat from Malonguane (the nearest city as the capital of Talaud region). The total land area is approximately 3.2 km^2. According to the data from Statistics Centre, the communities of Miangas island consists of 881 people in 324 households. The island of Miangas as the part of Talaud archipelagos is a northest border of Indonesia is shown in Figure 1.

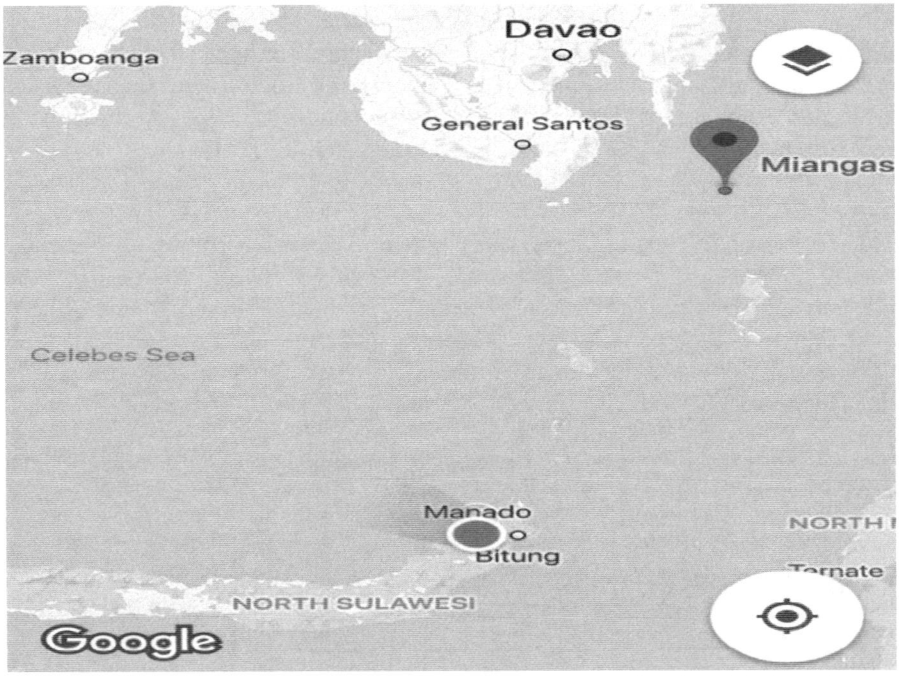

Figure 1. Location of Miangas Island as the northest border of Indonesia
Source: https://www.google.com/maps

Miangas island is facing the high cost of diesel fuel due to remoteness from Sulawesi mainland. This paper presents a techno economical study of PV-Diesel power system for the study of Miangas island in Indonesia.

2. Literature Reviews

This section presents literature review about the analysis works for power system infrastructure model for island communities by utilizing HOMER (Hybrid Optimization of Multiple Energy Resources) software from National Renewable Energy Laboratory (NREL).HOMER software has been used to perform the techno economic feasibility of possible models in developing the power system infrastructures. HOMER is an optimization software package, which can handle different technologies (including PV, wind, hydro, fuel cells) and evaluate design options for both off-grid and grid-connected power systems for remote, stand alone and distributed generations applications [1].

There are many studies has been conducted to study of HOMER utilization for analysing the model of power system generation. Dursun et al [2] studied a micro-grid wind-PV hybrid system for a remote community with 50 houses in order to find the optimal configuration and present a techno-economic analysis for the considered power generating system by the HOMER software. Bekel and Palm [3] presented a feasibility study for a stand-alone solar-wind based hybrid energy system for a model community of 200 families using HOMER software. Shaahid et al [4] evaluated the technical and economic potential of hybrid-wind-PV-diesel power systems to meet electrical energy demand of a remote village by using HOMER software. Sen and Bhattacharya [5] state a hybrid system design can overcome the problem of intermittency of renewable energies and increase reliability of supply. In their research for electrification of remote villages they carry out a case study of an Indian village using HOMER software and find that hybrid system ae technically and economically viable in many locations.

Many studies have been reported analysing renewable energy based on power generation using HOMER. Himri et al. [6] study of hybrid power system for a remote village in Algeria, while Nandi and Gosh [7] present a study of a Bangladesh village and Nfah et al [8] report case study of Ethiophia.

Homer's software capability for modeling has been demonstrated, through two experiments on small-scale systems by comparing HOMER modeling results with direct measurement results [9].
The location of the community is important to know as electricity demand patterns differ with geograhical site and cultural habits [10]. HOMER uses the optimizer propietary algoritm to search for the lowest net present cost system for the specific load and condition [11].

3. Methods

The method to conduct the techno economic study of hybrid power generation for communities in the remote island of Indonesia is using HOMER software. This software developed by the National Renewable Energy Laboratory (http://www.homerenergy.com) a division of the US Department of Energy used to design a hybrid power plant system using renewable energy.
The monthly load profil of electricity demand for Miangas island community for one year that used in HOMER simulation as shown in Figure. 2

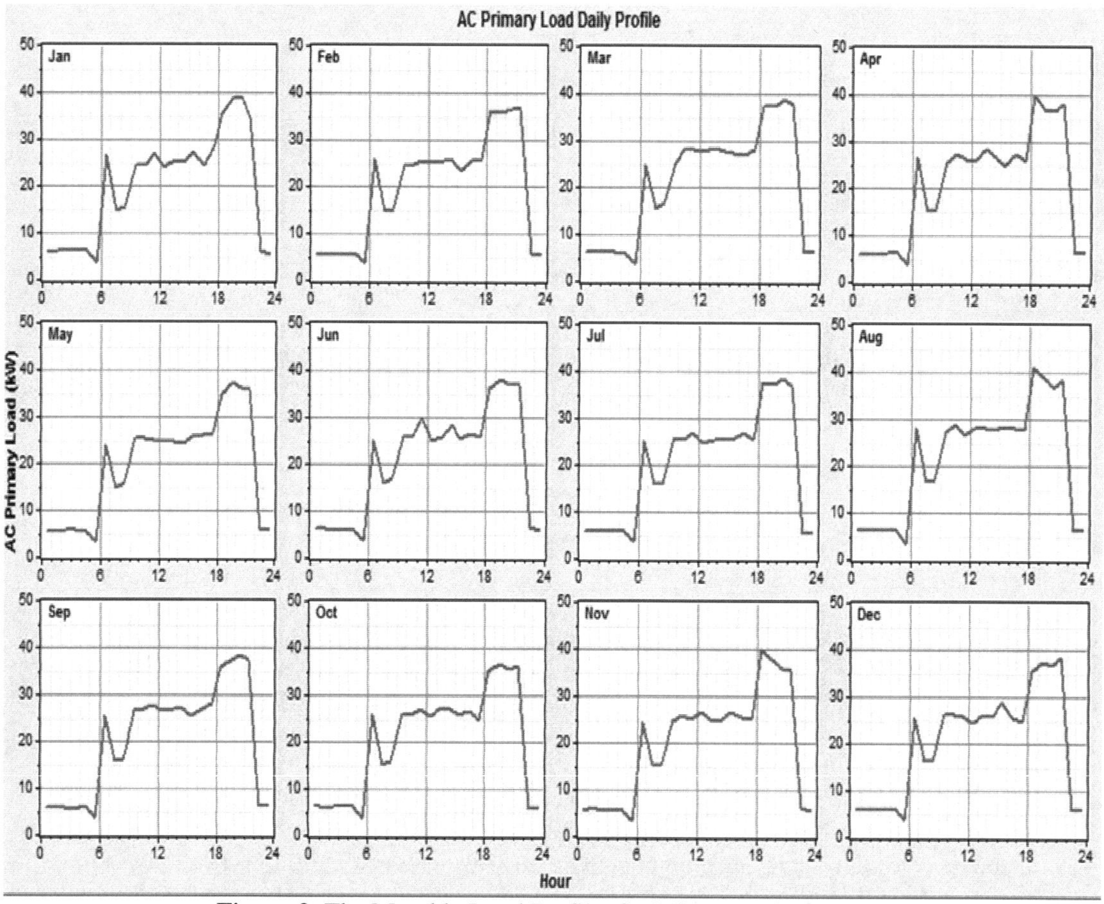

Figure 2. The Monthly Load Profile for Miangas Island

Feasibility of solar energy system mainly depends on solar radiation available at the specific location [12]. Data of solar energy sources in term of solar irradiation in Miangas island have been taken from NASA (National Aeronautics and Space Administration) website by input the latitude and longitude of the location. The average annual solar radiation equals to 5.53 kWh/m^2/day. HOMER also gives the clearness index data for Miangas island. Clearness index is the amount of global solar radiation on the surface of the earth divided by the extra-terrestrial radiation at the top of the atmosphere [1]. The data of solar radiation and cleaness index in Miangas island are presented in Table 1.

Table 1. Average Monthly Solar Irradiation in Miangas island

Month	Solar Irradiation ($KWh/m^2/day$)	Clearness Index
January	5.01	0.53
February	5.43	0.54
March	6.05	0.58
April	6.40	0.62
May	5.73	0.56
June	5.00	0.50
July	5.32	0.51
August	5.50	0.54
September	5.84	0.57
October	5.71	0.57
November	5.38	0.56
December	5.01	0.54

4. Results

The schematic of hibryd power generation model based on PV and diesel for Miangas island as shown in Figure 3.

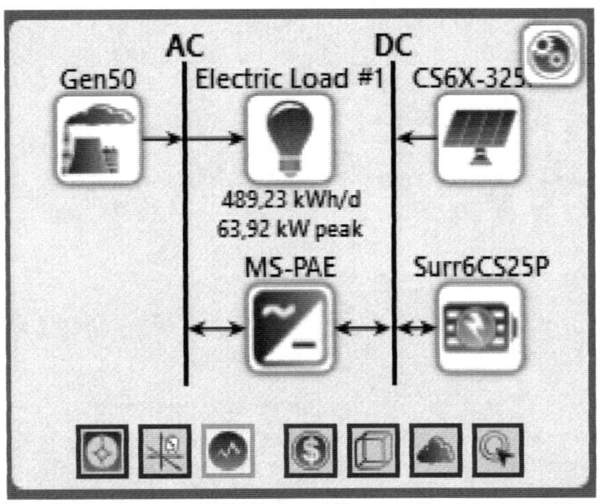

Figure 3. The Schematic of PV-Diesel Power System for Miangas Island

The design for PV-Diesel power systems uses Canadian solar Max Power C56x-325P of PV, battery Surette 6CS25P, Diesel Generator 50 kW and converter Magnum MS4448PAE. HOMER determines the value of the appropriate component capacity so as to produce a good and reliable power system in serving the load in terms of capacity of power plant component, yearly electric energy production, cost of energy (COE), total Net Present Cost (NPC), and operation cost. Techno Economic Analysis of PV Diesel Power System for Miangas Island are shown in Table 2.

Table 2. Techno Economic Analysis of PV-Diesel Power System for Miangas Island

Capacity of PV	150 kW
Capacity of Diesel	50 kW
Capacity of Battery	80 string
Capacity of Converter	25 kW
Yearly energy production from PV	249.688 kWh
Yearly energy production from Diesel	59.901 kWh
Energy cost per kWh (COE)	0.3118 US$
Net Present Cost (NPC)	719.673 US$
Operating Cost	36.857 US$

The simulation results of the electricity production of PV-Diesel power system and monthly average electricity generation are shown in Figure 4.

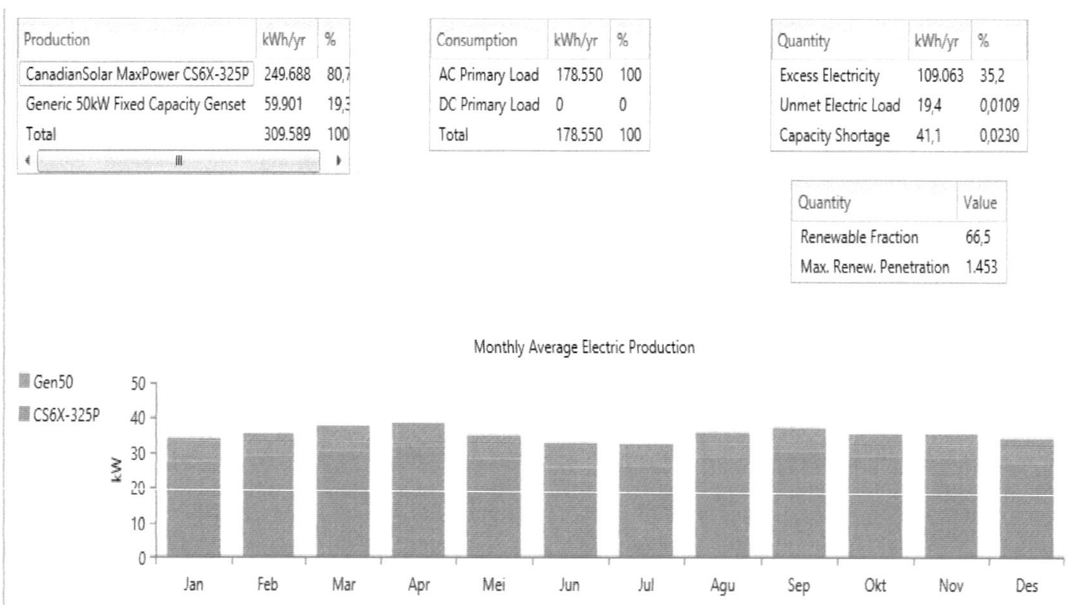

Figure 4. The Simulation Results for the PV Diesel Power System in Miangas Island

The annual electricity production from the PV Diesel system for Miangas island is 309.589 kWh in which 80.7% electricity comes from PV, 19.3% electricity comes from diesel with the 109.063 kWh excess electricity.

5. Conclusion

Based on the simulation result using HOMER software, the techno economic study of PV-Diesel power plant system based on energy demand in Miangas island can be obtained. The Cost of Energy (COE), Net Present Cost (NPC) and operating cost for proposed hybrid PV-Diesel power generation can be assessed for the design power systems uses Canadian solar Max Power C56x-325P of 150 KW PV, 18 string of Surette 6CS25P, Diesel Generator 50 kW and converter Magnum MS4448PAE 25 kW.

The cost of generating electrical energy with renewable energy is relatively high then there is a need for policies and strategies for the implementation of renewable energy in the archipelago, among others: cooperation with parties that have been successful with the implementation of renewable energy technologies, increased priority implementation of renewable energy-based energy infrastructure for potential locations. The adoption of a pro-island energy policy in the foremost island as a terrace that needs to be enriched for the sake of security, welfare and beauty as an added value in Indonesia's border region.

6. References

[1] Mamaghani A H, Escandon S A A, Najafi B, Shirazi A and Rinaldi F, 2016, *Techno-economic feasibility of photovoltaic, wind, diesel and hybrid electrification systems for off-grid rural electrification in Colombia*, Renewable Energy 97, pp 293-305.

[2] Dursun B, Gokcol C, Umut I, Ucar E and Kocabey S 2013 *Techno-economic evaluation of a hybrid PV wind power generation system*, Int. J. Green Energy 10, pp117-136.

[3] Bekele G and Palm B 2010 *Feasibility study for a standalone solar–wind-based hybrid energy system for application in Ethiopia* Applied Energy 87, pp 487–495.

[4] Shaahid S M, El-Amin I, Rehman S, Al-Shehri A, Ahmad F, Bakashwain J and Al-Hadhrami L M 2010, *Techno-economic potential of retrofitting diesel powerwith hybrid wind-photovoltaic-diesel systems for off-grid electrification of remote villages of Saudi Arabia*, Int. J. Green Energy 7, pp 632-646.

[5] Sen R and Bhattacharya S 2014 Off-grid electricity generation with renewable energy technologies in India: an application of HOMER. Renewable Energy, 62:389

[6] Himri Y, Stambouli A B, Draoul B and Himri S 2008 *Techno Economical study of Hybrid power system for a remote village in Algeria*. Energy, 33 (7), pp 1128-1136.

[7] Nandi S and Gosh HR 2010 *Prospect of Wind-PV-Battery hybrid System as an alternative to grid extentionin Bangladesh*. Energy 2, 35 (7): 3040-7.

[8] Nfah E M, Ngundam J M, Vandenbergh M and Schmid J 2008 *Simulation of off-grid generation options for remote villages in Cameroon* Renewable Energy ; 33 (5):1064-72.

[9] Brandon H. Newell. 2010. The Evaluation Of Homer As A Marine Corps Expeditionary Energy Pre-Deployment Tool. Thesis. Master Of Science In Electrical Engineering. Naval Postgraduate School.

[10] Naves D, Silva C and Connors S 2014 *Design and implementation of hybrid renewable energy systems on micro-communities: A review on case studies* Renewable and Sustainable Energy reviews 31, pp 935-936.

[11] Khare S, Nema P and Baredar P 2016 *Solar-Wind Hybrid Renewable Energy System: A Review* Renewable and Sustainable energy Reviews 58, pp 23-33.

[12] Chmiel Z and Bhattacharya SC 2015 *Analysis of off-grid electricity system at Isle of Eigg (Scotland): Lessons for developing countries*, Renewable Energy 81; pp 578-588.

Students Teach Pupils Environmental Issues and Renewable Energy

H Friman[1], I Banner[2], B S Tuchin[2] and Y Einav[1, 3]

1 Faculty of Engineering, H.I.T - Holon Institute of Technology, Holon, Israel
2 Dean of Students Office, H.I.T - Holon Institute of Technology, Holon, Israel
3 Dean of Students, H.I.T - Holon Institute of Technology, Holon, Israel
E-mail: henf@hit.ac.il

Abstract. Technological advances and accessibility to information on the internet have opened a new channel of pupils that are being taught by students throughout the country. Students, full of motivation and a will to learn and teach, have understood that this way is good for them – enabling them to profit from a side job and take advantage of the knowledge they have accumulated in their degree. Holon Institute of Technology ("HIT") developed a new program at the Faculty of Electrical Engineering. The Renewable Energy program gives the students technical and practical aspects of energy use (technology and methodology of the study) and energy efficiency. The program also deals with minimizing the environmental impacts of energy use, as well as with energy economy and environmental policy. The entrance of students to the field of teaching pupils while still in their studies brings many advantages, such as: fresh knowledge, motivation to teach, and innovative, out of the ordinary methods that arouse interest in the pupils and intrigue them.

1. Introduction

Technological advances and accessibility to information on the internet have opened a new channel of pupils that are being taught by students throughout the country. Students, full of motivation and a will to learn and teach, have understood that this way is good for them – enabling them to profit from a side job and take advantage of the knowledge they have accumulated in their degree.
The students are divided into two groups – students that teach classes of pupils and give lectures within their studies and students that teach private lessons to groups of pupils and individual pupils.
On first sight, it seems as though it would be better for a pupil who has difficulties to take private lessons with a qualified, veteran teacher that would surely be able to teach him well. However, this is not true. Professional teachers, especially veteran ones, normally use fixed teaching methods. Pupils that have had difficulties in school will continue to experience difficulties also in private lessons, if the teaching methods are similar. In contrast, students bring new, fresh teaching methods, which assist the pupils more. In addition, students are closer in age to pupils, therefore they may be more able to connect to them and understand them. As part of a social involvement project in which we participate in "Revivim" school in Holon, the students required to teach pupils about the fields of Environmental and Renewable Energy academic advisor Dr. Hen Friman.
We have decided to research teaching by students against teaching by teachers, and to examine it by means of the pupils' attention in class and their achievements in virtual quizzes we have conducted. We felt that we had succeeded to intrigue and arouse interest in the pupils by using different, unique teaching methods. The pupils, for their part, responded with excitement and participated throughout all

of the activity hours. In addition, significant improvement was seen in the knowledge they have accumulated from the initial quiz to the final quiz.
In conclusion, the entrance of students to the field of teaching while still in their studies brings many advantages, such as: fresh knowledge, motivation to teach, and innovative, out of the ordinary methods that arouse interest in the pupils and intrigue them. On the other hand, teachers have greater experience, a connection to the Ministry of Education and the required material, and teaching experience.

2. HIT – Holon Institute of Technology

HIT – Holon Institute of Technology was established in 1969 and became an independent public academic institution of higher education in 1999, certified by the Council of Higher Education of Israel. HIT focuses on the teaching of sciences, engineering, computer science and technology, management of technology and design. It also emphasizes multi-disciplinary theoretical and practical research of innovative technologies from a professional scientific, economic and cultural perspective. HIT trains highly qualified students in the realms of science, engineering, management and design. and plays an important role in their integration upon graduation into key positions within the industry. HIT aspires to quality and excellence in teaching and innovative research, and strives to introduce novel and unique cutting-edge teaching and research technologies. HIT also prides itself on its advanced academic achievements, application of innovative techniques and interdisciplinary professionalism that lead to creative teaching and new technologies. HIT aims to utilize the intellectual and professional potential of each and every student, so that they can fully integrate into the fast-paced technological world of today. Providing superior technological and scientific education enables HIT graduates to enter key leadership positions in both the private and public sectors.

2.1. Faculty of Engineering

The last decades have been dominated by the rapid changes introduced by the technology revolution, which has a tremendous influence on our daily lives. Today we are facing a myriad of new challenges. Technology-based industry has matured in many ways and the required skills for future engineers are much more complex in a world where "machines/computers" execute many of the engineering tasks. Most of all, we are facing a new generation of sophisticated students, who were born into the digitized/multimedia world. The mission of the study program is to encourage and initiate academic development, through the development of new study programs and methods, while being responsive to the rapidly changing trends in the field. The proper education of the undergraduate students must also be a function of market needs and predictions of how technology will develop in the foreseeable future. In order to ensure that our graduates are well qualified to meet the future needs of the market, meticulous attention must be paid to maintain a high standard in the fundamental courses and impart practical tools and skills. It is also important to introduce a wide variety of new subjects. The aims and goals of the Engineering faculty are to provide the students with a rich and comprehensive study program, and keep the study program updated to meet the ever-changing requirements for engineers of the future, enrich the student's theoretical knowledge as well as teach practical and design skills and knowledge; adapt its teaching methodologies and techniques, focusing on understanding as a goal; enable students to achieve skills such as self-learning and to acquire expertise via practice by understanding constantly update the teaching methods and the study program maintain relationships with the various relevant industry sectors introduce the students to state-of-the-art equipment and facilities, for conducting experiments that reinforce their understanding of the theoretical and practical issues studied in the courses promote research in the various fields; and explore cooperation with other institutes in Israel and abroad. Engineering is a practicing profession, a profession devoted to harnessing and modifying the three fundamental resources that humankind has for the creation of all technology: energy, materials, and information. The overall goal of engineering education is to prepare students to practice engineering and, in particular, to deal with the forces and materials of nature. Thus, from the earliest days of engineering education, instructional laboratories have been an essential part of undergraduate studies. Indeed, prior to the emphasis on engineering science, it could be said that

most engineering education took place in the laboratory. The emphasis on laboratories has varied over the years. While much attention has been paid to curriculum and teaching methods [1], [2].
The essential role of laboratories can be correlated with the fact that engineering is, in general, problem solving and analytical thinking. Well-designed laboratories during undergraduate engineering degrees can improve these skills for graduate engineers [3].

2.2. *Social Involvement Unit*

One of the many goals of the Social Involvement Unit, which is a part of Dean of Students Office, is to promote social involvement of students and staff in the community. It also promotes weak applicants and students at the institute by offering mentoring, tutoring, emotional support, guidance to learning, and adjustments in school. Over the years, the unit has worked in many education and welfare arenas to promote immigrants, youth, and more. The Social Involvement Unit serves as a professional centre to encourage and promote the social impact of students and staff and to leverage knowledge, expertise, and human capital for the benefit of the community through social involvement projects and course actions involving meaningful activities.

2.2.1. *Action learning course.*

The action learning course is an academic course which combines academic learning with social activities. These courses deal with processes and social challenges, reveal different ideologies, and develop critical thinking and pragmatic ideas. Students receive course credits and a grade for being part of such course. Participating students enrol in courses that involve action and activities to engage in the experiential learning process, thereby creating a dialogue and cross-fertilization between being taught in the classroom and experiencing the reality in the real world [4]. A learning experience includes meeting with social organizations, institutions, and state authorities and carrying out practical work with diverse populations. Through experience, students strengthen their academic skills, formulate ethical attitudes toward reality, develop professional and civilian perspectives, and realize how they can influence their surrounding in the present and hereafter.

3. Students Teach Pupils

Today, most schools in Israel and around the world are measured by achievements in national and international exams. Because the amount of theoretical material is large, and due to relatively short study time, it is almost impossible to deal with issues that are beyond the curriculum set by the Ministry of Education because of the teachers' lack of time and because of their erasure, most of the lessons in the school are monotonously boring and destructive in terms of curiosity and imagination [5]
The introduction of principles of environmental thought and energy efficiency in the early stages of childhood is very important, including the development of a sense of wonder, recognition of the beauty and mystery of the natural world [6]. The positions are formed at the very early stage of life [7]. The effective acquisition of environmental thought and energy efficiency in young children should be done in an integrated form of frontal learning that provides basic concepts along with experiential experience which [8], [9] This gives the pupils the motivation to learn, even for the weaker pupils, which allows the pupils to learn on several levels: 1. Hearing, 2. Touching, 3. Vision. These three levels give the pupils the application and understanding of the material, which at the end of the process the material learned is stored in a long-term memory for future use in the pupils daily life. Scientific and Technological Education in primary schools requires renewal and modernization [10]. There is some evidence that pupils claim that science and technology education is disconnected from reality and boring [11], [12]. Good teaching is context-dependent, and therefore what the outstanding teacher can do in his teaching may be different in different institutions with different populations of learners and different teaching objectives. Today, two definitions of good teaching are accepted when one argues that the ultimate goal of teaching is effective pupils learning [13], and the second argues that good teaching promotes high quality learning [14]. According to current teaching objectives, good teaching requires achieving multidimensional goals in addition to learning a particular material and developing related knowledge. For these purposes there is a cognitive dimension (developing

problem-solving abilities and independent learning skills, such as effective time management and coping with difficulties) and an effective / emotional dimension (development and change of attitudes, motivation and motivation for learning). The goal was that as a result of the teaching, in addition to the development of understanding of the learning material, there is also learning and development of a variety of skills and development and attitudes towards green thinking and energy efficiency.

3.1. The Methodology

The course sought to provide students with knowledge on a variety of environmental issues and experiential teaching skills for pupils. Multidisciplinary, environmental thinking, and simple ways to increase energy efficiency for young school pupils. Figure 1 shows the process steps. In order to evaluate the contribution and success of the course, a number of tools were used. We collected data on the perceptions of those involved, which we asked about the material being studied, the general experience in the classroom. The data collection tools were lesson observations. Interviews with pupils In these lessons, these interviews were conducted in a number of minutes, followed by lessons with about 5 pupils of each class, in order to understand the pupils experience and their positions. In addition, interviews were held with the class educators. In order to obtain numerical values that would indicate the success of the process, computerized surveys were conducted at the beginning of the activity, "entrance examiner", without prior knowledge, and a computerized survey at the end of the activity. These surveys asked five questions about the subjects learned in the session. At the end of the course, a comprehensive evaluation survey was conducted for all the systems to examine the impact and effectiveness of community intervention in the community [13]. A combination of visual teaching methods, including media in a school-run session, requires special effort and early preparation from pupils. Teaching through visual means and media requires the students who pass the lesson to reduce some of their control over the learning process. Pupils of the New Generation usually have more expertise, sophistication, and knowledge than the teachers in visual and media practice. This is actually the key to channeling their energy into learning [11], [12].

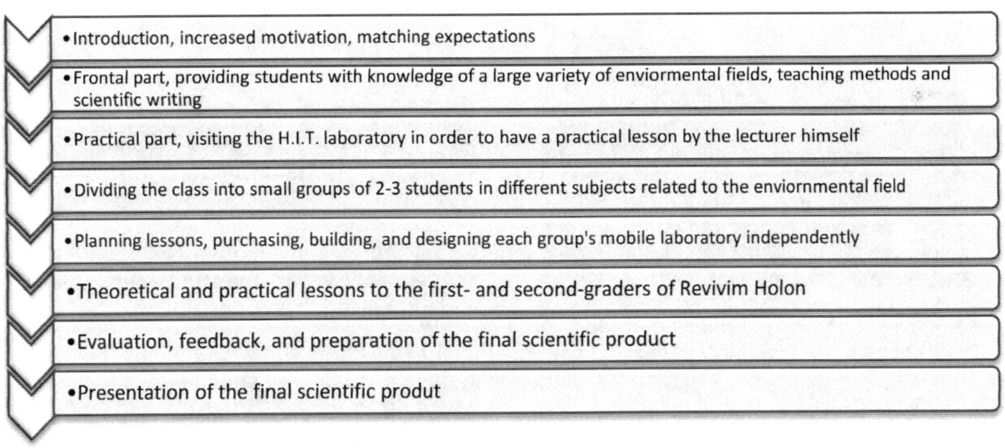

Figure 1. Students Teach Pupils methodology

4. Results

The course is a practical course. built of two parts: a frontal part and a practical part. 22 students from the faculties of Computer Science, Technology Management and Electrical Engineering at the Holon Institute of Technology chose to take part in the course, providing students with extensive environmental knowledge such as introduction to ecology, electricity generation, air pollution, The second stage of the practical course, the students were required to divide into work groups, to choose a specific subject in the field of knowledge, and to present to the pupils of the "Revivim" Elementary School in Holon, 90 minutes of exposure to one of the environmental issues Each group planned, integrated visual teaching methods, media and acquired and constructed models and experimental

IOP Conf. Series: Earth and Environmental Science **150** (2018) 012025 doi:10.1088/1755-1315/150/1/012025

systems that were used as instructional aids for the demonstration and experiential learning of scientific principles. In order to examine the effectiveness of the training, random interviews with pupils were conducted, interviews with pupils and educators were conducted at the beginning and end of each session "Trivia", virtual quizzes, and knowledge At the end of the course, after all the first and second classes (230 pupils) at "Revivim" School were exposed to the various workshops of the pupils, environmental higher.

4.1. The Frontal part

During the first six sessions, the students learned about the ability to produce existing electricity in Israel, which does not grow at the pace required by the increase in consumption. The State of Israel suffers from a shortage of electricity, The Ministry of Energy and Water is working to develop and integrate renewable energies into Israel's energy sector, as a result of the government's decisions to reduce energy consumption, subject, in order to promote security Energy independence and the strengthening of aspects of the environment In Israel and around the world awareness is raised of the importance of encouraging the transition to renewable energy - energy derived from the use of solar radiation, wind, biomass and other non-perishable sources that are not fossil fuels that pollute the environment. The management of the national water sector, while dealing with a resource in short supply, presents a complex challenge that requires optimal response The State of Israel is located in a region that suffers from water shortage, yet the state has achieved impressive achievements in the field of desalination, regular water supply of the highest quality in the world, The world's highest-level wastewater is low water. The course explained how this turning is made and what is required to be done in order to preserve the situation. The course dealt with another problem that exists in Israel is solid waste, the Ministry of Environmental Protection works to reduce the amount of waste generated and reduce the waste transferred for landfilling. One way to do this is through recycling, recycling is the process of removing materials from the waste stream and using them as raw materials, or creating new products, most of the recycling in Israel is done at sorting and separation plants. The students learned what effective teaching is and how to write lesson plans.

Figure 2. Convert kinetic energy (by blowing air) into electrical energy

Figure 3. created a "Grass Head" that performs photosynthesis

Figure 4. Speed of infiltration of water in different soil

Figure 5. Robots that avoided by Renewable Energy - sunlight

4.2. The practical part

In the practical part of the course, students were divided into work groups, 2-3 students per group. Each group wrote an experiential lesson plan that designed, purchased and built models that were integrated into a teaching system that exemplifies one of the scientific principles learned in the course. The students presented a teaching system to the first and second graders at the "Revivim" Elementary School, each week a different class. The group that dealt with energy efficiency, explained the concept, presented simple ways of making a preference for roasting laundry on a rope instead of turning on a clothes dryer, closing the light on the exit of the room, preferring sunlight to electrical lighting and activating a "Monopoly" game that included questions and activities related to saving electricity. The group that dealt with the water sector - in explaining the aquifer, which demonstrated the processes of water seepage in which the pupils created a model that describes the water cycle in nature. The energy conversion group presented the concept of energy "rolling" and held a competition between the pupils who were required to convert kinetic energy (by blowing air) into electrical energy - lighting the bulb (Figure 2). The group, which dealt with air pollution and the environment, explained to young pupils how the water value of water is and how the burning process affects this value. The children also created a "Grass Head" that performs photosynthesis (Figure 3). The group that dealt with soil in the environmental system taught how to characterize the soil and examined the speed of infiltration of different lands and the quiz: Which land is most suitable for agriculture? (Figure 4) The group that dealt with renewable energies presented the types of energy (wind, water, and sunlight) and the pupils experimented with vehicles and robots that avoided sunlight (Figure 5).

4.3. Effectiveness of teaching

In order to contribution to the learning process of pupils in "Revivim" elementary school. Each entry questionnaire and exit questionnaire consisted of five questions. When we examined the correct response rate in the entrance tests versus the correct answer in the exit tests, as shown in Figure 6, questionnaires entry 11-41% of pupils (depending on the question) Choose the correct answer. About 78-94% of the pupils chose the correct answer after studying in experiential material, experience and probably internalized the material.

Additional evidence of internalization of pupils' material and confidence in the new material they received was obtained when examining the time of response to questions of entrance exams and exit tests. As shown in Figure 7, the pupils were asked about 76 seconds on average to answer the various questions. On the other hand, in the summary questionnaires, the response time was reduced by an average of 31 seconds. Since the pupils were more confident in their answers and the material had a "fresh" in mind.

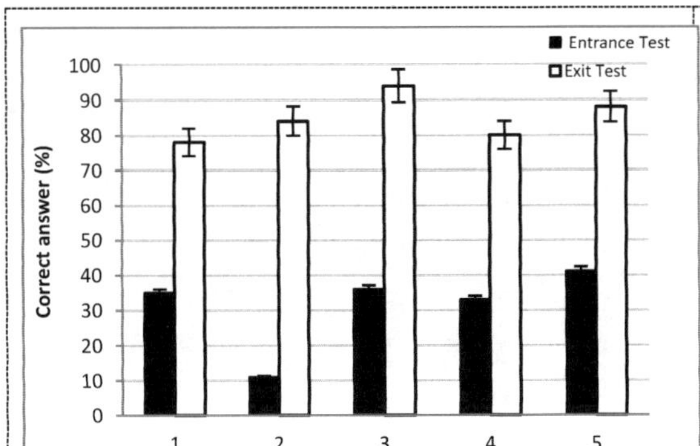

Figure 6. Correct Answers distribution

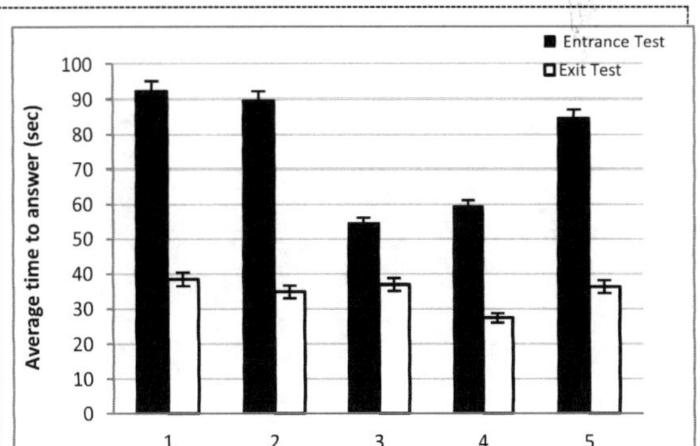

Figure 7. Average Time to Answer

When pupils were asked about their overall enjoyment of the course, all the sessions they performed (Figure 8) showed that 68% of the pupils enjoyed a great deal of activity during the course, 23% to a large extent, 8% 1% enjoyed very little.

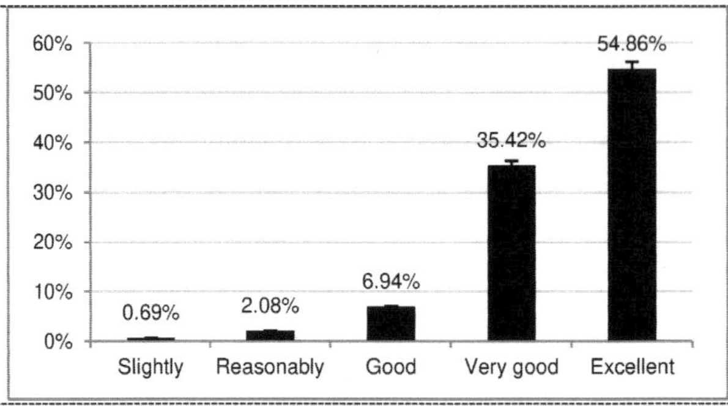

Figure 8. Pupil General Satisfaction

Data collection for the concepts involved, we asked them about the subject matter, the general experience in the classroom. Interview with pupils and classroom teachers: "I did not know that the wind could be turned on", "Now I understand why my mother keeps telling me to close the light in the room ..." said first grader pupil. The staff of the "Revivim" School also said in interviews that "the meeting of the young pupils with the students from HIT gave the pupils a different experience than what they are accustomed to on a daily basis", "The pupils were able to bring resources and formations we cann't display routine, diversity has contributed greatly to the learning process of the pupils". When students were asked about the extent of the personal contribution of the course, Maor Kakosh, a second-year electrical engineering graduate studies said that the institute, are mainly engaged in natural sciences: physics and mathematics. In this course, he received new tools dealing with the worlds of ecology and ecology. For Maor, this is the closing of a circle to return to "Revivim" School, where he studied in his youth. Amit Levy, a second-year computer science student, said he enrolled in studies to develop his personal career, but as part of the "Green Ambassadors" course, he taught first and second graders and was exposed to community issues. Dnial Srbi, second-year student in Management of Technology said that as part of a course "green ambassadors" She has the right to engage in teaching, whatever she thought she could get during the title. This experience was very experienced and meaningful for her and in her opinion may help her in the continuation of the professional path. An analysis of the results of the feedback shows that the students greatly benefited from the meetings with the students, learned a lot. The students noted that the pupils had a great commitment to the whole process. In addition, the institute noted that students are given the theoretical knowledge and the course was able to put it into practice and contribute to the community.

Figure 9. Pupils from the "Revivim" School received the "Green ambassador" Certificate

Figure 10. HIT students get Certificates of Appreciation

In March 2017, 230 pupils from the "Revivim" School gathered at an impressive ceremony held at HIT Holon Institute of Technology, where pupils received the green ambassador's certificate - a child with high environmental awareness (Figure 9). The HIT students get Certificates of appreciation and appreciation for their great contribution to the success of the course (Figure 10) During the ceremony, Jocelyn Koren, principal of "Revivim" School, said that "Revivim" is a "green" school. The meetings with the students were meaningful experience sessions and aroused a lot of curiosity among the pupils".

5. Conclusions

The Renewable Energy program gives the students technical and practical aspects of energy use and energy efficiency. The entrance of students to the field of teaching pupils while still in their studies brings many advantages. Based on the results, at the beginning of the lesson, pupils hesitated and took a long time before picking an answer. This hesitation can be explained by the fact that the pupils did not have extensive knowledge about the subject. On the other hand, on the exit test, there was a significant improvement as pupils quickly selected the correct answer, demonstrating their improved knowledge of environmental Issues and Renewable Energy.

6. References

[1] Feisel L D and Rosa A J 2005 The Role of the Laboratory in Undergraduate Engineering Education *Journal of Engineering Education* pp 121–130
[2] Wankat P C 2004 Analysis of the First Ten Years of Engineering Education *Journal of Engineering Education* 93 pp 13–21.
[3] Ma J and Nickerson J V 2006 Handson, simulated, and remote laboratories: A comparative literature review *ACM Computer Survey* 38 p7.
[4] Etkina E 2004 Developing and Assessing Scientific Abilities, Graduate School of Education. Rutgers University: http://paer.rutgers.edu/PT3
[5] Simmons B, McCrea E, Gay M, Herrmann L, Hutchinson L, Pistillo M B, Plevyak L H, Rivkin M, Williams S, Stenstrup A, Torquati J, Weiser B G and Wirth S 2010 Early Childhood Environmental Education Programs: Guidelines for Excellence, 2-8.
[6] Fernández C A, Primo M E, Lovatón C T and Olivvera J P 2016 Reflection in Action: A Key to Developing Competences for the Teaching of Science *Athens Journal of Education* pp 261-263.
[7] Schön D A 1983The Reflective Practitioner: How Professionals Think in Action. New York: Basic Books.
[8] Bamberger Y and Tal T 2008 Multiple outcomes of class visits to natural history museums: The students' view *Journal of Science Education and Technology* 17 pp 264–274.
[9] Falk J H and Dierking L D 2000 Learning from museums: Visitor experiences and the making of meaning. *Walnut Creek, CA* AltaMira Press.
[10] Osborne J and Dillon J 2008 Science Education in Europe: Critical reflections. Nuffield foundation.
[11] Lyons T 2006 Different countries same science classes: students' experiences of school science in their own words *International Journal of Science Education* 28 pp 591-613.
[12] Sjøberg S and Schreiner C 2006 How do learners in different cultures relate to science and technology? Results and perspectives from the project rose (the relevance of science education). Asia-Pacific Forum on Science Learning and Teaching 6 pp 1-17
[13] Donald J G 2000 Indicators of success: From concepts to classrooms. Paper presented at the American Educational Research Association, New Orleans.
[14] Ramsden P 1992 Learning to teach in higher education. London: Routledge.

Acknowledgments

We would like to thank the Higher Education Council for budgeting and supporting this course. To the Social Involvement Unit, thanks for their support and for allowing such a course to take place at the HIT, as well as helping us contribute to society and future generations. Last but not least, we want to thank "Revivim" School for the opportunity to take a part in the next generation of education for a better and cleaner environment.

An Economic Evaluation of Onshore and Floating Liquefied Natural Gas Receiving Terminals: the Case Study of Indonesia

M J Giranza[1] and A Bergmann[2]

1 MSc Candidate in the Centre of Energy, Petroleum, Mineral Law, and Policy, University of Dundee, Dundee, United Kingdom DD1 4HN
2 Energy Economist in the Centre of Energy, Petroleum, Mineral Law, and Policy, University of Dundee, Dundee, United Kingdom DD1 4HN
Email: eabergmann@dundee.ac.uk

Abstract. Indonesia has abundant natural gas resources, however the primary fuel used for electricity generation is coal and oil. Insufficient natural gas infrastructure with-in the country acts as a barrier to increased natural gas usage. In Indonesia LNG is the most efficient and effective method for distributing natural gas given the difficult geographical conditions, the world's largest archipelago and located in a deep sea area. The Government is planning to initiate natural gas imports by 2019 to meet the country's energy demands. In order to allocate adequate amounts of natural gas across the geographic regions Indonesia must build more LNG regasification terminals. The Indonesia government has not yet determined if the additional regasification terminals will be floating or land-based facilities. This paper assesses the two options and identifies which facility attains greater profitability. The financial analysis of investing in the Sorong LNG regasification terminal project is conducted using NPV, IRR, and sensitivity analysis. This analysis demonstrates that FSRU facilities have greater economic viability than onshore LNG regasification facilities. The FSRU project earns greater than a 12% IRR as compared to a negative IRR earned by an onshore project. The government can make the onshore projects viable by increasing the sales fee from US$10.00/MMBTU to US$10.60/MMBTU.

1. Introduction

1.1. Background

Indonesia is well-known as an international exporter of natural gas (NG) since the 1970s.[1] In order to export NG to international markets which may be located several thousands of miles away the most efficient transportation is by liquefied natural gas (LNG) tanker ships. LNG infrastructure is required to convert the gas phase to a liquid phase at a liquefaction gas terminal, a liquefaction terminal is the same as a shipping terminal.[2] In 2016 Indonesia had four liquefaction terminals; Arun, Bontang, Tangguh, and Donggi Senoro. At least ten countries located in Asia, the Americas, and Africa became LNG importers of Indonesian NG in 2015. Indonesia has not been a net oil exporter since 2003, but is still a significant exporter of NG. Globally Indonesia is ranked tenth for NG exports and fifth for LNG exports specifically.

Indonesia has abundant NG resources and reserves, but the principle energy source used for electricity generation in Indonesia consists principally of coal and oil. In 2016, 31 % and 46 % of the power plants in Indonesia are supported by coal and oil, respectively. [3] Approximately 18 % of power plants in Indonesia use NG as fuel. Indonesia exhibits a preference for oil as a fuel for power generation, which is imported from abroad, as compared to NG which is abundant in Indonesia. This situation has developed as the result of an inadequate number of LNG receiving terminals. [4] A LNG receiving terminal is the same as a LNG regasification terminal. To fulfill the nation's demand for NG, the government is planning to import natural gas for the first time by 2019. [5] In order to allocate an adequate supply of NG to all regions and areas, Indonesia must build more LNG regasification terminals. Indonesia has had relatively flat production during the past 5 years. With an increasing amount of NG retained for domestic consumption. Figure 1 below shows a decline in domestic LNG production of 13.2% (2,900,000 tonnes) between 2011 and 2015.

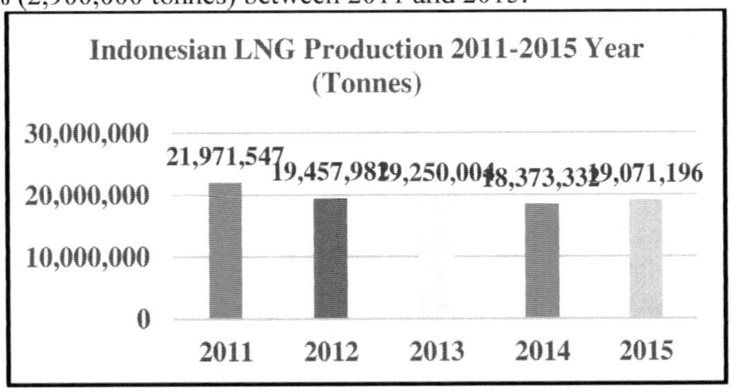

Figure 1. Indonesia LNG Production 2011-2015 [1]

LNG is the superior resource option for Indonesia because of complicated geographical conditions, which consist of several thousand islands, with some separated by a deep sea [5]. As can be seen in Figure 2 below the Indonesian archipelago is located between two oceans, the Pacific and Indian Oceans. The eastern region mainly consists of small islands relatively evenly distributed. [6] with seabed's ranging from 1,000 to 7,000 meters in depth. This configuration of small islands separated by a deep-sea has created a significant challenged to Indonesian government for developing infrastructures in the area. NG pipelines cannot be used in this depth of water for such long distances. It is not technologically appropriate economically efficient to deliver NG to the eastern region of Indonesia. [7] As a result, the Indonesian Government has decided to use a virtual pipeline, which is LNG shipping, to transport NG from areas with excess NG to areas experiencing a shortage of NG.

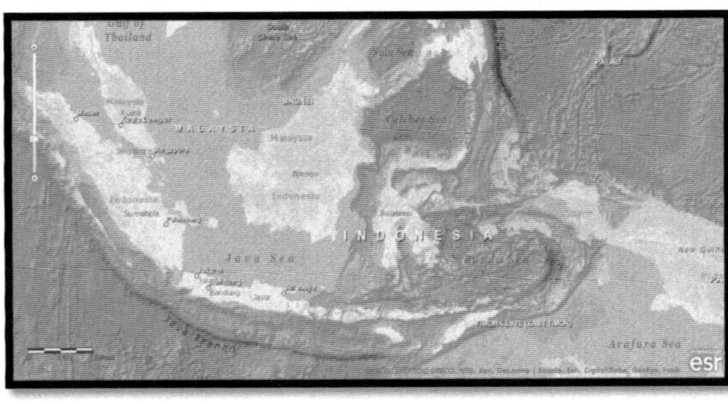

Figure 2. Map of Indonesia [6]

A total required investment of US$1,484 million [5] has been estimated to develop all the requested LNG infrastructure to support this virtual pipeline. The financial analysis conducted by [5] is based on the total cost and total revenue estimates from all associated projects, and is not based on site specific facilities. The Indonesian government has not yet decided whether the LNG receiving terminals will be built in the form of floating or land-based facilities. The motivation for this research paper is develop a better understanding of the financial impact of each type of facility.

1.2. Indonesian Gas Allocation and Monetisation

In 2016, the Minister of Energy and Mineral Resources (MEMR) issued the Minister Regulation No. 6/2016 [8] regarding the allocation, monetisation, and market price of NG. In this regulation, the government arranges the priority and order of NG allocation. Exporting of NG is now the lowest ranking allocation after all domestic demand is fulfilled.

In order to supply any power plant with NG, the Indonesian State Electric Company (PLN), must propose an allocation of NG to the MEMR through SKK Migas. SKK Migas is an institution established by the Indonesian Government to manage the country's upstream oil and gas activities. According to MEMR the proposal consists of the expected gas price and the method of transporting the natural gas. The final gas price is than agreed between PLN and the NG contractor, through a legal contract. PLN has the option of selecting NG supplier to transport and process the NG or LNG. The MEMR will evaluate the PLN proposal and determine its decision with assistance from SKK Migas.

Prior to 2009 PLN was the monopoly power company of Indonesia and controlled both transmission and sales of electricity to households and industry. A major regulatory change occurred in 2009 when the Indonesian Government enacted the Law of National Electricity No.30/2009, [9] from which time private parties were allowed involvement in the electricity industry.

1.3. FSRU vs Onshore LNG Receiving Terminal

Capital expenditures for floating storage and regasification units (FSRU) may be lower than onshore, due to potential construction cost efficiencies, through repeatability gains (iteration of design) where there is a generic design concept, with minor adaptations for each specific project. [7] It is assumed that costs will be better controlled with construction/assembly activities that are organized through established shipyards and supply chains using established labour and materials resources. Finally, the lack of offshore pipelines and near/onshore facilities will reduce costs and there is no need to purchase or lease waterfront land for facilities.

However, certain key issues need to be addressed when considering floating liquefied natural gas (FLNG) facilities. FLNG are ships that transport LNG and are not designed for regasification. Concerns of host governments about the economic impacts on local employment opportunities and skills development and supporting industries may or may not arise from the FLNG ships being sourced domestically or internationally. [7] However, long-term local jobs will be created during the operational phase (the operation of an FLNG facility is similar to the operation of a conventional LNG liquefaction facility). Moreover, there will be a significant increase tax revenues for the host government, and significant expenditure on goods and services in the host country during both the construction and operations phases. [7]

1.4. How to Evaluate the Investment Profitability of a Project

Discounted Cash Flow (DCF) analysis offers the simplest approach and is the most common method used to value oil and gas projects. [10] There is little disagreement in the literature on the appropriate evaluation criteria. In contrast, there is much debate over the appropriate modelling methodology. The DCF method remains the preferred tool, particularly as managerial flexibility is less of an issue in an individual investment project. [11]

In order to validate the ratio of profitability change due to uncertainty of certain characteristics in the investment, a sensitivity analysis can be conducted on a proposed project. [12]-[14] A sensitivity analysis identifies the value of more accurate information and the influence of input parameters on the objective criteria for investment decision making. [13]

2. Data and Methodology

2.1. Data Collection

The numerical data analysed was obtained from research carried-out by IGN Wiratmaja Pudja (General Director of Oil & Gas, Ministry of Energy and Mineral Resources, Republic of Indonesia 2014-2019) entitled *Virtual Pipeline to Support Natural Gas Infrastructures Development in Eastern Indonesia Region* presented and published as part of the 2016 Australian Pipelines & Gas Association

(APGA) Annual Convention & Exhibition. [5] This is the preferred data and was selected as the project investment assumptions are based on Pertamina's "rule of thumb", and Pertimina's extensive experience in such Indonesian projects. The case study is developed based on the LNG receiving terminal planned for deployment in Sorong, West Papua Province, Indonesia. According to *Roadmap of Indonesian Natural Gas Policy 2014-2030*, the project is planned for development to start by 2020. [4]

2.1.1 Investment Cost Assumptions

The investment cost is partially or completely irreversible. [15] In other words, the initial cost of investment is at least partially sunk, or irretrievable. Also, there is uncertainty over the future rewards from the investment. Consequently, there is need to assess the probabilities of alternative outcomes that can result in greater or lesser profits (or losses) for the venture.

According to Griffin, [7] the common capital cost characteristics of each regasification terminal can be classified as follow:

- Onshore LNG Receiving Terminal: LNG carrier, pipeline, and land-based regasification unit.
- Floating LNG Receiving Terminal: LNG carrier, Floating Storage and Regasification Unit (FSRU), and pipeline.

The capital expenditure for this project is assumed only for the storage and regasification unit. It means that the company purchases the LNG and transports it by vessel chartered to the LNG receiving terminal. According to one of Pertamina's rule of thumb, the FSRU and land-based regasification project investment is assumed to be US$0.6 million/MMSCFD and US$2.1 million/MMSCFD respectively. [5]

2.1.2 Other Costs Assumption

There are two central operational activities for a LNG receiving business; shipping of LNG and operation of the regasification terminal. [7] It is assumed that the company charts the LNG carrier and pays for shipping costs. And the company pays for operational costs and maintenance of the regasification unit, which it owns.

The data presented in Table 1 below is from PLN, the anchor buyer of NG, and other parties that have contributed to LNG development. [5] PLN has investigated/estimated possible costs based on information from various liquefaction operator sources. Table 1 shows the cost of distribution (shipping and regasification) of LNG from Bontang to Sorong is US$3.20/MMBTU.

Table 1. Estimated Distribution Cost from LNG source [5]

Origin	Primary Delivery	Demand	Capacity		Distance	Days 1-way	Round Trip	Ship Capacity	Storage Capacity	Shipping + Regas Cost
			Weight	Volume						
		[MMSCFD]	[ton/day]	[m3/day]	[km]	[day]	[day]	[m3]	[m3]	[$/MMBTU]
Bontang	Sorong	27.9	642	1,335	1,600	3.2	10.1	13,468	17,509	3.20
	Jayapura	28.9	664	1,380	2,400	4.8	13.9	19,235	25,005	3.20
Salawati	Ambon	19.1	439	914	440	0.9	4.5	4,125	5,363	4.00

With regard to other related gas distribution costs, Perusahaan Gas Negara (PGN), Indonesia's state-owned gas company and the country's largest NG utility, has identified other costs: trading fees of US$0.03/MMBTU; operating and maintenance costs of US$1.38/MMBTU; and taxes of US$0.41/MMBTU (Pudja et al., 2016). The feed price for NG is assumed to be US$4.72/MMBTU. Interest rate on debt is assumed at be 4.75% per annum.

2.1.3 Revenue Assumption

It is assumed that PLN does not have to construct or commission ancillary infrastructures other facilities for purchasing and receiving NG directly to its power plant gates. Therefore, revenue for the NG supplier will be the purchase price paid by PLN.

There is no direct fee, fixed or variable, paid by PLN to the company that provides the LNG infrastructure and facilities. The fee is established and contracted based on a negotiated agreement of the NG seller and purchaser in a contract. According to Pudja (2016), the NG supply company is able to purchase NG at maximum price of US$13.00/MMBTU from domestic NG producers for delivery to PLN.

2.1.4 Financial Calculation and Economic Evaluation

Project investment analysis seeks to analyse projected cash inflows (benefits) and projected cash outflows (costs) than compares the results with established decision-making criteria. [16] There are several techniques for analysing project investments, common methods include payback analysis, Net Present Value (NPV), Discounted Pay Back (DPB) and Internal Rate of Return (IRR). [17] This research evaluates LNG receiving terminal investments based on two of these financial analysis techniques; NPV and IRR.

a. Net Present Value

In order to calculate NPV, we calculate at the net cash flow in each period and then discount this stream back to the present time period. [18] The general mathematical equation for deriving the NPV is:

$$NPV = \sum_{t=0}^{N} \frac{Yt}{\prod_{j=0}^{t}(1+ij)} \tag{1}$$

Where NPV = the net present value
Y_t = the net cash flow at the end of period t, i.e., at time t
i_j = the interest (discount rate) for period j
N = the life of project
j = points in time prior to t
t = the point in time under consideration [19]

The investment should be purchased if and only if the NPV positive. [15] [18]

b. Internal Rate of Return

The IRR is another method in capital investment appraisal which uses the time value of money but results in an answer expressed in percentage form. [20] The internal rate or return is found by solving for the value of K from the following formula [21]:

$$I_0 = \frac{FV_1}{1+K} + \frac{FV_2}{(1+K^2)} + \cdots + \frac{FV_n}{(1+K^n)} \tag{2}$$

The investor determines whether the project is acceptable or not based on the calculated Internal Rate of Return. If the IRR is meets the required threshold, then the proposed project/investment is deemed acceptable and vice versa. [22]

2.1.5 Methodology

This paper conducts a quantitative financial analysis of two types of LNG receiving terminals; an onshore LNG receiving facility and a FSRU. The facility with the greater IRR would be considered the superior investment and the more viable project. A sensitivity analysis of service fees paid by PLN is conducted. The minimum fee level required to be paid by PLN must provide the NG supply company a 12% IRR.

3. LNG Receiving Terminal Project Investment Valuation

Given the information provide in section two the analysis shows that the FSRU Sorong investment is more profitable as compared to a land-based regasification facilities with a NPV if US$16.95 million. Table 2 below reports a higher IRR value which is 13.77 %. The main determinant is that the total capital investment cost of the FSRU is lower than a land-based terminal. The total investments cost of the FSRU is only US$16.74 million compare to US$58.95 million for a land-based regasification unit.

[5] The lack of offshore pipelines and near/onshore facilities significantly reduces the cost of FSRU investments. [7] Moreover, there is no requirement to purchase or lease shoreline land for facilities.
On the other hand, this analysis indicates a negative impact on project economics for the land-based regasification facilities. Table 3 below shows the simulation results. The IRR for land-based regasification facilities is -0.27 %. In additional to the higher capital costs, land-based regasification facilities have another economic disadvantage, that of stranded assets. At termination of the NG supply contract the assets of the company, pipeline and regasification unit, cannot be relocated as flexibly, if at all, as a FSRU. [7], [23] This supports the assumption that shutdown costs for land-based regasification facilities will be substantially greater than for FSRU facilities.

Table 2. Economic Evaluation of Investment Profitability in FSRU LNG Receiving Terminal Project

Year	Cost Component (Million USD)					LNG Delivered (MMBTUD)	Revenue (Million USD)	Net Benefit (Million USD)
	Capital	LNG	Shipping & Regas	Other	Total			
1	16.74				16.74			-16.74
2								
3		51.53	34.93	19.87	106.33	29,908.80	109.17	2.84
4		51.53	34.93	19.87	106.33	29,908.80	109.17	2.84
5		51.53	34.93	19.87	106.33	29,908.80	109.17	2.84
6		51.53	34.93	19.87	106.33	29,908.80	109.17	2.84
7		51.53	34.93	19.87	106.33	29,908.80	109.17	2.84
8		51.53	34.93	19.87	106.33	29,908.80	109.17	2.84
9		51.53	34.93	19.87	106.33	29,908.80	109.17	2.84
10		51.53	34.93	19.87	106.33	29,908.80	109.17	2.84
11		51.53	34.93	19.87	106.33	29,908.80	109.17	2.84
12		51.53	34.93	19.87	106.33	29,908.80	109.17	2.84
13		51.53	34.93	19.87	106.33	29,908.80	109.17	2.84
14		51.53	34.93	19.87	106.33	29,908.80	109.17	2.84
15		51.53	34.93	19.87	106.33	29,908.80	109.17	2.84
16		51.53	34.93	19.87	106.33	29,908.80	109.17	2.84
17		51.53	34.93	19.87	106.33	29,908.80	109.17	2.84
18		51.53	34.93	19.87	106.33	29,908.80	109.17	2.84
19		51.53	34.93	19.87	106.33	29,908.80	109.17	2.84
20		51.53	34.93	19.87	106.33	29,908.80	109.17	2.84
21		51.53	34.93	19.87	106.33	29,908.80	109.17	2.84
22		51.53	34.93	19.87	106.33	29,908.80	109.17	2.84
	Present Value of Net Benefit (Million US$)						US$16.95	
	Internal Rate of Return						13.77%	

Table 3. Economic Evaluation of Profitability in Onshore LNG Receiving Terminal Project

Year	Cost Component (Million USD)					LNG Delivered (MMBTUD)	Revenue (Million USD)	Net Benefit (Million USD)
	Capital	LNG	Shipping & Regas	Other	Total			
1	58.59				58.59			-58.59
2								
3		51.53	34.93	19.87	106.33	29,908.80	109.17	2.84
4		51.53	34.93	19.87	106.33	29,908.80	109.17	2.84
5		51.53	34.93	19.87	106.33	29,908.80	109.17	2.84
6		51.53	34.93	19.87	106.33	29,908.80	109.17	2.84
7		51.53	34.93	19.87	106.33	29,908.80	109.17	2.84
8		51.53	34.93	19.87	106.33	29,908.80	109.17	2.84
9		51.53	34.93	19.87	106.33	29,908.80	109.17	2.84
10		51.53	34.93	19.87	106.33	29,908.80	109.17	2.84
11		51.53	34.93	19.87	106.33	29,908.80	109.17	2.84

12	51.53	34.93	19.87	106.33	29,908.80	109.17	2.84
13	51.53	34.93	19.87	106.33	29,908.80	109.17	2.84
14	51.53	34.93	19.87	106.33	29,908.80	109.17	2.84
15	51.53	34.93	19.87	106.33	29,908.80	109.17	2.84
16	51.53	34.93	19.87	106.33	29,908.80	109.17	2.84
17	51.53	34.93	19.87	106.33	29,908.80	109.17	2.84
18	51.53	34.93	19.87	106.33	29,908.80	109.17	2.84
19	51.53	34.93	19.87	106.33	29,908.80	109.17	2.84
20	51.53	34.93	19.87	106.33	29,908.80	109.17	2.84
21	51.53	34.93	19.87	106.33	29,908.80	109.17	2.84
22	51.53	34.93	19.87	106.33	29,908.80	109.17	2.84
Present Value of Net Benefit (Million US$)						-US$23.00	
Internal Rate of Return						-0.27%	

Sensitivity analysis performed on investment profitability was conducted using different sales fees. Table 4 below shows that the IRR of the NG supplier can be increased to 12.94% if PLN would pay at least US$10.60/MMBTU. The profitability of the project is highly sensitive to the sales fee as a US$0.60 increase from US$10.00/MMBTU to US$10.60/MMBTU results in an IRR increases from -0.27% to 12.94%.

Table 4. Sensitivity Analysis of Sales Fee paid PLN versus the Profitability of Onshore LNG Receiving Terminal Project

Description	**Unit**	**Sales Fee (US$/MMBTU)**			
		10	**10.3**	**10.6**	**10.9**
NPV	million US$	-23.00	1	52.99	90.99
IRR	%	-0.27	7.38	12.94	17.66

Figure 3 illustrates the increase in both the NPV and the IRR that results from an increase in the sales fee. Less than a $0.30 increase in sales fee changes the NPV from negative to positive. The same results are seen with the IRR.

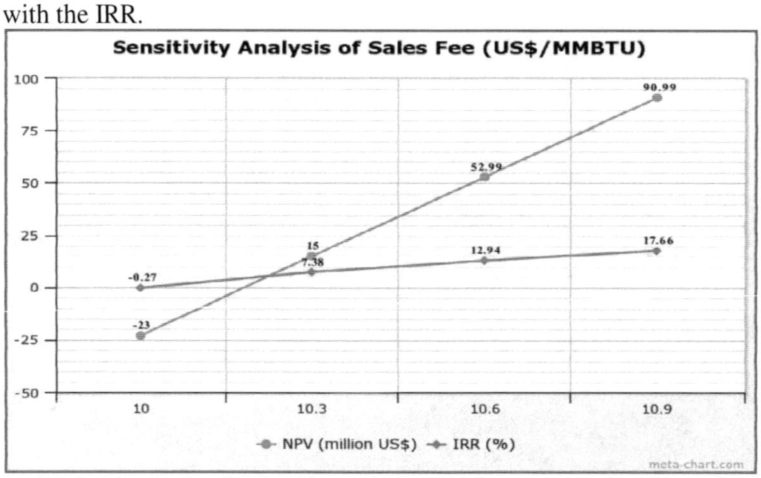

Figure 3. Sensitivity Analysis of Sales Fee paid by PLN versus Profitability of Onshore LNG Receiving Terminal Project

4. Conclusions and Recommendations

A FSRU is more economically viable than an onshore LNG receiving terminal. The FSRU project produces a greater IRR (12.94%) as compared to a negative IRR (-0.27%) for the onshore project in this case study, the Sorong LNG receiving terminal project. The primary reason for this result is the substantially lower initial investment cost required for FSRU projects. In addition, termination and

closer (shutdown) costs of onshore projects will be greater as storage and regasification facilities cannot be relocated as FSRU are capable.
If the Indonesian Government wants to promote investment in and deployment of onshore LNG facilities and infrastructure an increased sales fee will be required, an increase from US$10.00/MMBTU to US$10.60/MMBTU. This would raise the IRR up to 12.94%.

5. Reference

[1] The Ministry of Energy and Mineral Resources of the Republic of Indonesia 2015 *Statistik Migas*
[2] Paltsev S, O'Sullivan F, Lee N, Agarwal A, Li M, Li X and Fylaktos N 2013 *Natural Gas Monetization Pathways for Cyprus: Interim Report - Economics of Project Development Options* MIT Energy Initiative, Massachusetts Institute of Technology, Cambridge, MA
[3] PT PLN (Persero) 2016 *National Electric Generation Plan 2016-2025*
[4] The Ministry of Energy and Mineral Resources of the Republic of Indonesia 2014 *Roadmap* of Indonesian Natural Gas Policy 2014-2030
[5] Pudja I G N W, Dwiantoro D, and Simarmata M R 2016 Virtual Pipeline to Support Natural Gas Infrastructures Development in Eastern Indonesia Region October 1–16
[6] Research and Development Center for Marine and Coastal Resources 2016 *Map of Indonesia*
[7] Griffin P 2012 *The Law and Business of LNG* Second Edition Globe Law and Business
[8] Minister of Energy and Mineral Resources of the Republic of Indonesia 2016 *Minister Regulation No. 6/2016*
[9] Law of the Republic of Indonesia 2009 *Law of National Electricity No.30/2009*
[10] Emhjellen M and Alaouze C M 2003 A comparison of discounted cashflow and modern asset pricing methods – Project selection and policy implications *Energy Policy* 31(12) 1213–20 https://doi.org/10.1016/S0301-4215(02)00181-7
[11] Daniel P, Keen M and McPherson C 2010 *The Taxation of Petroleum and Minerals* The Taxation of Petroleum and Minerals: Principles, Problems and Practice Oxon United Kingdom: Routledge https://doi.org/10.4324/9780203851081
[12] Iloiu M and Csiminga D 2009 *Project Risk Evaluation Methods – Sensitivity Analysis* Annals of the University of Petroşani Economics 9(2) 33–38
[13] Jovanović P 1999 Application of sensitivity analysis in investment project evaluation under uncertainty and risk *Int J Project Management 17*(4) 217–22 https://doi.org/10.1016/S0263-7863(98)00035-0
[14] Martin R, Lazakis I, Barbouchi S and Johanning L 2016 Sensitivity analysis of offshore wind farm operation and maintenance cost and availability *Renewable Energy* 85 1226–36 https://doi.org/10.1016/j.renene.2015.07.078
[15] Dixit A and Pindyck R 1994 Investment Under Uncertainty *Investment Under Uncertainty* 1–17 Princeton University Press https://doi.org/10.2307/1909571
[16] Arnold G 2006 *Essentials of Corporate Financial Management* Harlow: Prentice Hall
[17] Tennent J 2008 *Guide to Financial Management* The Economist John Wiley & Sons
[18] Varian H R 2010 *Intermediate Microeconomics: A Modern Approach* New York: W.W. Norton & Co
[19] Bussey L E 1978 *The Economic Analysis of Industrial Projects* Englewood Cliffs N J:Prentice-Hall
[20] Weetman P 2010 *Management Accounting Second Edition* Pearson Education Limited
[21] Drury C 2015 *Management and Accounting 9th Edition* Cengage Learning EMEA.
[22] Anthony R N, Hawkins D F and Merchant K A 2011 *Accounting, Text and Cases 13th Edition* New York, NY : McGraw-Hill/Irwin
[23] Mokhatab S, Mak, J Y, Valappil J V and Wood D A 2014 *Handbook of Liquefied Natural Gas* https://doi.org/10.1016/B978-0-12-404585-9.01001-1

The Effects of Domestic Energy Consumption on Urban Development Using System Dynamics

M D Saryazdi[1], N Homaei[2] and A Arjmand*[3]

1 Department of Computer Engineering, Vali-E-Asr University of Rafsanjan, Rafsanjan, Iran
2 Department of Industrial Engineering, Mazandaran University of Science and Technology, Mazandaran, Iran
3 Department of Industrial Engineering, College of Engineering, Alzahra University, Tehran, Iran
E-mail: a.arjmand@alzahra.ac.ir

Abstract. In developed countries, people have learned to follow efficient consumption patterns, while in developing countries, such as Iran, these patterns are not well executed. A large amount of energy is almost consumed in buildings and houses and though the consumption patterns varies in different societies, various energy policies are required to meet the consumption challenges. So far, several papers and more than ten case studies have worked on the relationship between domestic energy consumption and urban development, however these researches did not analyzed the impact of energy consumption on urban development. Therefore, this paper attempts to examine the interactions between the energy consumption and urban development by using system dynamics as the most widely used methods for complex problems. The proposed approach demonstrates the interactions using causal loop and flow diagrams and finally, suitable strategies will be proposed for urban development through simulations of different scenarios.

1. Introduction

Energy is one of the primary requirements for industrial and commercial goals of each country and has an important role in the country's rapid socio-economic growth and improvement of the quality of life [1], [2]. Also it is considered as a determining factor for employment, economic competitiveness and social welfare [3]. In 2007 energy consumption in the world was 495 quadrillion and it is predicted that it will reach 739 quadrillion by 2035. This represents 1.4 percent of annual growth for energy consumption in the world [4]. Undoubtedly, depending on the type of consumption pattern, level of development and economic growth, energy consumption is varied in different countries [5], [6].
In 2007 energy consumption in non-OECD countries first surpassed as much as 1.5 percent by OECD. It is expected that the difference in energy consumption between OECD and non-OCED will increase to 32% by 2035. Today, the rapid growth in demand and consequently the decline of non-renewable resources and increasing pollution, led to the environmental and energy crisis because, along with consumption, production and supply methods of energy are of effective factors in greenhouse gas emissions and environmental pollutants [7]-[9]. Researchers, managers and politicians are trying to provide appropriate programs to limit energy consumption. Today, two-third of total world energy consumption and CO2 gases emissions is related to urbanization [10]. It is observed that challenges such as lack of energy sources as a geophysical concern and warming the climate through greenhouse

IOP Conf. Series: Earth and Environmental Science **150** (2018) 012027 doi:10.1088/1755-1315/150/1/012027

gas emissions as an environmental concern, along with increased urbanization and urban development are growing and it is daily added to global problems [11].
Therefore this paper examines the interaction between energy consumption and urban development. It should be noted that the amount of energy consumption in different countries is different depending on patterns of their urban development. As a result the different policies and strategies need to meet the challenges in the future.
According to the above mentioned cases, the problem of domestic energy management is too complex. Therefore it is necessary to analyze energy management in urban residences systematically. Thus the present study examines the effects of domestic energy consumption and urban development.

2. Theoretical framework of system dynamics

Demonstrating nonlinear and feedback effects of decision-maker variables in complex systems is very difficult for human. One can predict through a linear relation between variables while this could lead to incorrect decisions. System dynamics are approach to analyze such nonlinear and complex socio-economic systems. To explain system dynamics, there are five basic elements such as system, feedback, flow variables, relevant diagrams and modeling processes.

3. Modeling process

3.1. *Main variables*

The names of model variables are presented in Table 1 to create a dynamic model.

Table 1. Main variables.

Variables
Energy consumption, average consumption per household, total energy consumption per household, level of energy production, total cost of energy production for domestic applications, budget, allocation of budget to raise the awareness, invest to increase awareness, amount of energy consumption by buying efficient devices, bill of energy consumption per household, decrease energy consumption by modifying consumer patterns, municipal fund, urban development, population, residences, air pollution, green spaces, cost paid by energy consumption.

R1. Reinforcing loop for government budget:

Figure 1 shows that as the government budget increases, its percentage to raise the level of public culture and awareness increase too and government is ready to invest more. Instead, high the level of public awareness, their consumer patterns varied and they become more mature, on the other hand they desire to buy efficient consumer devices according to their financial capacity and because of a time distance, efficiency of energy consumption increases. It can be observed that the budget and level of energy consumption in Iran has a bilateral relationship.

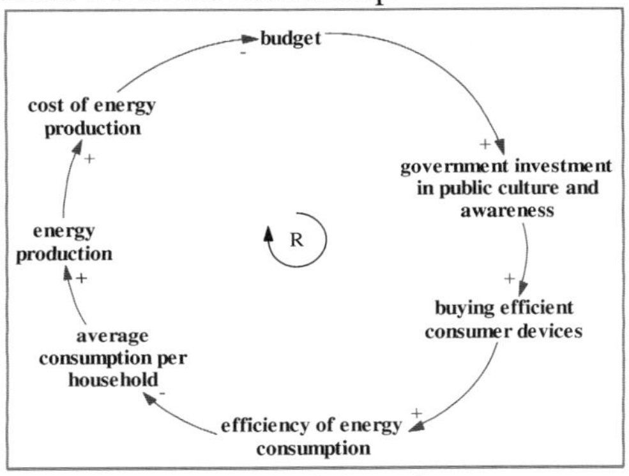

Figure 1. Reinforcing feedback loop for government budget.

B1. Balancing loop for energy consumption

Figure 2 shows that as average energy consumption increases, it leads to increase bill and through time distance causes to reduce the energy consumption. Since high energy bill is considered as an economic challenge for households, they forced to modify their consumer patterns to cope with these challenges. Thus high energy bill plays a balance role in reducing consumption.

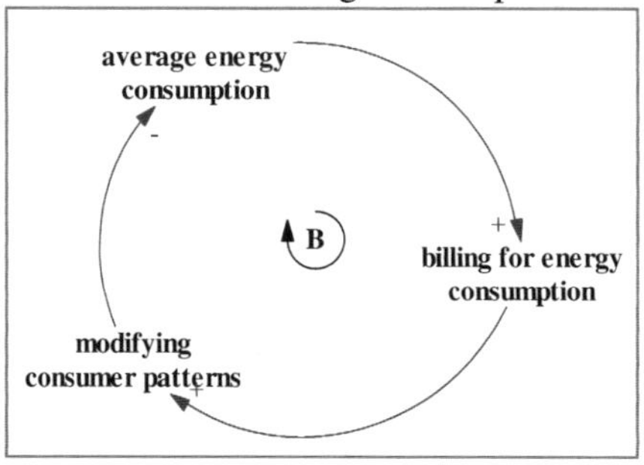

Figure 2. Balancing feedback loop for energy consumption.

B2. *Balancing feedback loop for urban development*

According to Figure 3, as the government budget increases, municipal budget also increases and it leads to develop more because cities need enough budgets to implement their programs. When a city develops, its population increases because residents of rural and other areas immigrate to developing cities and this leads to reduce the mortality rate and increase medical and health care. In addition pregnancy rate because of high education reduces but the population in urban areas has a direct relationship with the rate of development.

High population causes to increase rate of construction in the city, which increases energy consumption. This increases the difference between energy supply and demand and government had to produce more energy so it decrease the government budget and affects negatively on urban development.

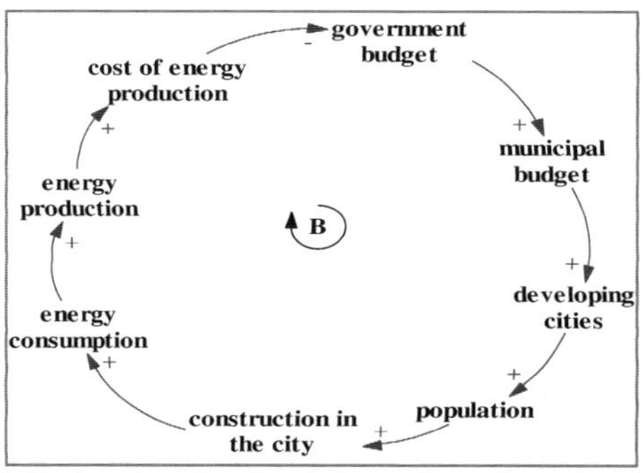

Figure 3. Balancing feedback loop for urban development.

R2. Reinforcing loop for pollution reduction

According to Figure 4, high urban development, greater the amount of green spaces will be because the green space is considered as one of the criteria to estimate the rate of development. By increasing green spaces, air pollution reduces thus pollution reduction is one of the criteria to increase urban development.

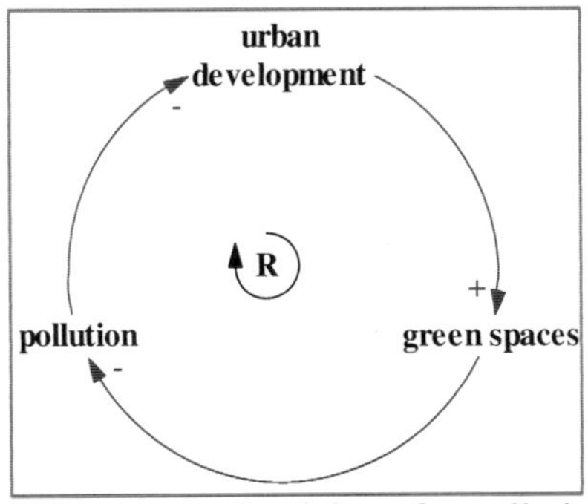

Figure 4. Reinforcing feedback loop for pollution reduction.

B3. Balancing loop for urban development and air pollution

Figure 5 shows that as urban development increases, population and the number of residences increase, so the level of energy production and consumption increases. Since domestic energy production such as electricity, oil and gas depends on fossil fuels, more energy production leads to increase pollution.

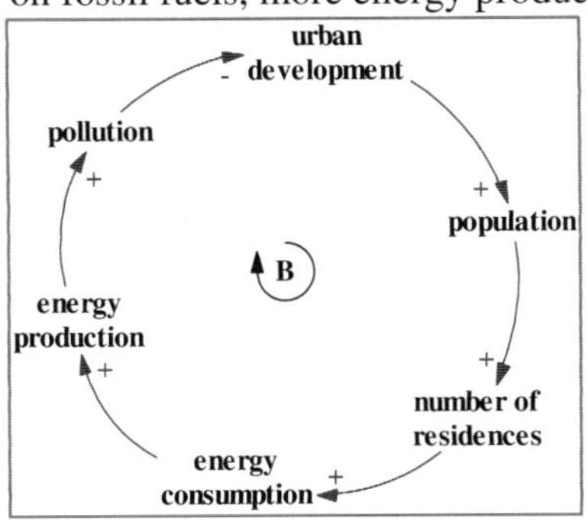

Figure 5. Balancing feedback loop for urban development and air pollution.

R3. Reinforcing loop for Urban Development and budget

Figure 6 shows that as government budget increases, the municipal budget will increase too and this leads to develop more urban areas. When a city develops, the number of residences and energy consumption increase. By increasing energy consumption, cost of bill paid by households will increase so that it leads to increase the government revenues and budget.

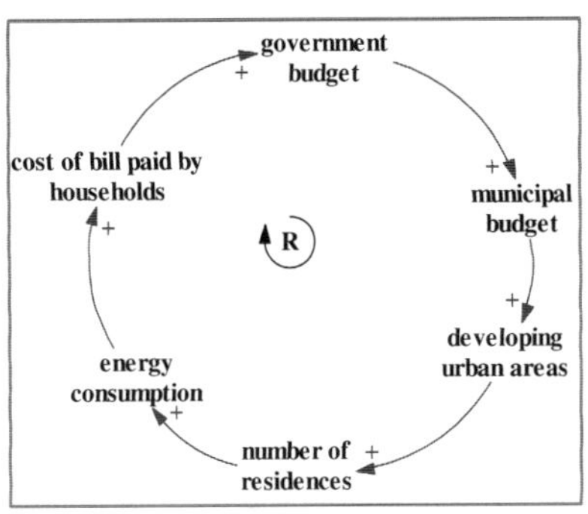

Figure 6. reinforcing feedback loop for urban development and Budget.

3.2. Modeling flow diagram

According to the causal diagram presented in the previous section, flow diagram shows the impact of domestic energy consumption on urban development. Measurement for time variable is year. This model is implemented using software Vensim. It should be noted that this model was simulated for a

100-year period from 2011. For the purpose of building model and decision-making, levels of budget, urban development, population as well as residences were determined that after the simulation stage, the behavior of these variables were analyzed. We defined 4 levels in the model including Government budget, Developing urban, Population and Number of residences.

4. Model simulation

By using dynamic model and relations between variables, the model was simulated and results were analyzed based on the predetermined levels. Figure 7 shows that urban development increases for a 100-years period. In contrast budget decreases initially and then increases. It happens because of high urban development and low budget and since the interests is obtained by urban development, revenues increase relative to the cost and it leads to increase the budget (according to reinforcing loops in the model). Given that urban development process rapidly progresses and then slows down and desires to equivalent level therefore budget reaches normal condition and fluctuates around it. (According to equivalent loops embedded in model). Urban development leads to increase population and consequently, residences increase too (according to causal relationships in the model).

5. Testing dynamic model performance

Based on the relationships shown in the attachment, the relationships among model variables, initial and fixed values of problem were defined. Different tests were used to verify the simulated model:

Testing adequacy scope focuses on adequacy of parameters and causal loops based on the purpose, which this problem by modeling, eliciting the results and focus on model framework was verified.

Testing compatibility of units emphasizes their homogeneity and it was verified based on matching all units by unit Cheque option in Vensim software.

The cumulative error focuses on the lack of sensitivity relative to time scale, that is, if the measurement of time equals to a year, the similar results can be obtained by changing it to 6 months. The test was conducted on the model and results were entirely consistent with previous results. For example, urban development is presented in Figure 8.

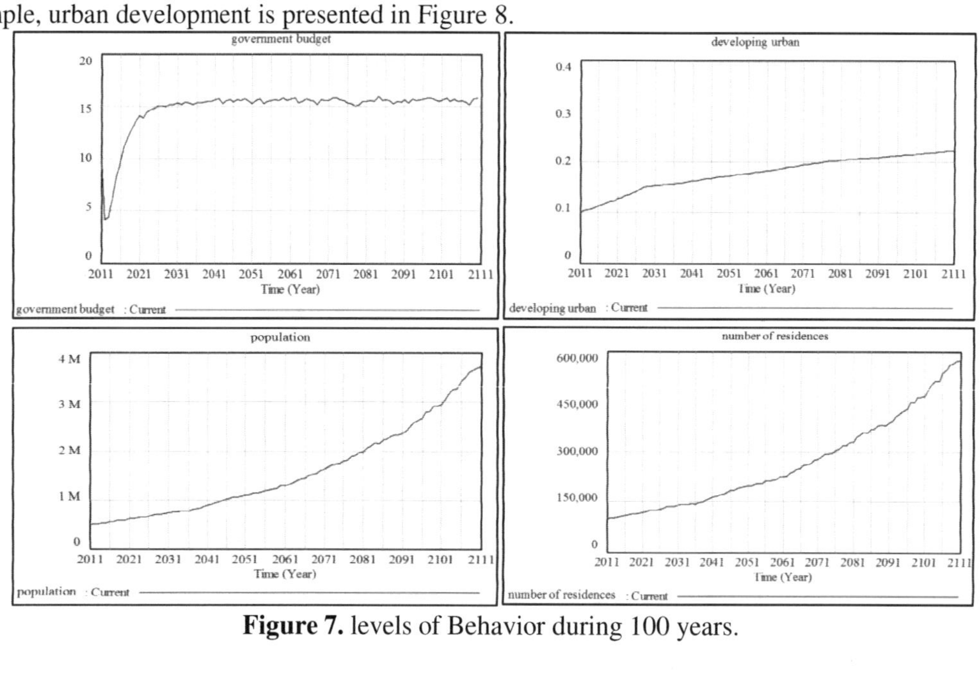

Figure 7. levels of Behavior during 100 years.

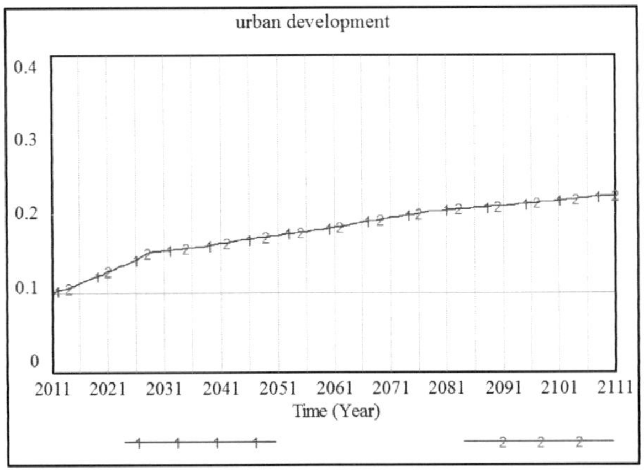

Figure 8. Cumulative error on the urban development variable.

Structure assessment determines compatibility of model behavior with its structure. It is necessary that variables behavior in a simulated model with negative feedback must be purposeful. Figure 5 shows that the slope of the urban development process slow down over time and has a purposeful behavior.

Boundary conditions focus on model resistance, that is, model must show the expected behavior under the conditions of changing systems and input values. Model was verified in terms of boundary conditions.

To perform this test, the "percentage of budget awarded to the city" under boundary conditions 0 was tested and their results on the urban development were considered. It is expected that the urban development should be reached zero over time when government does not allocate fund to the city. The results are shown in Figure 9.

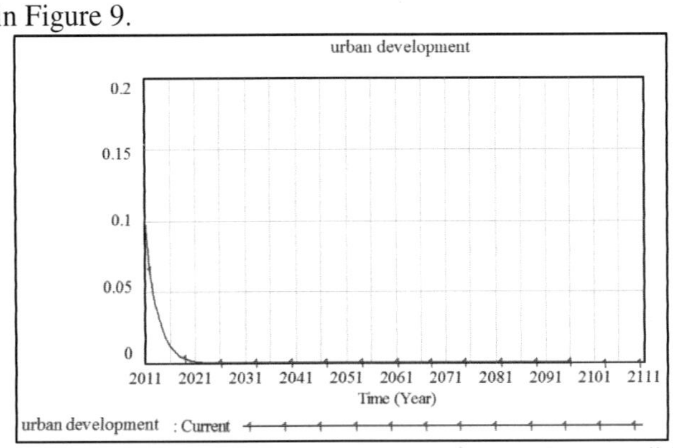

Figure 9. Behavior of urban development in boundary conditions 0.

6. Scenarios

After testing the model performance, dynamic model was evaluated under the different scenarios and the best scenario was selected. In regard to limited resources and the types of policies, the model needs to be done such a way that it shows the most effective performance to develop urban areas.

These 3 scenarios are simulated as follows:

Scenario 1: this scenario called "high allocation of budget without growing green space" focuses on issue of allocation of budget, but it does not include growing green space. Thus this scenario increases "rate of the budget awarded to the city "by 25 %.

Scenario 2: this scenario called "high growth of green space by allocating low budget" focuses on green space, but the allocation of budget will not increase so this scenario improves "rate of green space growth in terms of urban development" by 20%.

Scenario 3: this scenario called the "moderate allocation of the budget with low growth of green space" focuses on both of budget and green space. Thus this scenario improves "percent of the budget allocated to city" and "the rate of green space growth in terms of urban development" by 10 % respectively. The results of each scenario for urban development are as follows:

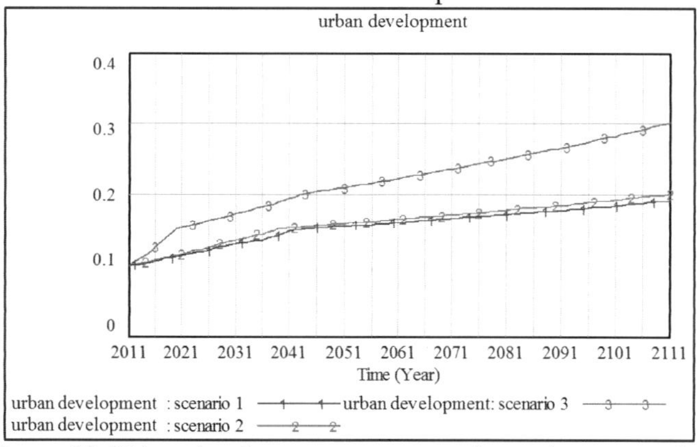

Figure 10. Urban development process under various scenarios.

Figure 10 shows that among three scenarios, the scenario 3 "moderate allocation from the budget with low growth of green space" is better than the other two scenarios.

7. Conclusion

Deeply real improvements in business thinking and practice emerge when managers are responsibility for modeling systems in functional areas. This should be based on personal involvement of managers in the development of a system perspective and define the goals, ways to achieve and success criteria in applying the methods. Given that dynamic system is one of the most widely used method in complex problems, this paper tries to examine the interaction between urban development and energy consumption using causal loops and flow diagram. Finally, the suitable strategies for urban development are proposed by model simulation and different scenarios. Among these scenarios, senario3 "moderate allocation of the budget with low growth of green space" is better than the other scenario. It is showed that that the high allocation of budget without growing green space and vice versa does not affect on urban development and managers need to focus on both of them equally and if necessary, they continue to work more. This model provides the important advantages:

Considering the temporal dimension between cause and effect. Time distance between the cause and effect was considered through delay.

Do analysis "what happens- if". This will reduce the risk of action plans before implementing. In this model, three scenarios were examined and analyzed and then the best one was selected.

Areas could be considered for future research:

It is necessary to be discussed more detail in further research to improve and develop relations and the relevant equations.

More simulations implemented by different policies under various scenarios and analyze the sensitivity of their results can provide more accurate conclusions.

8. References

[1] Abdelaziz E A, Saidur R and Mekhilef S 2010 *A review on energy saving strategies in industrial sector* (Renewable and Sustainable Energy Reviews)

[2] *U.S. Energy Information Administration International Energy Outlook* 2010 World Energy and economic Outlook

[3] Azam S A, Nabi Z R and Godin K *Investigate the relationship between energy consumption in the central campus of Tehran University and environmental emissions arising from it* (Journal of Health and Environment, Environmental Health Association) vol 2, Issue 2, summer 1388, p 150-159

[4] Bhushan Energy Management Cell 2000 Save Energy *Energy Management*
[5] The Organization for Economic Co-operation and Development (OECD) 2002 *Urban Development*, www.oecd.org/gov/regional/innovation
[6] United Nations Educational, Scientific and Cultural Organization (UNESCO).
[7] You F, Hu D, Zhang H Tao, Guo Z, Zhao YH, Wan B N and Yuan Y 2011 *Carbon emissions in the life cycle of urban building system in China—A case study of residential buildings*, Ecological Complexity, vol 8, Issue 2, p 201-212.
[8] Dhakal S 2009 *Urban energy use and carbon emissions from cities in China and policy implications*, Energy Policy, vol 37, Issue 11, p 4208-4219.
[9] Hepbasli, Arif, *Development and restructuring of Turkey's electricity sector: a review*, Renewable and Sustainable Energy Reviews, Volume 9, Issue 4, August 2005, Pages 311-343.
[10] Halsnæs K and Garg A 2011 *Assessing the Role of Energy in Development and Climate Policies—Conceptual Approach and Key Indicators* World Development, vol 39, Issue 6, p 987-1001.
[11] Ren H B, Zhou W S, Nakagami K, Gao W J and Wu Q 2010 *Feasibility assessment of introducing distributed energy resources in urban areas of China*, Applied Thermal Engineering, vol 30, Issue 16, p 2584-2593.

External costs as a measure of environmental impact in the generation of electricity in Poland

W Cel[1], A Czechowska-Kosacka[1], Justyna Kujawska[1] and H Wasąg[1]

1 Lublin University of Technology, ul. Nadbystrzycka 38D, 20-618 Lublin, Poland
E-mail: w.cel@pollub.pl

Abstract. The depletion of natural resources, rising fossil fuel prices and growing environmental awareness, are leading to an increase in the popularity of renewable energy sources. In Poland, the share of energy derived from renewable sources continues to grow and now stands at 12.9% of the country's gross electricity consumption. Energy from renewable sources in Poland is 60€ more expensive per MWh than energy from conventional sources. According to the European Climate and Energy Package, Poland is committed to increasing its share of renewable energy to 15% in 2020, and a further 5% by 2030. It is very important to ensure that the increase in the share of renewable energy will increase the price of energy for the end users. To convince the public of the need to incur greater costs in the purchase of "green" power, we should put forward arguments showing the benefits of its use. The aim of this paper is to demonstrate the viability of support through a system of certification for renewable energy sources and also to estimate the potential increase in energy prices caused by raising RES contribution.

1. Introduction

Widely circulating opinions say that the greenhouse effect is a major threat to our planet. Not many people realize that the rapidly disappearing non-renewable fossil fuels constitute a far greater threat to our civilization [1]-[3]. According to the latest available data, the world's oil reserves will run out by 2047 whilst natural gas by 2068. Coal is in a better situation being exhausted by 2140 and an estimate for uranium is 2144 (Europe's Energy Portal, 2011). Unfortunately, in the now-dominant democracies, there are strong tendencies to favour the present without looking into the future. Successive generations are rarely taken into account because, as they do not exist they have no impact on present decision-making processes. It is vividly seen that the modern world is developing in an unsustainable way. We have seen tremendous growth in the techniques and technologies which are able to transform the modern world and potentially lead to its destruction [4], [5]. Development of human capability to transform the world is beyond the social sciences, and consequently there is no clear criteria for establishing the values to which these changes should serve. Rapid depletion of natural resources requires a new definition of development rules for the modern world even more urgent and necessary [6]-[8].

The "Polish Energy Policy until 2030" project, prepared by the Ministry of Economy provides, in accordance with the guidelines of the European Commission, for increasing the share of renewable energies in the gross electricity consumption to 15% in 2020 and to 20% ten years later. Energy providers are required to provide part of their energy from renewable sources, which will steadily increase in the coming years. Energy production from renewable sources is encouraged by strong economic subsidies. The cost of renewable energy in Poland consists of a basic price for energy and the cost of Renewable Energy Certificates (Green Certificates). Today, one certificate costs 60€/MWh.

Thus, by increasing the proportion of renewable energy in our consumption, this will lead to higher energy costs for the end user. The situation is so serious that, according to a Eurostat report, published in June 2010, the Poles pay the highest price for energy among all EU citizens when revenue per capita is compared.
However, for a comprehensive analysis of the costs and benefits of using energy from renewable sources, compared to the energy from the combustion of fossil fuels, we have to take into account economic, environmental and social aspects. The combination of these issues is consistent with the principles of sustainable development. In order to present these aspects in a quantified manner and compared on a common scale we use ExternE - the methodology for estimating external costs which was proposed by the European Commission. External costs in this methodology include the cost of the health, welfare, environmental damage and moreover, costs associated with the greenhouse effect. These costs are not included in the price of electricity, but are paid by society in an indirect way. Assessing the costs requires taking into account the whole cycle of energy production in accordance with the principles of LCA. The LCA Methodology contains extraction of raw materials, the production phase, decommissioning and waste disposal.

2. The ExternE methodology

Most people believe that the actions taken to increase energy production from renewable sources are caused by the need to avoid the greenhouse effect. Developing renewable energy sources can significantly reduce the greenhouse gas emissions, indicated by the carbon footprint. The carbon footprint shows all the greenhouse gas emissions from the energy production process and allows us to make comparisons using a common scale known as the carbon dioxide equivalent. As a result, we can determine a given process's potential for global warming. The value of carbon footprint for renewable energy sources is significantly lower compared to energy from conventional sources. Electricity derived from coal produces emissions with a value around 1100 gCO_2e/kWh, whilst from renewable energy sources the value is below 100 gCO_2e/kWh [9].
Yet, carbon footprint does not take into account the depletion of fossil fuels. This is also dangerous, because in the long term we will run out of fossil fuels. In addition, it also turns out that the methodology for estimating costs is incomplete because it does not include damage to the environment caused by the various methods of producing energy. However, subsidizing renewable energy under the green certificate system increases the price of energy.
In the 1970s and 1980s, in world literature, many different methods and techniques were described in the assessment of the environment, both in cognitive and practical aspects. The development of these methods was driven by assessments of the environment without which it is impossible to understand the basic laws of nature and man's relationship with the environment [10]. The concept of external costs was introduced in the theory of economy by Pigou at the beginning of the last century [11]. The development of the ExternE methodology started in 1991 and was first planned as a joint venture between the European Commission and the US Department of Energy. The ExternE methodology is a bottom up methodology assessing the whole life cycle and fuel cycle of a specific plant. It employs an impact pathways approach tracing the emissions from the source to the impact. A wide range of impacts is considered ranging from the impact on human life and health through to visual amenities [12]. One impact which the ExternE cost methodology takes into account is climate change. The impact of global climate change is diverse and possibly immense. The interactions between the global climate system, the ecosystem and the socio-economic system are very complex. Intensive research over the past few years has increased our understanding of the various areas and leads to continual improvement in the corresponding models. The ExternE study by the European Commission shows that specifically, the increase in the mortality risk arising from chronic exposure to particulate matter leads to substantial external costs. The current approach of the ExternE study (ExternE Methodology update 2005) is also used to quantify external costs arising from ill-health. Other external factors arising from reduction of agricultural yield and material damage are quantified according to the ExternE method, but these are very small. Up to now there has been no satisfactory means of expressing, in monetary terms, the impact on biodiversity and ecosystems of SO2 and NOx emissions

due to acidification and eutrophication [13]. As the external costs are normalized to the electricity output of the power plants, the unit is therefore €/kWh.

This paper attempts to estimate the external cost of electricity production in Poland. It turns out that if they were included in the cost of generating electricity from various sources, the final price would differ significantly from that at present. In other words, the external costs are a measure telling us how much we can change the environment for present and future generations.

3. Energy balance and external costs in Poland

Production of electricity in Poland is mainly based on the combustion of coal and lignite. In 2008, the combustion of coal produced 85,8 TWh, whereas the combustion of lignite produced 53,1 TWh (Table 1), which was over 90% of the overall energy production in Poland. As a result of such a large proportion of coal in comparison to other fuels, the weighted mean external cost of electricity production from coal in Poland reached 52,7 €/MWh (Table 2)

Table 1. Electricity production by fuel type in Poland (TWh)

Fuel Type	Year		
	2006	2007	2008
Coal	93,4	92,8	85,8
Lignite	53,4	51,0	53,1
Gas	4,6	4,5	4,1
RES	5,3	5.4	6.4
Others	5,1	5,1	5,0
Total	161,7	158.8	154.4

External costs, dependent on the type of fuel are presented in Table 2. The table contains in detail pollutant quantities. Note the significant decrease in external costs, which are due to running the desulphurization systems [14]. The costs referring to CO2 emissions reflect any losses caused by the greenhouse effect. The smallest part of the total external cost of burning coal and lignite in Poland is dust and NMVOC emissions.

Table 2. External cost of electricity production in Poland (€/MWh) [15]

	Coal		Lignite	
	2007	2008	2007	2008
SO2	28,59	18,52	47,52	31,94
NOx	10,69	10,08	9,12	8,45
PM 2,5 - 10	0,11	0,06	0,11	0,10
PM 2,5	1,24	0,67	0,66	0,60
NMVOC	0,06	0,07	0,05	0,05
CO2	20,27	20,25	24,48	24,60
Total	60,98	40,67	81,94	65,73

4. Renewable energy sources in Poland

4.1. The potential of renewable energy in Poland

In the development strategy of renewable energy sources in Poland the dominant role is played by biomass [16] whose potential is estimated at 152,8 TWh [17]. It should be emphasized that the biomass potential has a wide range of values depending on the source mentioned in the literature. Currently, biomass is mainly used for heat production in the generators of low and medium power, and in the combustion process for electricity generation in power stations.

The potential of hydropower is 12 TWh, and its main resources are focused in the Vistula basin 9270 GWh (77.6%), the Oder 2400 GWh (20.1%) and coastal 280 GWh (2,3%). Recently a significant increase in the use of small hydropower plants with a capacity below 2 MW was noted, due to the simpler administrative procedures. But in the longer term the development of the hydropower sector as a whole should be supported, along with power plants with greater capacity [18].

The real economic potential of wind energy in Poland is about 123,3 TWh, of which 107,4 TWh wind power is located on-shore, and 18,6 TWh off-shore [18]. The current pace of development of wind energy should ensure 24% of the entire share of renewable energy sources in 2020 and reduce its production costs.

The potential of solar energy in Poland is about 23 TWh, of which about 0,2% is currently utilized. This will however increase over the coming years. The market potential for solar energy usage in 2020 is estimated at 5,35 TWh, mainly for heating [19].

4.2. External costs of RES compare to fossil fuels

In Poland, the proportion of energy derived from renewable sources continues to grow and now stands at 12.9% of the country's gross electricity production. The greatest dynamics of growth in recent years has been wind power, however the largest share in electricity production, at about 50%, is from biomass. In the overall RES balance, biomass reached up to 85.8% in 2009. The share of renewable energy sources in electricity production is shown in Table 3.

Table 3. Proportion of renewable energy in Poland [20]

Year	2007		2008	
	GWh	%	GWh	%
Hydro Power	2350.9	43.3	2099.6	32.6
Wind Power	521.2	9.6	817.9	12.7
Biomass	2361.8	43.5	3278.2	50.9
Biogas	195,4	3.6	244.7	3.8
Total	5429,3	100	6440,4	100

Such a large proportion of biomass in the gross electricity production indicates the need for greater support for other renewable sources, while it should be noted that biomass burning is characterized by the largest external costs of all RES as shown in the Figure 1. In Europe the primary energy production share of the various sources of renewable energy varies. However, biomass still has the dominant position, but its percentage is smaller at only 65.6% (compare to 85.8% in Poland), whilst that of water is 21.7%, wind 6,6% and geothermal 5,2% [21].

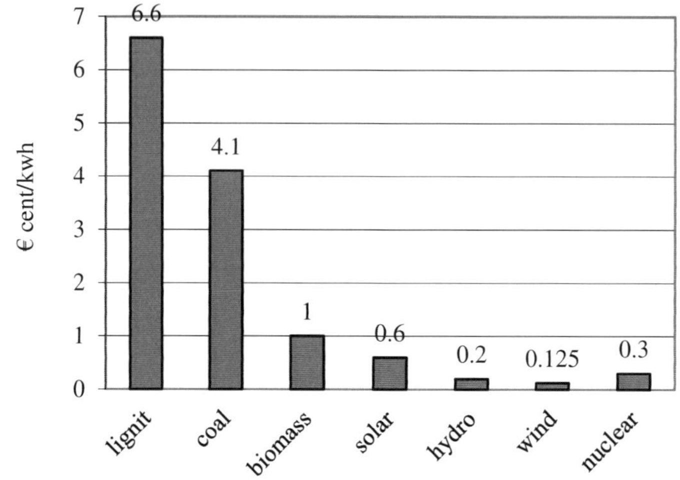

Figure 1. External cost of renewable energy sources and fossil fuels [11], [12], [22]-[24].

4.3. External costs avoided in Poland

The system of green certificates allows for the development of renewable energy sources. In 2008, the annual production of electricity from renewable sources was 6,44 TWh representing 4,1% of the entire electricity production in Poland. Replacing conventional energy with renewable energy sources allows us to avoid environmental losses and as a result leads to economic benefits. Use of renewable sources in 2008 resulted in a saving, in monetary value, of 339,4 million euros in external costs. At the same time the energy from renewable sources, received an additional 386,4 million euros under the green certificate system. Figure 2 demonstrates that the additional price of energy from renewable sources is compensated by a saving of over 87%, in the external costs which should be considered by the final recipient of the energy as a satisfactory level.

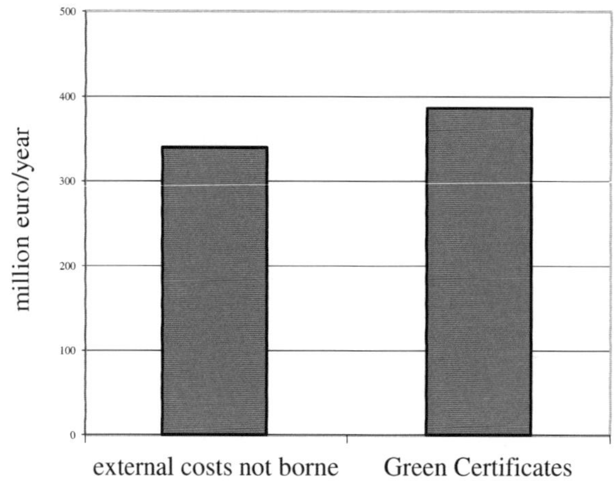

Figure 2. Comparison of Green Certificate amounts with external costs not borne

4.4. Forecast for developing of RES

According to the accepted European Climate and Energy Package, Poland is committed to increasing its share of RES to 15% in 2020. Detailed target forecasts for electricity derived from RES is 10.4% in the years from 2010 to 2012, 12.9% in 2017 and 15% in 2020. Long term forecasts predict a 20% share of RES in 2030 [25]. A new Energy Law was introduced in 2005, which created a support system to assist the development of renewable sources. Under this system, green certificates have been introduced, which subsidize the energy produced from RES.

Table 4. Forecast for the final energy price due to increase of RES share (€/MWh)

Year	2017	2020	2030
Share RES (%)	12.9	15	20
Price of Green Certificates (€)	60	60	60
Amount of increase (€) (share RES x price of Green Certificates)	7.74	9	12
Final energy price (€)	38.25[a]	40.25	43.25
Percentage increase	–	**5,22%**	**13,07%**

[a] current price of energy in Poland 38.25 €/MWh

As a result, energy from renewable sources in Poland is 60 €/MWh more expensive compared to energy from conventional sources. It is important to notice that the increased proportion of renewable energy sources in the gross electricity consumption will increase the final cost of energy. Table 4 shows, for a particular RES percentage a corresponding increase in the cost of energy for the end user. It should be also be mentioned that the amount of subsidies is fixed at the same level for all types of renewable energy sources.

5. Conclusion

One of the most frequently repeated opinions regarding energy from renewable sources, is that it is more expensive than energy from conventional sources. In Poland, as a result of the introduction of green certificates, RES energy is 60€/MWh more expensive. However, that increase is offset by a saving of over 87% of external costs, which society would pay indirectly.

One of the priorities of the Polish energy sector is fulfilling its commitments under the European Climate and Energy Package, including an increase in the share of RES in the overall balance of 15% in 2020, this growth will increase the price of electricity to end users, but despite fears, growth will be modest and will be only about 5.22% whilst the share of renewable energy sources will be 15%. According to growth forecasts for RES of up to 20%, the electricity prices will rise by 13.07% for the end user. However, we should not be concerned about increasing support for RES, which causes an increase in the price of energy, as the current level of support for renewable energy in Poland is one of the highest in Europe.

Insufficient rate of development of RES in Poland could therefore be caused by systemic barriers, rather than by not enough support. In order to accelerate the development of renewable energy sources, it should also take into account the diversity of support depending on the type of renewable energy source. The analysis indicates that the external costs of renewable energy sources are characterized by a wide range of costs. Energy derived from biomass and solar power costs more than 3 times in external costs compared to the cost of energy from water or wind. External costs could be one of the factors taken into account for estimating levels of support under green certificates for each RES. This would mean different levels of support for the different RES.

Current projections indicate that Poland will be dominated by lignite-fired and coal-fired power stations for the foreseeable future. Significant development of renewable energy sources in Poland is

expected between 2010-2020. Analysis of external costs in determining the future directions of energy generation in Poland indicates the need to reduce electricity production from lignite. At the same time technology to remove sulphur dioxide should be developed in order to reduce external costs. Constructions in 2008, of the flue gas desulfurization system in Poland, helped to reduce the external cost of the coal and lignite combustion by more than 20% per MWh.
During the analysis of external costs it is also worth considering obtaining energy from nuclear power stations, whose operation involves very low external costs which are comparable with renewable energy sources. However, nuclear power still causes a lot of social fears in Poland. An information campaign is needed to convince the public. In this case, presenting the results of the analysis of external costs could also be very helpful.

6. References

[1] Lindzen R S 2010 Global warming: the origin and nature o the alleged scientific consensus. *Problemy Ekorozwoju* [Problems of Sustainable Development] **5** 13-28
[2] Paryjczak T 2008 Promoting Sustainability through Green Chemistry *Problemy Ekorozwoju* [Problems of Sustainable Development] **3** 39-44
[3] Paryjczak T 2008 Promoting Sustainability through Green Chemistry *Problemy Ekorozwoju* [Problems of Sustainable Development] **3** 45-51
[4] Pawlowski A 2008 The role of social sciences and philosophy in the shaping of the of sustainable development concept *Problemy Ekorozwoju* [Problems of Sustainable Development] **3** 3-7
[5] Pawlowski A 2009 The sustainable development revolution *Problemy Ekorozwoju* [Problems of Sustainable Development] **4** 65-76
[6] Laszlo Ch 2008 Sustainable value *Problemy Ekorozwoju* [Problems of Sustainable Development] **3** 25-29
[7] Pawłowski A 2009 Sustainable energy as a sine qua non condition for the achievement of sustainable development *Problemy Ekorozwoju* [Problems of Sustainable Development] **4** 3-7
[8] Borys T 2010 Decade of education for sustainable development *Problemy Ekorozwoju* [Problems of Sustainable Development] **5** (1): 59-70
[9] Cel W, Pawlowski A and Cholewa T 2010 Slad weglowy jako miara zrownowazonosci odnawialnych zrodel energii [Carbon footprint as measure of sustainabilityof renewable sources of energy] *Scientific Conference in Bydgoszcz Poland* Bydgoszcz Poland
[10] Jozwiak M 2000 Zintegrowane wskazniki stanu srodowiska przyrodniczego [Integrated indicators of the state of the natural environment] *Regionalny Monitoring Środowiska Przyrodniczego* **3** 25-27
[11] Hamacher T, Cabal H, Hallberg B, Korhonen R, Lechon Y, Saez R M and Schleisner L 2001 External costs of future fusion plants *Fusion Engineering and Design* **54** 405-411
[12] Hamacher T, Saez R M, Aquilonius K, Cabal H, Hallberg B, Lechon Y, Lepicard S, Schleisner L, Schneider T and Ward D 2001 A comprehensive evaluation of the environmental external costs of fusion power plant *Fusion Engineering and Design* **56** 95-103
[13] Krewitt W and Schlomann B 2006 External costs of electricity generation from renewable energies compared to electricity generation from fossil energy sources [online] Available on the Internet: www.bmu.de/erneuerabare_energia/ dowlands/do c/37085.php
[14] Strupczewski A 2010 Koszty *zewnetrzne wytwarzania energii elektrycznej w Polsce* [External costs of electricity production in Poland] Available on the Internet: www.elektroonline.pl/a/1032, Koszty zewnetrzne wytwarzania energii elektrycznej w Polsce Energetyka
[15] Radovic U 2009 Porownanie wplywu na zdrowie człowieka i srodowisko naturalne roznych zrodeł energii – wyniki badan w programie externe [Comparison of effects on human health and the environment of different energy sources - the results of research in the externe] *Energy Market* Warsaw Agency Warsaw Poland
[16] Panoutsou C, Eleftheriadis J and Nikolaou A 2009 Biomass supply in EU27 from 2010 to 2030 *Energy Policy* **37** 5675-5686

[17] Szyszlak-Bargłowicz J and Piekarski W 2009 Wykorzystanie biomasy jako elementu bezpieczenstwa energetycznego [The biomass as an element of energy security] *Wiadomości Elektrotechniczne* **3** 22-23

[18] Steller J 2009 Energetyka wodna w Polsce [Hydropower in Poland] *International Scientific Conference* Lomza Poland 17-18 April 2009

[19] Wisniewski G and Michalowska-Knap K 2008 Stan obecny i potencjal energetyki odnawialnej w Polsce [Current status and potential of renewable energy in Poland] *International Conference - renewable energy in the EU and Poland by 2020* Warsaw Poland

[20] Main *Statistical Office in Poland* 2010 Energia ze zrodel odnawialnych 2009 [Renewable energy 2009] Available on the Internet: www.stat.gov.pl. ISSN 1898-4347

[21] *European Commission [online]* www.ec.europa.eu

[22] Nisan S and Benzarti N 2008 A comprehensive economic evaluation of integrated desalination systems using fossil fuelled and nuclear energies and including their environmental costs *Desalination* **229** 125-146

[23] Owen A D 2006 Renewable energy: Externality costs as market barriers *Energy Policy* **34** 632-642

[24] Streimikiene D, Roos I and Rekis J 2009 External cost of electricity generation in the Baltic States *Renewable and Sustainable Energy Reviews* **13** 863-87

[25] Duda M, Mikolajuk H and Okrasa S 2009 Prognoza bilansu energetycznego Polski do 2030 r. [Forecast of Poland's Energy Balance to 2030] *In the 23rd Conference of energy raw materials and energy in the national economy* Zakopane Poland 2009 ISBN 978-83-60159-57-4

IOP Conf. Series: Earth and Environmental Science **150** (2018) 012029 doi:10.1088/1755-1315/150/1/012029

Sustainability aspects of biofuel production

L Pawłowski[1], W Cel[1] and K Wójcik Oliveira[1]

1 Faculty of Environmental Engineering, Lublin University of Technology, Nadbystrzycka Street 40B, Lublin 20-618, Poland
E-mail: l.pawlowski@pollub.pl

Abstract. Nowadays, world development depends on the energy supply. The use of fossil fuels leads to two threats: depletion of resources within a single century and climate changes caused by the emission of CO_2 from fossil fuels combustion. Widespread application of renewable energy sources, in which biofuels play a major role, is proposed as a counter-measure. The paper made an attempt to evaluate to what extent biofuels meet the criteria of sustainable development. It was shown that excessive development of biofuels may threaten the sustainable development paradigms both in the aspect of: intergenerational equity, leading to an increase of food prices, as well as intergenerational equity, resulting in degradation of the environment. The paper presents the possibility of sustainable biofuels production increase.

1. Introduction

The idea of sustainable development was created by Brundtland. It calls for such development that meets the needs of the present generation without compromising the ability of future generations to meet their own needs. The main goals to be reached in the frame of sustainable development are:

- Ensuring the availability of basic goods which are essential for life of the present and future generations, including an equitable access to raw materials [1]-[5].
- Ensuring the caring ability of the environment to sustain the existence of human civilisation, especially in term of the environment quality [6]-[8].

Although this definition of sustainable development was formulated in 1987, the roots of the idea can be found in the works of Hans-Carl von Carlowitz, who in 1713 suggested the trees feeling should only be carried out on such a degree that makes it possible to regrow them in the foreseeable future.
The development of human civilisation was based on the use of natural resources in an uncontrolled manner, without taking into account degradation of the environment and what is as well important without considering an interest of the future generation. Generally, sustainability is based on the integration of environmental management social and economic goals, mainly reduction of poverty. The main aim of sustainability is to interlink the ecological and socio-economical border in pathway of developments to assure the ecological, social and economic stability [5], [8]-[12]. Only the development that is based on the consumption of resources which does not lead to their exhaustion and offering their use for all people constitutes an important paradigm of the global justice. Undoubtedly, energy supply is and will be one of the key factors governing the development or even survival of the human civilisation. Its consumption has grown over the last decade by 15,4% (see Table 1).
In the case of Poland, the consumption of primary energy grew rapidly from 1950 to 1980 due to the development of mostly heavy industry, similarly to other socialist countries. After the transitioning from the socialist to capitalist system, which started in 1990 through the introduction of so-called Balcerowicz Plan, a number of industrial plants went bankrupt, resulting in a significant decrease in

the consumption of primary energy in 1990-2015 from 100.6 Mtoe to 94.6 Mtoe/year, i.e. by 5.7% (see Table 2).

Table 1. Characteristic of global primary energy consumption 2006-2016.

Year	Total primary energy	Total renewable energy		Total biofuels	
	Mtoe	Mtoe	%	Mtoe	%
2006	11 267	93	0,83	28	0,25
2007	11 627	107	0,92	38	0,33
2008	11 784	123	1,04	50	0,42
2009	11 602	144	1,24	56	0,48
2010	12 170	170	1,40	64	0,53
2011	12 455	204	1,64	66	0,53
2012	12 621	239	1,90	67	0,53
2013	12 866	280	2,18	72	0,56
2014	12 989	317	2,44	80	0,62
2015	13 105	367	2,80	80	0,61
2016	13 276	402	3,03	82	0,62

Table 2. Total consumption of primary energy in Poland (Mtoe).

Year	Mtoe
1950	28.6
1960	54.4
1970	84.7
1980	124.5
1990	100.6
2000	89.8
2006	94.1
2007	92.7
2008	95.4

Table 3. Characteristic of primary Energy consumption and CO_2 emissions in Europe and selected countries (Listed from the biggest emitters of CO_2 per capita).

Europe	$MgCO_2$/capita/year	Primary energy consumption, Mtoe/year
Czech Republic	10.7	43.4
Netherlands	10.5	77.4
Belgium	9.9	59.1
Germany	9.1	311.8
Austria	8.1	33.0
Poland	7.8	101.9
United Kingdom	7.1	188.1
Italy	6.5	167.4
Spain	5.9	125.6
France	5.0	252.8
Sweden	4.8	49.0
	Other biggest emitters of CO_2	
Australia	17.4	122.9
USA	16.9	3191.2
Canada	15.4	251.9
Korea	11.8	260.4
Russian Federation	11.7	731.8
China	5.9	2727.7
Mexico	4.0	186.2

The best parameter characterizing the impact of primary energy consumption by individual countries is the amount of annual CO_2 emissions per capita. The selected EU countries and other biggest emitters of CO_2 are depicted in Table 3. The data shows that the highest CO_2 emissions per capita come from Australia, the USA and Canada, while the lowest come from Mexico, Sweden, France, Spain and China, although the total CO2 from China is the highest (10.6 mln Mg CO_2/year, i.e. 29.31% global emissions in 2015), due to the largest population. For example, the CO2 emissions from Poland per capita per year account for 7.8 Mg CO_2 per capita. Burning primary energy sources is responsible for 3 main air pollutants in Poland, i.e. SO_2, NO_x and CO_2. Due to the economic changes caused by the transition from the socialist to capitalist system in 1990, a substantial decrease in SO_2, NO_x and CO_2emission occurred, mainly resulting from the bankruptcy of large industrial plants in 1990-2000 (see Table 2).

As was mentioned, the problem with energy is not only its effect on climate change but as well exhaustion fossil fuels reserves.

According to the forecasts, the estimated reserves of coal are approximately 1140 billion Mg and will suffice for about 114 years, the reserves of crude oil amount to 1701 billion barrels and will be enough for roughly 51 years, whereas the reserves of natural gas in amount of 190 trillion m3 which will be sufficient for 52 years [13]. Therefore, developing the renewable energy sources is necessary not only from the point of view of climate protection, but also to ensure new energy sources for our civilisation. The development of renewable energy may mitigate these threats. Global use of renewable energy grows from 93 Mtoe = (0.83%) in 2006 to 402 Mtoe (3.03%) in 2016 (see Table 1 and 2)

One of primary energy resources are biomass of which biofuel becomes the most important. Its consumption is growing from 28 Mtoe (0.25%) in 2006 to 82 Mtoe (0.62%) in 2016 [13]. Application of biomass as a primary energy source varies greatly between the developing and the developed countries. The developing countries mainly utilize biomass – especially wood – as one of the basic sources of primary energy, whereas the developed countries are developing advanced methods of biomass use for energy purposes, primarily including the production of biogas and liquid biofuels, i.e. ethanol and biodiesel. These advanced biofuels are usually produced with biomass from special plantations [13]-[17].

While having a discussion about the environment, we usually think about the natural environment and ignore the human environment, which includes an equally important component of socio-economic relations. The development of biofuels exerts a significant influence on the functioning of societies [18], [19].

2. Socio-economic consequence of biofuel development.

Focusing the energy policy only on mitigation of the CO_2 emission threatens the sustainable development paradigm i.e. intra-generational equity, which demands equitable access to food for all people. The food demand will continue to increase for two reasons: growing human population and greater number of better-fed people. At present, 250 thousand people starve to death each day, while 780 million people in developing countries and 27 million people in developed countries are malnourished. In such situation, devoting large areas of land for the biomass used for fuel production raises moral concerns. This is especially relevant in the context of biofuels from agricultural crops. According to the European Commission decision from 2009, as much as 10% of energy used in transportation should be obtained from biofuels. In order to make biofuels a viable solution, European governments subsidize powerful industrial and agricultural lobbies. For instance, by 2020 each person in Great Britain will pay approximately £35 per year (£1-2 billion in total) for biofuel subsidies, while in Germany – roughly €30 (€1.4–2.2 billion in total).

Also in the United States, the production of ethanol (mainly from corn) which is used as fuel additive developed rapidly thanks to subsidies. In 2011, as much as 127 million tons of corn, i.e. 40% of annual production was devoted for bioethanol production. Allocating such

a great amount of corn for ethanol production in 2007–2012 caused a 100% increase in the prices of corn. Large import of crops for biofuels by the European Union caused a dramatic 2.5-fold increase in the food price index. The increase in food prices is especially severely felt by poor people, who spend most of their income on this basic commodity.

In the light of the above-mentioned statements, utilizing biomass from agricultural crops as a source of energy endangers the implementation of sustainable development, as it violates the intra-generational justice paradigm by limiting the access to food for the poor.

3. Environmental consequences of biofuel development.

The promotion of biofuels is based on wrong assumption that their combustion releases equal amount of CO_2 to the one absorbed earlier by plants. This oversimplification does not take into consideration the entire biofuel production cycle. The effects of land use change and energy outlays for the cultivation and processing of biomass used in fuel production are omitted in the calculations [17].

A pressure put on the usage of biofuel in transportation, exerted especially by the European Union policy, resulted in the deforestation of tropical forests to clear land for cultivation of biofuel plants – especially in the developing countries [18]. According to the research conducted by Danielsen [13], the absorption of CO_2 by tropical forests is much greater than the one of plants grown in their place. Consequently, the biofuel production decreases CO_2 absorption on the lands of felled tropical forests. Transforming tropical forests for the cultivation of biofuel plants leads to an additional emission of 55 Mg CO_2 annually per hectare for the period of 120 years.

Moreover, in order to produce biofuel, e.g. corn bioethanol, it is necessary to provide energy for the cultivation, fertilizer production, plant collection, as well as fuel processing through fermentation and distillation. By using the life cycle assessment technique, it was shown that the amount of the emitted CO_2 per unit of energy obtained from corn ethanol is 60% greater in relation to the CO_2 amount emitted in combustion of the equivalent petroleum-based fuels. Even in the case of sugar cane ethanol produced in Brazil, where the industry is most advanced and where the leftover biomass is entirely used, e.g. sugar cane stalks are burned for heat energy; it was not possible to decrease emission of CO_2 to a lower level than the one of oil-based liquid fuels. However, in the case of Brazil, the development in sugar cane ethanol helped to create approximately 700 thousand new jobs, which can be considered a positive effect increasing the social sustainability. It allowed Brazil to become independent from the liquid fuels import, and the price of ethanol is competitive to the one of petrol. In Brazil, it was possible to achieve sustainable access to liquid fuels for transportation.

The full analysis of biofuel production cycle showed that cultivation of some plants for biofuel production, such as rape, requires application of large amount of fertilizers which increase the emission of another greenhouse gas; namely, nitrous oxide, thus contributing to the greenhouse effect. In the case of rape, the nitrous oxide may increase the greenhouse effect by as much as 70%. The development of biofuel plant cultivation destroys habitats and biological diversity. The forests in south-eastern Asia, which provide habitats for a plethora of various organisms, are the most endangered by the biofuel plant plantations. It is estimated that the number of species living in the area decreases fivefold when forests are cut and turned into biofuel plantations.

Production of liquid biofuels for the transportation also has a negative impact on the aquatic environment due to a high consumption of water both for watering plants and during their processing to biofuels. Moreover, the processing involves production of large quantities of hazardous wastewater, e.g. production of 1 litre of ethanol generates to 6–12 litres of highly contaminated wastewater. Meanwhile, water shortage in that regions is already negatively influencing food production.

Generally, approximately 2500 litres of water are used in order to obtain 1 litre of biofuel. Such amount of water is enough to produce food for one person. In order to irrigate 30.000.000 hectares of land for biofuel cultivation, roughly 180km3 of fresh water will need to be used. One must bear in mind that due to the growth of population to 8.3 billion in 2030 (from 7.2 billion in 2012), food, water and energy demand will increase by 35%, 40% and 50%, respectively. Large monocultures, which are usually used in the case of biofuel plants, require wide application of herbicides and pesticides, which then infiltrate to ground waters – contaminating them. Soybean farming in Brazil is an example of a negative effect of pesticides. Along with herbicides, they are used on a large scale in the Pantanal wetland, which constitutes one of the most important habitats for hundreds of birds, mammals and reptiles. Another example includes 20,000 hectare sugar cane plantation, intended for ethanol production, which is located in the delta of Tana River in Kenia. With the planned water uptake of

1680 m^3 water/min, which equals 30% of river flow rate, it seriously threatens the local ecosystem which is a habitat to 345 species of water and marsh birds.

4. Conclusions

1. Attention was drawn to the fact that the development of civilization is threatened not only by the excessive CO2 emission from the combustion of fossil fuels, but also by their depletion.
2. In such case, it is necessary to develop other energy sources, including renewable ones.
3. There is no doubt that the development of renewable energy sources may mitigate climate changes on the one hand and ensure the adequate supply of energy for the proper functioning of our civilization on the other.
4. It was pointed out that the cultivation of energy crops for the purpose of biofuel production may also have negative consequences, because:
- rain forests, which absorb significant amounts of CO2, are cut down in order to acquire land for the cultivation of energy crops,
- increasing the area of subsidized energy crops cultivation will limit the food production, resulting in the increased food prices.

5. References

[1] Velkin V I and Shcheklein S E, 2017, Influence of RES integrated system on energy supply improvement and risks, *Problemy Ekorozowju/Problems of Sustainable Development*, **12**, 123-9.
[2] Pawłowski A, *Sustainable Development as a civilizational revolution. Multidimensional approach to the challenges of the 21st century*, CRC Press, Taylor & Francis Group, Balkema Book, Boca Raton, London, New York, Leiden, 2011, 240.
[3] Dłużewska A, 2017, Wellbeing versus sustainable Development- conceptual framework and application challenges, *Problemy Ekorozowju/Problems of Sustainable Development*, **12**, 2, 89-97
[4] Palanisamy K and Parthasarathy K, 2016, Urbanization, food insecurity and agriculture – challenges for social sustainable development, *Problemy Ekorozowju/Problems of Sustainable Development*, **12**,1, 157-162.
[5] Rydzewski P, 2017, General Social survey and sustainable development. Methodological and empirical aspects. *Problemy Ekorozowju/Problems of Sustainable Development*, **12**, 1, 21-26
[6] Berkes F, Folke C and Colding J, 2000, *Linking social and ecological systems*, Management practices and social mechanisms for building resilience, Cambridge, University Press
[7] Pimentel D, 2012, *Global Economic and environmental aspects of biofuels*, CRC Press.
[8] Mason L R, 2015, Environment and Social development, Cen[16] ter for Social Development, CSD Working Papers, no. 15-33
[9] Adams W M, 1990, Green Development 2ng edition, *Environment and sustainability in the third world, The dilemma of sustainability*. Taylor & Francis.
[10] Meadows D H, Randers J and Meadows D L, 2004, *The limits to growth: The 30-years update*, Chelsea green Publishing, White river Jct., Vermont.
[11] Konarski W, 2014, Mineral energy source and political activities: introduction to mutual dependencies and their selected exemplification, *Problemy Ekorozowju/Problems of Sustainable Development*, **9**, 1, 63-70
[12] Cao Y and Pawłowski A, 2013, Life cycle assessment of two emerging sewage sludge to energy systems: Evaluating energy and greenhouse gas emissions implications, *Bioresource Technology*, **127**, 81-91,
[13] Danielsen F, Beukema H, Burgess N D, Parish F, Brühl C A, Donald P F, Murdiyarso D, Phalan B, Reijnders L, Struebig M and Fitzherbert E B, 2009, Biofuel plantations on forested lands: double jeopardy for biodiversity and climate, *Conserv Biol.*, **23**,2, 348-58
[14] Report 2015, *The impact of biofuels on transport and the environment, and their connection with agricultural development in Europe*, Directorate-General for Internal Policies,
[15] Searchinger T, Use of U. S. Croplands for biofuels Increases Greenhouse Gases Through emissions from Land-Use change, 2008, *Science*, **319**, 1238

[16] Fargione J. et al., 2008, Land Clearing and the Biofuel Carbon Debt, *Science*, **319,** 1235
[17] Young ChA, 2009, Biofuels boom could fuel rainforest destruction, Stanford Report
[18] Gibbs H K and Johnston M, 2008, Carbon payback times for crop-based biofuel expansion in the tropics: the effects of changing yield and technology, *Environmental Research Letters*, **3**, 3
[19] Dragomirescu H and Bianco L, 2017, Tackling Sustainability from a Systemic Perspective: A Contextualized Approach, *Problemy Ekorozowju/Problems of Sustainable Development*, **12**, 1, 31-39

A Comparative of Life Cycle Assessment of a Conventional Van and a Battery Electric Van for an Online Shopping System in Thailand

J Koiwanit[1*] and C Hamontree[1]

[1] Faculty of Engineering, King Mongkut's Institute of Technology Ladkrabang, Bangkok 10520, Thailand
E-mail: Jarotwan.ko@kmitl.ac.th
*Corresponding author

Abstract. The transportation sector is responsible for one of the main emitters of large quantities of pollutions to the atmosphere, which impacts local, regional or global environment receptors. In Thailand, many retail chains have been trying to launch many campaigns and projects to reduce GHG emissions together with offering the best convenience services to serve customers' needs. By promoting an online shopping system for the workplace, this will mitigate even more of the air pollutants than the conventional online shopping system, where the products are delivered to customer's doorsteps. This study aims to investigate and compare the impact of different vehicle technologies for an online shopping using Life Cycle Assessment (LCA) methodology especially in the vehicle use phase. The observed results showed that the electric van has the potential of reducing emissions and consequently showed lower impacts in most impact categories.

1. Introduction

In recent decades, climate change has had the strongest and most comprehensive impact to natural systems [1]. Greenhouse gas (GHG) emissions are known to contribute to the climate change and carbon dioxide (CO_2) is a major GHG [2]. In Thailand, transportation was ranged as the second largest contributor to CO_2 emissions [3]. Not only does road transportation contribute to the high costs of fuel consumption, but it will also lead to high GHG emissions and other environmental impacts. In 2015, there were approximately 39.3% of Thailand's population using the internet and this results in an increase in online shopping system [4]. To cope with this issue, the use of an online shopping system has become an approach in ensuring lower emissions and thus alleriate traffic congestion. Furthermore, this consequently will benefit its users via lowered travel costs and time consuming [5]. However, instead of delivering products individually to each customer, retailers can deliver to workplaces at a certain time. Employees can take their groceries from their workplaces directly home instead of having to take individual trips to the grocery stores [6]. Previous research [6] has been done on evaluating and comparing the environmental impacts of the vehicle's fuel consumption of the online shopping system with the traditional shopping system. In this study, the use phase of two types of vehicles, which include 1) battery electric vans and 2) conventional vans has been modelled to compare the life cycle impact using an LCA methodology.

As complexity of a product especially vehicles have a significant impact on the complexity of the LCA study, LCA on both conventional and battery electric vans (BEVs) are a challenging task. In recent decades, EVs using the LCA methodology have been studied [7]-[11], while [10], [12]-[15]

investigated electric vehicles (EVs) on specific components, especially Li-ion batteries. Most LCA studies focus on EVs production and their raw material acquisition, there is a few focusing on the use phase [16], [17]. In contrast to the previous LCA studies, this study attempts to model the use phase scenarios based on real life previous research data [6]. This study is a gate-to-gate study focusing mainly on culminating in the creation of a use phase database for conventional and battery electric vans on an online shopping system for workplace delivery at King Mongkut's Institute of Technology Ladkrabang (KMITL) in Bangkok, Thailand. Center of Environmental Science of Leiden University (CML) 2001, the life cycle impact assessment (LCIA) method, is used to convert Life Cycle Inventory (LCI) data into 11 environmental impacts. This database is currently unavailable and few consider life-cycle impacts. As a result, this study will fill the gap and a useful database will be produced.

2. Implementation of LCA methods

2.1 Goal and scope definition
The goal of this LCA is to study environmental performance over the life cycle of an electric designed van compared to a conventional designed van. The main objectives of this study are:

- To have a complete understanding of the emissions, wastes, resource use and energy consumption in the use phase of the delivery vehicles.
- To improve our understanding of all related processes in road transportation for emission reductions.
- To encourage the increased use of an electric designed van as a solution for CO_2 emissions reduction for any retailer chains or logistic companies.

2.2 Functional unit
The most common way to compare the life cycle impact of one vehicle to another is to measure emissions per kilometer driven over their lifetimes. This is a common functional unit for comparing different forms of transportation that aims to carry a different number of passengers. Therefore, the functional unit selected in this comparative study is 1 km travelled. Even though running fuel is computed differently during on and off-peak times, this study assumed that the average electricity consumption rate is constant at any period of time.

2.3 System boundaries
This study is a gate-to-gate study that takes into account fuel type and fuel consumption through the vehicle use. The system boundaries of the electric and conventional designed van. The reference case (conventional van) system boundaries include production and usage of fossil fuel. The electric van system boundaries include production and usage of electricity from battery.

2.4 Temporal and geographical boundaries
In this study, a vehicle's lifetime of 12 years and a lifetime driven distance of 230,500 kilometers were assumed. These number are the default expected life span values for a typical vehicle and car used in most research [9], [18], [19].

2.5 *Technological boundaries*
This study evaluates modern road-vehicle technologies and their emission impacts. There are two systems analyzed, which include: (i) the electric designed van intended for use in urban areas, and (ii) the conventional designed van.

3. Life cycle inventory analysis (LCI)
The majority of data are specific to Bangkok/Thailand. The assumptions about energy consumption and emissions emitted from the electric designed van are mainly derived from [6], [20], and Ecoinvent. Nevertheless, if there is still not enough data available to conduct the study of the use phase of the vehicles in Thailand, the data is taken from several literature reviews shown in Table 1.

Table 1. Main operating parameters of electric vans.

Parameter	Value	Unit	Reference/comment
Driving behavior	-	-	Cautious
Type of road	-	-	City
Slope of the road	-	-	Flat
Acceleration	0	m/s^2	
Total weight (kg)	3500	kg	www.targetmovethailand.com
Rolling friction of the tires	0.2	-	www.engineeringtoolbox.com
Temperature at time of the travel	35	c	Expert's estimate
Density of air at 35 °C	1.15	$kg.m^{-3}$	www.engineerstudent.co.uk
Aerodynamic drag coefficient	0.38		[21]
Frontal area of the vehicle	1.8 × 2.2	m × m	Expert's estimate
Velocity	16.7	m/s	Expert's estimate
Energy consumption	**0.29**	**kWh/km**	**Calculation based**

3.1. Description of e-shopping system for workplace delivery using EVs

Online shopping system in Thailand continues to increase because of the Thai consumer new lifestyles, high competition, and information technology. Many retailers launch their online shopping system at customers' doorsteps but none have done the online shopping system for a workplace delivery to reduce more number of kilometres travelled, compared to a conventional online delivery system. In order to be environmentally friendly, different vehicle technologies can be considered. Electric vehicles (EVs) are one of the alternative options to reduce the amount of fossil fueled usage in today's conventional vehicles. EVs also reduce noise and eliminate tailpipe emissions.

As the range of EV's energy consumption depends on many parameters, it is variable. However, the energy consumption of EVs can be decided into 3 main aspects [7], [20], [22]:

- The driving resistances must be overcome to achieve and put the vehicle into movement.
- The use of auxiliaries such as light, air conditioning, radio, navigation, ventilation, etc.
- Energy losses in the process of converting electric energy of the battery into mechanical energy at the wheels of the vehicle.

In this study, in order to model delivery vehicles, a designed van was chosen as to drive their way from Srinakarin storage location to KMITL and driving resistances were taken into account.

3.2. Description of e-shopping system for workplace using conventional delivery vans

Based on the previous work [6], the conventional designed vans are based on Tesco Lotus's vans in which there are 3 separate compartments. These compartments are for frozen (15 baskets), chilled (15 baskets) and non-perishable products (45 baskets) and this in total can carry 75 baskets per trip as shown in Figure 1.

Figure 1. Tesco Lotus's designed vans (shoponline.tescolotus.com)

Based on the interviews in the previous work, it can be concluded that each faculty member will get around one and half basket each week and most of them are non-perishable products. The study assumed that if all new members which are 23 buy products for 2 baskets each week, the designed van can deliver their products to the institute one time per week. The travel distance between the closet storage location and the institute is 34 km or 68 km in return. According to [23], the average is 22 mile per gallon which is equal to the fossil fuel usage at 0.107 liters per km.

4. Life Cycle Impact Assessment (LCIA)

Since this study aims to evaluate the overall environmental impact associated with each unit process of an electric designed van and a conventional designed van specific to Bangkok/Thailand, the accurate evaluation should be represented by the life cycle impact assessment methodology best suited for the Thai environment. This study employs CML 2001 since this method is well-established, comprehensive, and commonly used. In addition, GaBi 7, an LCA software product of PE International, Germany, is used in this study since this software is an easy-to-understand structure and friendly user interface.

5. Results

The two scenarios were compared based on these environmental impact categories and the results were analyzed. In terms of the electric vans, because no emissions emitted during the use stage, all contribution from any impact categories were derived from electricity generation. The percentage change (from the conventional van) in environmental impacts categories for the electric van is shown in Figure 2. The analysis of the results of LCA study is presented below.

5.1. Global Warming

The impact to the category of global warming was decreased by 57.39% in the electric designed van compared to the conventional designed van. The main reduction is mainly derived from using electricity instead of fossil fuel. The main substances that contributed to the global warming category include CO_2, methane (CH_4), and nitrous oxide (N_2O), which accounted for 95.1%, 2.48, and 1.08%, respectively. In the "conventional vehicle" scenario, around 83% of the contribution was derived from the car using stage and only 17% from gasoline mix at refinery.

5.2. Abiotic Depletion (ADP elements)

The results showed the impact to abiotic depletion considering ADP elements in the electric designed van is 7.83E-08 kg Sb-Equiv., which is a 32.5% increase for the electric designed van system. The main substances that caused this abiotic depletion impact category include Copper (Cu), silver, and Lead (Pb) which accounted for 24.3%, 21%, and 18.2%, respectively of the impact in this category. The increased impact of abiotic depletion (ADP elements) is mainly due to the electricity needed for operation of the vehicle system. In the "conventional van" scenario, gasoline mix contributed 100% to this impact category.

5.3. *Abiotic* Depletion (ADP fossil)
A total number in the environmental impact category of abiotic depletion considering ADP fossil is 1.73 MJ in the electric van system. The results showed that the impact to the abiotic depletion (ADP fossil) is reduced by approximately 66.4% for the electric designed van system compared to the baseline scenario. The gasoline mix for the conventional van and the electricity generation system are the primary sources for abiotic depletion. The dominant contributors to the abiotic depletion were lignite and hard coal, which accounted for 47.3% and 33.3%, of the contribution in this impact category, respectively. In the "conventional van" scenario, gasoline mix contributed 100% to this impact category.

5.4. *Acidification Potential (AP)*
The results showed that the total impact of the electric designed van system is 2.52E-04 kg SO_2-Equiv. and this has better performance compared to the conventional system. The impact to the acidification potential of the electric designed van scenario is reduced by 96.3% compared to the baseline scenario. The main substances contributing to the acidification potential impact category are hydrogen sulphide (H_2S), ammonia (NH_3), and hydrogen chloride (HCl), which accounted for 79%, 25.9%, and 24.9%, respectively. In the "baseline" scenario, around 53.2% of the contribution was from vehicle using stage and 46.8% was from operation of the gasoline.

5.5. *Eutrophication Potential (EP)*
The results showed that the total impact to the eutrophication potential is 4.01E-05 kg Phosphate-Equiv. The impact is decreased by 97.9% for the electric designed van system compared to the baseline scenario. The impact is mainly due to less amount of NO_x and N_2O emissions emitted into the air, which accounted for 57.2% and 4.84%, respectively. In the "baseline" scenario, the vehicle using stage and gasoline mix contributed 52.9% and 47.1% respectively to the impact in the acidification category.

5.6. *Freshwater Aquatic Ecotoxicity Potential (FAETP)*
In the electric designed van scenario, the impact to the freshwater aquatic ecotoxicity is decreased by 93.3% mainly due to heavy metals emitted into the fresh water. The total impact was 2.34E-04 kg DCB-Equiv. The main substances that contributed to the freshwater aquatic ecotoxicity were Nickel (Ni) and Vanadium (V), which accounted for 40.6% and 22.2%, respectively. In the "baseline" scenario, 97% of the contribution was derived from vehicle using stage and 3.04% from the gasoline mix at refinery.

5.7. *Human Toxicity Potential (HTP)*
The electric designed van system shows the total impact in human toxicity potential which is 5.82E-03 kg DCB-Equiv. The total impact is reduced by 98.7 % compared to the baseline scenario. It was observed that the heavy metals (e.g. As, Cd, Cu, VOCs), and different inorganic emissions (e.g. hydrogen fluoride (HF) and NO_x) are emitted to atmosphere. In the "capture" scenario, 59.3% of the contribution was derived from vehicle using stage and 40.7% from the operation of gasoline mix.

5.8. *Marine Aquatic Ecotoxicity Potential (MAETP)*
The online shopping system using the electric designed van showed a total marine aquatic ecotoxicity potential impact of 19.9 kg Ethene-Equiv. Compared to the baseline scenario, the potential for marine aquatic ecotoxicity in the electric vehicle scenario increases by 235.2%. Electricity generating system of the electric van contributed 100% to the marine aquatic ecotoxicity impact category. HF from electricity generating station for the electric designed van accounted for 98.1% of the contribution in this impact category. In the "baseline" scenario, almost 100% of the eutrophication potential was derived from vehicle using stage.

5.9. Ozone Depletion (ODP)
Compared to the conventional designed van system, the results showed the impact to ozone depletion is increased by 205.02% for electric vehicle system. The total result of this impact category is 1.05E-11 kg R11-Equiv. The increased impact of ozone depletion is mainly due to the higher impact on electricity generation system compared to the operation of the gasoline. The dominant contributors to this impact category were dichlorotetrafluoroethane (R114) and dichlorodifluoromethane (R12) and, which accounted for 97.8% and 0.01%, respectively. In the "baseline" scenario, gasoline mix contributed 100% to this impact category.

5.10. Photochemical Ozone Creation Potential (POCP)
The results from the CML 2001 method indicated that the total impact of the photochemical ozone creation impact category is 1.88E-05 kg Ethene-Equiv. Compared to the conventional vehicle system, the impact is reduced by 102.9% for the electric vehicle system. The decreased impact is attributed to the less amount of emissions emitted into the air from electricity generation system compared to gasoline mix. The main substances that caused photochemical ozone creation include dichlorotetrafluoroethane (R114) and chlorodifluoromethane (R22) which accounted for 97.8% and 2.97%, respectively of the impact in this category. In the "baseline" scenario, the photochemical ozone creation impact was due to gasoline mix at refinery and vehicle using stage, which accounted for 54.4% and 45.6% of the impact, respectively.

5.11. Terrestrial Ecotoxicity Potential (TEP)
The results showed that the total potential for terrestrial ecotoxicity in the electric van system is 1.31E-04 kg DCB-Equiv. The impact to this impact category is increased by around 101% for the electric van system compared to the baseline scenario. The impact is derived from the trace elements collected from the electricity generation system. Heavy metals, which include mercury (Hg), chromium (Cr), As were the main substances that contributed to terrestrial ecotoxicity, which accounted for 78.5%, 5.31%, and 3.36%, respectively. In the "baseline" scenario, the processes of vehicle usage and gasoline mix contributed 81.4% and 18.6% respectively to the impact on the terrestrial ecotoxicity impact category.

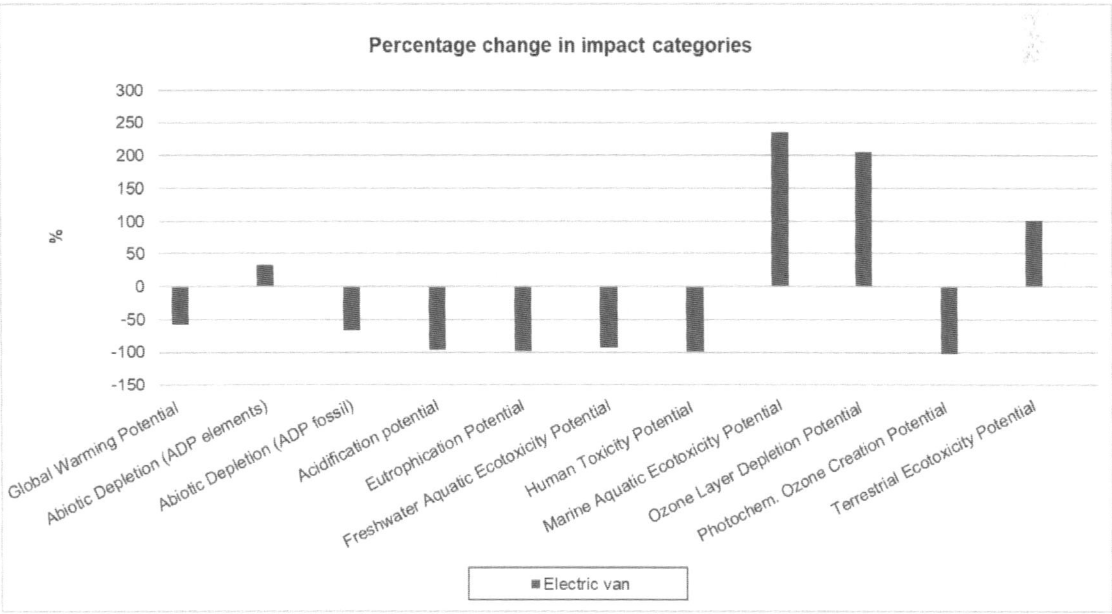

Figure 2. Percentage change in impact categories for electric vehicle scenario.

6. Conclusions

Over a decade, e-commerce has seen significant growth in a new and more efficient era as the changing lifestyles of today's shoppers. Together with undergoing changes to be more environmentally friendly, retailers have been launching many campaigns and project to reduce GHG emissions. Electric cars can be one of the alternative ways to help boost business by building retailers' "green" image. The objective of this study was to examine the environmental effects of an electric designed van compared to a conventional van. An electric vehicle has become an increasingly more attractive option by reducing or eliminating the downsides of using a conventional fossil fueled vehicle. In big cities such as Bangkok, which suffer from severe air pollution, the use of electric vehicles is very valuable. This study recognizes gate-to-grate impacts of alternative modes of vehicles. Results suggest that the reductions in concentrations of CO_2 and other emissions accounted for the benefits in the environment impact categories of GWP, ADP elements, ADP fossil, AP, EP, FAETP, HTP, and POCP. This shows that the electric van contributes to a reduction in a number of emissions compared to the today's designed vans, which means that using electric van was significantly better than using a designed van, especially in the use stage. In this study, percentages are used rather than absolute quantities; consequently, the impact categories, particularly marine aquatic ecotoxicity, appear to see significant increases. In addition, there are some weaknesses in the current version of the LCA analysis in which there are many other processes (e.g. parts production, transportation of parts production, end-of-life products), which are not taken into account. These will definitely result in different environmental impacts. It is recommended for future work that this LCA study should be extended to include other processes.

7. References

[1] IPCC 2014 Summary for policymakers in climate change 2014: Impacts, adaptation, and vulnerability. Intergovernmental Panel on Climate Change. New York, U.S.
[2] Koiwanit J, Manuilova A, Chan C, Wilson M and Tontiwachwuthikul P 2014 *Int. J. Greenhouse Gas Control* **28** 257
[3] Mangmeechai A 2016 *Int. J. of Environ. Stud.* **73**(5) 778
[4] Ngamprasertkit S 2016 *Thailands retail foods 2016.* USDA Foreign Agricultural Service. Thailand
[5] Chen T D and Kockelman K M 2016 Carsharing's life-cycle impacts on energy use and greenhouse gas emissions. *Transport. Res. D Tr. E.* **47** 276
[6] Koiwanit J and Suksangpanya N 2017 Proc. *Int. Symp. on Multimedia and Communication Technology: ISMAC 2017,* Ayuttaya, Thailand
[7] Kuçukay F 2014 *Proc. 11th Symp.* Braunschweig p 9-45
[8] Hawkins T R, Singh B, Majeau - Bettez G and Strømman A H 2013 *J. Ind. Ecol.* **17**(1) 53
[9] Messagie M, Boureima F, Coosemans T, Macharis C and Mierlo J V 2014 *Energies* **7**(3) 1467
[10] Notter D A, Gauch M, Widmer R, Wager P, Stamp A, Zah R and Althaus H 2010 *Environ. Sci. Technol.* **44**(17) 6550
[11] Faria R, Moura P, Delgado J and de Almeida A T 2012 *Energy Convers. Manage.* **61** 19
[12] Zackrisson M, Avellan L and Orlenius J 2010 *J. Clean. Prod.* **18**(15) 1519
[13] Majeau-Bettez G, Hawkins T R and Stromman A H 2011 *Environ. Sci. Technol.* **45**(10) 4548
[14] Ellingsen L A, Majeau - Bettez G, Singh B, Srivastava A K, Valoen L O and Stromman A H 2014 *J. Ind. Ecol.***18**(1) 113
[15] Sullivan J L and Gaines L 2012 *Energy Convers. Manage.* **58** 134
[16] Faria R, Marques P, Moura P, Freire F, Delgado J and de Almeida A T *Renew. Sustainable Energy Rev.* **24** 271
[17] Geyer R, Stoms D and Kallaos J 2013 *Environ. Sci. Technol.* **47**(2), 1170
[18] Boureima F, Messagie M, Matheys J, Wynen V, Sergeant N, Van Mierlo J, De Vos M and De Caevel B 2009 *WEVJ* **3**(1) 1
[19] FEBELAUTO 2012 *Yearly Report* Febelauto: Brussels, Belgium. Retrieved from http://issuu.com/ jaarverslagfebelauto/docs/jvs_febelauto_final (accessed August 18, 2017)

[20] Ayoubi M, Eilemann A, Mankau H, Pantow E, Repmann C, Seiffert U, Wawzyniak M and Wiebelt A 2013 Fahrzeugphysik. In: Bress H and Seiffert U (eds) *Vieweg-Handbuch Kraftfahrzeugtechnik* 7th edn. Springer Vieweg, Wiesbaden pp 47-118

[21] Pannone G 2015 *Technical analysis of vehicle load reduction potential for advanced clean cars* California Air Resources Board and the California Environmental Protection Agency. California. U.S.

[22] Del Duce A, Egede P, Ohlschlager G, Dettmer T, Althaus H, Thomas Butler T and Szczechowicz E 2013 *Guidelines for the LCA of electric vehicles.* European Union Seventh Framework Programme.

[23] The commonwealth of Massachusetts 2016 *Fuel efficiency standards for state fleet.* The commonwealth of Massachusetts. Massachusetts, U.S.

Acknowledgements
We are grateful for the required information from Dr. Ronnapee Chaichaowarat for this project.

IOP Conf. Series: Earth and Environmental Science **150** (2018) 012031 doi:10.1088/1755-1315/150/1/012031

A Comparative Assessment of Life-Cycle Greenhouse Gas Emissions from Hypothetical Electric Airport Transportation Services in Thailand

J Koiwanit[1*]

1 Faculty of Engineering, King Mongkut's Institute of Technology Ladkrabang, Bangkok 10520, Thailand
E-mail: Jarotwan.ko@kmitl.ac.th

Abstract. Global warming is an increase of average temperature in the atmosphere, which causes adverse effects on the environment. Carbon dioxide (CO_2) from transportation sector is one of the main contributors of the overall greenhouse gases (GHG). To cope with this issue, electric car services are increasingly seen as popular alternative modes of green transportation especially for urban cities as it is more flexible, more environmentally-friendly, and less expensive than the use of conventional vehicles. The study analyses and compare the hypothetical electric car systems from airport transportation services. Center of Environmental Science of Leiden University (CML) 2001, the Life Cycle Impact Assessment (LCIA) method, is applied to convert life cycle inventory data into environmental impacts. The observed results showed that the electric shuttle bus had the highest impact in global warming potential (GWP) compared to other transportation types. Alternatively, this Life Cycle Assessment (LCA) study that evaluated different transportations provided important information for decision makers on quantifying the differences between each scenario.

1. Introduction

Innovation and creative industries are the core components to help Thailand to move towards 'Thailand 4.0' policy, ideally on a low carbon economic foundation. Its low-carbon economy roadmap can be consequently achieved by clean technologies. As transportation was ranged as the second largest contributor to greenhouse gas (GHG) emissions in Thailand [1], green vehicles are an important transportation option to reduce the amount of fossil fueled usage in today's conventional vehicles [2]. At the end of March 2017, the Thai government launched its new electric vehicle policy to encourage the automotive industry to produce their next generation vehicles in Thailand [3]. As electric public vehicle services continue to rise, this study presents a life cycle assessment study on hypothetical electric airport transportation services in Bangkok, Thailand. This study aims to evaluate the environmental impacts of different types of the hypothetical electric airport transportation services using the Life Cycle Assessment (LCA) methodology. The electric cars and shuttle bus have been designed to run on the same route as the conventional cars and bus services. In recent decades, electric cars have been studied using the LCA methodology [4]-[7]; however, most of them focus on electric cars production and their raw material acquisition [8], [9]. There is a few focusing on the use phase; consequently, this study attempts to model the electricity use based on real life previous research data. This study is a gate-to-gate study focusing mainly on the use phase of public vehicles from Bangkok's Suvarnabhumi airport to Don Mueang airport, Bangkok, Thailand. Center of Environmental Science of Leiden University (CML) 2001, the Life Cycle Impact Assessment (LCIA) method, is used to convert Life Cycle Inventory (LCI) data into Global Warming Potential (GWP) impact. This database is currently unavailable. As a result, this study will fill the gap and useful results will be produced.

2. Implementation of Life Cycle Assessment (LCA) method

2.1. Goal and scope definition
The goal of this study is to evaluate the GWP impact category over the use phase of different types of electric airport transportation. The LCA results are specific to AOT transportation services in Thailand. This study uses the LCA which provides information and database on the GHG effects of the process or product. This helps to improve understanding of the GWP implications in the use of hypothetic transportation services in Thailand.

2.2. Functional unit
One passenger-km road transportation (one passenger-kilometer-travelled (PMT)) was chosen as a functional unit for this study. This unit allows a significant comparison to be made since a shuttle bus normally uses much more energy to produce and operate while showing higher capacity in carrying passengers than a normal small car. Therefore, the functional unit selected in this comparative study is PMT.

2.3. System boundaries
Though there are different transportation options, especially for the transit passengers from Suvarnabhumi International airport to Don Mueang airport, this study focuses one three types of hypothetical electric transportation options, which include: (i) the taxi, (ii) the limousine, and (iii) the shuttle bus. This study is a gate-to-gate study that takes into account electric consumption through each vehicle use.

2.4. Temporal and geographical boundaries
The lifespan of the vehicles in this study is assumed to be 12 years and an assumed lifetime driven distance is 230,500 kilometers. These are the default expected life span values for a typical shuttle bus and car used in most research [5], [10], [11]. However, it is possible that the vehicles will be able to operate more than their design lives depending mostly on maintenance and sympathetic use [12].

2.5. Technological boundaries
This study evaluates road-vehicle types and their effects of greenhouse gas (GHG) emissions. EV covers various vehicle types such as hybrid EVs and battery EVs. Hybrids have a combustion engine and an electric while battery EVs have an electric motor and a battery [13]. In this study, a battery EV in use phase have been chosen and investigated for its GHG impacts.

3. Life Cycle Inventory *(LCI)* analysis
In this study, most of the data included in the system boundaries are specific to Bangkok/Thailand. The assumptions about electricity consumption and emissions emitted from different types of vehicles are mainly derived from [14], [15], and Ecoinvent. Nevertheless, the data is taken from several literature reviews.

3.1. Description of airport transportation service using electric taxi

The automobile is generally considered as the most convenient commuter transport and consumes inexpensive cost of fuel compared to other transportation forms. In Bangkok, Thailand, there is an overwhelming number of taxis and the country's best-selling car model for taxi drivers is Toyota Corolla Altis. The airport provides taxi service to various destinations in Thailand and the most widely used alternative fuels are Liquefied Petroleum Gas (LPG) and Compressed Natural Gas (CNG) [16]. The average number of passengers per riding is 2.21 [17]. In this study, the number of passengers is assumed to be 1 and 2 at a time. The main operating parameters of the electric taxi is shown in Table 1.

Table 1. Main operating parameters of electric taxi.

Parameter	Value	Unit	Reference/comment
Driving behavior	-	-	Cautious
Type of road	-	-	City
Slope of the road	-	-	Flat
Acceleration	0	$m \cdot s^{-2}$	
Total weight (kg)			www.sgcarmart.com
- with 1 passenger	1380	kg	
- with 2 passengers	1430	kg	
Average passengers per round	1-2	passengers	[17]
Rolling friction of the tires	0.02	-	www.engineeringtoolbox.com
Temperature at time of the travel	35	c	Expert's estimate
Density of air at 35 °C	1.15	$kg \cdot m^{-3}$	www.engineerstudent.co.uk
Aerodynamic drag coefficient	0.32		[18]
Frontal area of the vehicle	1.775 × 1.46	m × m	www.sgcarmart.com
Velocity	22.2	$m \cdot s^{-1}$	Expert's estimate

3.2. Description of airport transportation service using electric limousine

Airports of Thailand Public Company Limited (AOT) offers the official limousine service for passengers at the Suvarnabhumi International airport. Though the airport transfers services across Thailand with various types of vehicles, which include Isuzu MU-7, Toyota Camry, BMW Series 7, Mercedes-Benz E 300, and Toyota Commuter [19]. This study focuses on serving passengers from Suvarnabhumi International airport and Don Mueang airport using Isuzu MU-7 as the hatchback size vehicles tend to fall in to most people price points and the average number of passengers per riding is between 1 and 3 based on the information from the AOT's staff. The energy use for the conventional limousine is from the combustion of diesel. The operating parameters are summarized in Table 2.

Table 2. Main operating parameters of electric limousine.

Parameter	Value	Unit	Reference/comment
Driving behavior	-	-	Cautious
Type of road	-	-	City
Slope of the road	-	-	Flat
Acceleration	0	$m \cdot s^{-2}$	
Total weight (kg)			www.cardekho.com
- with 1 passenger	2040	kg	
- with 2 passengers	2090	kg	

- with 3 passengers	2140	kg	
Average passengers per round	1-3	passengers	www.rotmaethai.com
Rolling friction of the tires	0.02	-	www.engineeringtoolbox.com
Temperature at time of the travel	35	c	Expert's estimate
Density of air at 35 °C	1.15	kg·m^{-3}	www.engineerstudent.co.uk
Aerodynamic drag coefficient	0.328		[20]
Frontal area of the vehicle	1.8 × 1.805	m × m	www.cardekho.com
Velocity	22.22	m·s^{-1}	Expert's estimate

3.3. *Description of airport transportation service using electric shuttle bus*

Suvarnabhumi Airport offers a free shuttle bus service for passengers. The shuttle buses are operated by an internal provider named AOT and runs on a schedule. The shuttle bus service operates everyday between 5am to midnight. The buses used for this route are the Euro II buses.
As a carsharing, all the fuel consumptions and emissions are distributed across the number of passengers each round and at a given time. However, this study assumed that 35 passengers will be on the bus each round. The assumptions and operating parameters of the shuttle bus are summarized in Table 3.

Table 3. Main operating parameters of electric shuttle bus.

Parameter	Value	Unit	Reference/comment
Driving behavior	-	-	Cautious
Type of road	-	-	City
Slope of the road	-	-	Flat
Acceleration	0	m·s^{-2}	
Total weight (kg) (with 1 to 3 passengers)	16750	kg	[21]
Average passengers per round	35	passengers	www.rotmaethai.com
Rolling friction of the tires	0.01	-	www.engineeringtoolbox.com
Temperature at time of the travel	35	c	Expert's estimate
Density of air at 35 °C	1.15	kg·m^{-3}	www.engineerstudent.co.uk
Aerodynamic drag coefficient	0.65		www.clubcobra.com
Frontal area of the vehicle	2.447 × 3.44	m × m	[22]
Velocity	22.22	m·s^{-1}	Expert's estimate

Taking steps to convert the bus fleet to all electric will be challenges for AOT; as a result, effective planning is required to make the change nearly invisible to passengers [23].

4. Life Cycle Impact Assessment (LCIA)

The CML2001 life cycle impact assessment methodology was applied in this study. The CML2001 methodology produced assessment results on the midpoint impact categories; however, GWP is taken into account in this study. In addition, GaBi 7, an easy-to-understand structure and friendly user interface software tool produced by PE International, Germany, is used in this study.

5. Global Warming Potential (GWP) Results

In this study, the GWP is 0.097 and 0.048 kg CO_2-Equiv. in the taxi services with 1 and 2 passengers. The airport transportation service system using limousine showed a total GWP impact of 0.133, 0.067, 0.048 kg CO_2-Equiv for 1, 2, and 3 passengers, respectively. A total number in the environmental impact category of GWP is 0.018 kg CO_2-Equiv in the electric bus system. The main substances that contributed to the global warming category include CO_2, methane (CH_4), and nitrous oxide (N_2O), which accounted for 95.13%, 1.08%, and 3.79%, respectively. The GWP impact was highest for the limousine without sharing the ride compared to the taxi and bus services. However, the limousine with 2 passengers was shown to require less electricity consumption and showed lower GWP impact compared to the taxi without sharing the ride. In addition, there was the least electricity consumption, net total CO_2, and GHG emissions in shuttle bus transportation services since there was a carsharing service system sharing more number of passengers than other airport transportation services. This helps lower energy output and lower CO_2 emissions emitted to the atmosphere. The study showed that GHG emissions varied depending on loads and types of transportation services. The transportation services' impacts on the GWP impact category are shown in Figure 1.

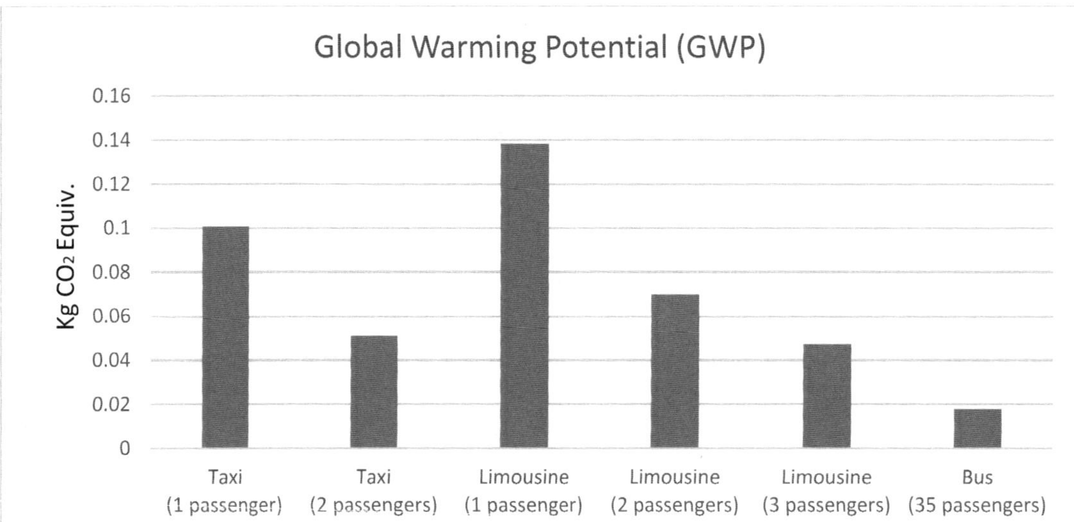

Figure 1. Global warming potential of different types of airport transportation services.

6. Conclusions

As changing the entire vehicle systems to electric vehicles would result in a reduction of emissions, the objective of this study was to examine the GHG effects of different types of the hypothetical airport transportation services. Carsharing is becoming an increasing attractive approach by allowing users to share travel miles and costs and also save in travel-time because of less vehicles on the road. Therefore, reducing traffic density and simultaneously reducing fuel consumption per person per kilometer. In this study, the services ran from Suvarnabhumi airport to Don Mueang airport and it was found that this action decreased an impact in all impact categories compared to a private car. The reductions in concentrations of CO_2 accounted for the benefits in the environment impact category of GWP. This shows that shuttle bus served during the day allows for an efficient car-sharing and consequently contributes to a reduction in a number of emissions compared to today's private cars.

This means that using carsharing is significantly better than using a private car service. However, electricity consumption will vary based on variables such as car model, load, climate, etc. It is recommended for future work that this study should be extended to take sensitivity analysis into consideration.

7. Acknowledgements

I am grateful for the required information from Dr. Ronnapee Chaichaowarat for this project.

8. References

[1] Mangmeechai A 2016 An economic input-output life cycle assessment of food transportation in Thailand. *Int. J. Environ. Stud.* 73(5) 778
[2] Chen T D and Kockelman K M 2016 Management of a shared autonomous electric vehicle fleet: Implications of pricing schemes. *Transp. Res. Rec.* 2572 37
[3] Larkin C 2017) *The Thai Government's latest gamble: Electric vehicle policy.* Retrieved from https://frontera.net/news/asia/the-thai-governments-latest-gamble-electric-vehicle-policy/ (Accessed 25 October 2017)
[4] Hawkins T R, Singh B, Majeau - Bettez G and Strømman A H 2013 *J. Ind. Ecol.* 17(1) 53
[5] Messagie M, Boureima F, Coosemans T, Macharis C and Mierlo J V 2014 *Energies* 7(3) 1467
[6] Faria R, Moura P, Delgado J and de Almeida A T 2012 *Energy Convers. Manag.* 61 19
[7] Notter D A, Gauch M, Widmer R, Wager P, Stamp A, Zah R and Althaus H 2010 *Environ. Sci. Technol.*44(17) 6550
[8] Faria R, Marques P, Moura P, Freire F, Delgado J and de Almeida A T 2013 Impact of the electricity mix and use profile in the life-cycle assessment of electric vehicles. *Renew. Sus.t Energ. Rev.* 24 271
[9] Geyer R, Stoms D and Kallaos J 2013 *Environ. Sci. Technol.* 47(2) 1170
[10] Febelauto 2012 *jaarverslag2012. (Yearly report)* (Brussels, Belgium: Febelauto)
[11] Boureima F S, Messagie M, Matheys J, Wynen V, Sergeant N, Mierlo J V, De Vos M , De Caevel B 2009 Comparative LCA of electric, hybrid, LPG and gasoline cars in Belgian context, WEVJ 3 469
[12] Rose D 2016 *What's the average life expectancy of a car battery – and how can I extend it?*Retrieved from http://www.telegraph.co.uk/cars/advice/whats-the-average-life-expectancy-of-a-car-battery--and-how-can/ (Accessed 31 October 2017)
[13] Egede P, Dettmer T, Herrmann C and Kara S 2015 *Procedia CIRP* 29 233
[14] Koiwanit J and Suksangpanya N 2017 Environmental analysis of a hypothetical online shopping system in Thailand. *ISMAC 2017,* Ayuttaya, Thailand.
[15] Ayoubi M, Eilemann A, Mankau H, Pantow E, Repmann C, Seiffert U, Wawzyniak M, Wiebelt A. (2013). Fahrzeugphysik. In: Braess H. Seiffert U (eds) *Vieweg-handbuch kraftfahrzeugtechnik, 7th edn.*(Germany, Weisbaden: Springer Vieweg) p 47
[16] Transport Research and Innovation Monitoring and Information System (TRIMIS) 2016 *Sharing opportunities for low carbon urban transportation.* Retrieved from https://trimis.ec.europa.eu/project/sharing-opportunities-low-carbon-urban-transportation (Accessed 18 October 2017)
[17] European Conference of Ministers of Transport (ECMT) 2007 *(De)regulation of the taxi industry.* (MA, USA: Organisation for Economic Co-operation and Development (OECD))
[18] Katz J 2016 *Automotive aerodynamics.* (New Jersey, US: John Wiley & Sons, Inc.)
[19] Airports of Thailand Public Company Limited (AOT) 2011 AOT limousine. Retrieved from http://suvarnabhumiairport.com/en/115-suvarnabhumi-limousine (Accessed 15 October 2017)
[20] Pannone G 2015 *Technical analysis of vehicle load reduction potential for advanced clean cars.* Retrieved from https://www.arb.ca.gov/research/apr/past/13-313.pdf (Accessed 22 October 2017)
[21] Isuzu Automotive Dealership, Inc. 2007. *Isuzu FTR33P bus.* Retrieved from https://2549-ph.all.biz/isuzu-ftr33p-bus-g11223 (Accessed 25 September 2017)

[22] Isuzu Motors Limited 2003 *Isuzu LV423R*. Retrieved from http://jbus.0fees.net/jbus/isuzu/lv423r/index.html?i=1 (Accessed 25 September 2017)
[23] Aber J 2016 *Columbia University*. Retrieved from http://www.columbia.edu/~ja3041/Electric%20Bus%20Analysis%20for%20NYC%20Transit%20by%20J%20Aber%20Columbia%20University%20-%20May%202016.pdf (Accessed 12 October 2017)

Use of ICT for sustainable transportation

P Agarwal[1] and M A Alam[2]

1 Assistant Professor, Department of Computer Science and Engineering, Jamia Hamdard, New Delhi, India
2 Professor, Department of Computer Science and Engineering, Jamia Hamdard, New Delhi, India
E-mail: pagarwal@jamiahamdard.ac.in, aalam@jamiahamdard.ac.in

Abstract. The world is experiencing an unfettered growth in terms of development but shrinking the way these developments are leading to the societal, economic and environmental changes and chaos. Sustainability is the answer and needs to be addressed effectively. ICT (Information and Communications Technology) has revolutionized the way things can change. This paper deals with sustainable transportation. Sustainable transport system is a sub topic of a bigger issue "Sustainable Development". So what does this imply? ICT, can indeed enable the designing of smarter cities that offer a better quality of life for their residents while being more sustainable and cost effective. It is not just the citizens of a country but also the government that can gain benefit from initiatives and meet the objectives faster. This paper digs into the traditional transport systems and the sustainable transport system which we thrive for. Green vehicles/ Electric Vehicles/ Driverless cars/ Hybrid vehicles are the need of the hour. This paper extensively explores the issues and inventions that can lead to sustainable transportation. It further explores the problems associated with them and their solutions. These solutions cover the major aspects of sustainability like meticulous planning, correct usage of ICT and a well drafted and implemented governance framework.

1. Introduction

In terms of economy, sustainable transport system should be cost effective and lead to an economic growth of the country as a whole. In terms of society, it should be the one that results in less accidents, less commuting time (so that time to meet the family on to and fro from job to home is reduced), be safe and should provide comfort and convenience to the commuter and the driver, thus leading to health benefits. Traffic congestion incurs cost and reduces the economic growth by wasting people's time and energy and by slowing the delivery of goods or services. As far as environment is concerned, since transport systems have significant impacts on the environment, accounting for between 20% and 25% of world energy consumption and carbon dioxide emissions [1], so Sustainable transport system should minimize harmful emissions, should be environment friendly and thrive to use resources that are renewable and inexhaustible.

2. Background

Sustainability in the transport sector implies finding a proper balance between social, economic and environmental qualities [2] and is a major concern throughout the world. The major stakeholders are the travellers, the technologists and the government. The major concern related to non-sustainability is the traffic volume resulting in traffic congestion where [3] states that congestion is a situation that arises when addition of a vehicle to a traffic flow that might have the power to increase the trip time of others, greenhouse gas emissions – a major factor leading to climate change, vehicle accidents resulting in human loss. The concern and aim of this paper is to first highlight the problems related to the above issues in major and then discussing the ICT innovations that can help us to achieve sustainability. [4] discusses various definitions of ICT in various contexts but broadly associates it as use of communication technologies by people or organizations for accessing information. This paper concludes by examining the challenges in achieving sustainability and concludes by examining the fact that ICT can prove to be a major driving force if the issues are well addressed and implemented owing to the potentials that ICT possesses.

3. Literature Survey

This section provides an insight to the work that has been done in the related area.

In [5], the authors discuss the influence of ICT by studying the relation between ICT and transport and examines the applications of ICT and the transport sector. They also discuss the existing and the futuristic aspects of ICT role in making transportation sustainable. They further compare the differences in adoption of ICT applications between U.S. and Europe. In [6], the author discusses about the data collection from various sources about the traffic congestion and the ways it can be improved. [7] sneaks into how extensible applications can be extended beyond the obvious e commerce applications or the information sites. In [8], the authors discuss about how the society is changing and the use of technology particularly mobile phones is widespread across the world and in all societal layers. The paper discusses about the increased urbanization resulting in environmental problems. It thus explores the potentials and possibilities of ICT in achieving sustainability particularly in the urban context. The prime focus of this paper is the concern about climate change and how ICT can tackle the environmental issues and in order to do so they identify about 35 applications that are related to transportation, housing, consumption and urban exploration. The authors in [9] The authors in this paper discuss about the changing definitions of sustainability and its relation with transportation. The primer [10] provides an overview of the key elements of congestion pricing and illustrates the various aspects and skill sets necessary to analyze and thus implement congestion pricing. This paper further examines the effectiveness of both primary and advanced technology in tolling and congestion pricing. The focus is in identification in the current scenario and also to explore the future aspects to make it work better in the future. [11] Discusses various solutions that IBM provides in the transportation sector. They focus on resource optimization solutions from IBM that offer tools and environment for developing short term as well as long term planning and scheduling applications. These solutions are flexible enough to be adjusted with the changes in regulation, infrastructure, business and operations and can help us to take the right decisions related to transportation. In addition, IBM also provides digital video surveillance and analytical solutions on video data collected from multiple sources like video cameras and biometrics which can be used to identify risks. These solutions open an avenue for wide range of applications with respect to surveillance systems. IBM further illustrates its traffic prediction tool that can provide real time traffic predictions and support route guidance. The paper [12] discusses about the how the physical infrastructures can be interpreted as electronic ones and the changes that telecommunications can bring in structuring of the cities. In [13] the author discusses about the Intelligent Transport systems (ITS) that can use IT in enabling advanced traveler information system which can provide information related to navigation, possible accidents, weather conditions, electronic toll collection, real time data to passengers about arrival and departure of trains or buses and enable communication among various assets of the transportation sector. These changes come with a baggage of innumerable benefits which are further discussed. The paper [14] focuses on the relationships between transport and telecommunications. The paper [15] focuses on telecommuting and its impact on traffic reduction. It also talks about the social aspects of work, technological advances, management and legal issues. In [16], the paper focuses on relation between urban mobility and sustainability in terms of travel time, travel costs and travel patterns.

4. Issues, Problems and their solutions

Various inventions and concepts related to transport sustainability are discussed and their associated problems and solutions are presented in this section.

4.1 The Autonomous/ Driverless cars

Imagine a scenario. You get up in the morning, a sleep hangover and a pending presentation at office. You get into the car, take a nap for about half an hour. Post that you prepare your presentation. Aloha!!!! How is that possible? Well, it is indeed possible today. We are talking about the driverless car. A rotating LIDAR(Light Detection and Ranging) technology, radar systems on front and rear side, an Artificial Intelligence (AI) software[17] connected to all the sensors, a simulation of human perception using AI to control the steering and the brakes, "THE AUTONOMOUS VEHICLE" is one of the best examples of use of embedding ICT in the automotive industry. It is not only the comfort it provides but autonomous cars can truly minimize accidents by increasing the precision and reducing human intervention (one of the causes of traffic accidents), reduce congestion thus improve the traffic flow, and also reduce vehicle carbon emissions. This can only be achieved by deploying the "platooning" technology [18] wherein the driverless car shall travel to another driverless car at high speed so as to mitigate traffic. But it comes with its own baggage of bottlenecks. Are the things as simple as they seem? Will driving in real time traffic be capable enough to detect an anomaly? Driving in real time traffic is a challenging task for machine intelligence. A precision of as high as of milliseconds of reaction time is required to avoid unforeseen obstacles. A "Human Sense" too is necessary. Thus an extensive computing power needs to be built in to the car to minimize the latencies both application and network. It does not hold today due to space constraints, heat dissipation, and many more. IOT (internet of things), High speed networks, Sensor technology, Cloud Computing are creating huge opportunities for sustainability. As you can see a "**driverless car**" **is not** just devising "***a car on wheels with no driver***", but uses all the technologies this paper mentioned above. A cloud to drive a car? Yes, an intelligence can be build into the cloud rather than a car [19]. This storage power in cloud infrastructure can be used effectively to intelligently drive a car. But this requires a paradigm shift from the centralized clouds to a local cloud and then to a mini cloud which is closer to the car. This will enable "**Human Sense**" and "**Machine Intelligence**" to work together.

4.2 Car sharing

Car sharing has a great impact on traffic control. In India, it has been observed that shared services being offered by Ola and Uber have extensively reduced the usage of personal cars as it provides comfort and moreover is economical too. These services are successful only because of smart applications installed on the phone that enables him/her to request a taxi passing by nearby location of the commuter. In addition, the application on driver's mobile (Google Maps to name one) helps him to track the commuter for pick up fast if its confirmed booking. Further, mobile application also allows the driver to choose the best route in terms of time duration and traffic so that the commuter reaches the desired destination in best case time. taxi service, but one that is full digitally enabled. Rather than maintaining a fleet of cars, mechanics, drivers and the taxi service providers permit the cars to return to the legal car owners when not in use for a job.

4.3 Car pooling

We are sure that there must have been instances when you step out of your locality and see two persons known to you who are heading to a nearby destination as yours. This is where car sharing could work. The end user enters the travel requirements and there is a car for you. You share your travel cost and travel to reach the respective destinations. Great, isn't it?

4.4 Eco – Driving

The concept of eco- driving has already been introduced. Though it can be achieved by keeping speed in control, syncing the speed and the gear, decreasing the car speed at turns and so on but ICT can enhance the effectiveness of these. In [20] the authors discuss the technical acquisition of data by using modern ICT tools and technologies.

In so as to collect vehicle trajectory and eco-driving data, a fuel consumption sensors and a portable Global Positioning System (GPS) receiver were installed in test vehicles. But a possible drawback is a lack of satellite signal which could prevent vehicle positioning. In [21] it has been identified that eco driving can reduce GHG (Green House Emissions) by a great amount. It works through use of ICT built into a car or through smart phone that reduces fuel consumption by updating drivers when they are accelerating too fast or their car having air pressure that is suboptimal.

4.5 *Congestion Relief*

Congestion Relief is related to decreasing the density of traffic. Certain technologies that decrease congestion need to be formulated. Various forms of ICT like Advanced Traveller information systems, Video surveillance and response are necessary to control congestion, Congestion free lanes need to be developed. In addition, when a vehicle enters the toll area it is automatically directed to a particular lane. Relating to toll fees, users set up an account and transponders initiate withdrawal of funds based on accounts.

4.6 *GPS/ Navigation software*

As discussed in car sharing, GPS helps the commuters, taxi drivers in identifying the best route. But this has still not been adopted by private vehicle owners and truck drivers. Since truck is responsible for logistics and transportation of goods between states so a software that measures truck-route-efficiency is also needed. Not only shall it find best route in terms of time and congestion but also take the tolls and other barriers into account to determine which route will require the lowest amount of fuel for a particular trip and even how much profit that trip would generate. These fuel savings translate into lower GHG emissions.

4.7 *Automatic license plate recognition (ALPR) Technology*

This technology [22] is based on taking the images of the number plates of the vehicle crossing a particular area which helps in the identification of the vehicle and the driver. Front and rear cameras are installed at several locations to improve the identification rate. But the problem associated with it is the reliability of images. This can occur due to damaged or dirty license plates, light reflection on the image to name a few.

4.8 *E-world*

Use of several E-services [5] for banking, shopping, entertainment, government services, booking for hotels, holidays, trips, e-networking etc reduces travel significantly. But they may cause an addition of travel time as new travel demand, successful social networking and all. Though the concept of "Work from Home" is quite prevalent in many countries including India, where employees complete the assigned work or do chats with clients using chat applications, video-conferencing which again reduces travel time significantly.

4.9 *Informational signaling(Variable Message Signs/ Electronic signboards)*

VMS gives information on changing road network conditions ahead and have assisted in communication with drivers regarding major congestion areas ahead. Electronic boards have displays that broadcast messages on the screen about the rules to be observed for driving. In addition to directional guidance, traffic rules of a city, these boards also can also contain weather updates so that a person can re route or avoid pending work enroute if a hailstorm or heavy rain forecast is seen by the driver. This could save him from resulting congestion and time wastage that may occur otherwise.

The figures: Figure 1, Figure 2 are a few Electronic signboards which have been captured on roads of Delhi, India.

Figure 1. It shows a signboard

Figure 2. It shows a signboard depicting a traffic rule.

4.10 *Video Surveillance and Response*

Well, you may agree that cars may break down in the middle of the road any day, any time. We all must have faced this problem. Or let us visualize another situation particularly in India. Why we said so is because, Roads are narrow and vehicles are many. Many new flyovers have come up in metropolitan cities like Delhi. Though these infrastructures have improved the road condition but in case of break down, interrupts the traffic flow too causing a cascading effect several kilometres far off too. So it is necessary to continuously monitor locations to identify whether traffic is moving or encountering congestion. This can be achieved by strategically placing sensors or television cameras. In case of perceiving a disabled vehicle being the cause of congestion, a a repair/assistance vehicle is dispatched to the location. Upon arrival, the obstacle is removed and appropriate assistance could be offered.

4.11 *Advanced Driver's Assistance(ADAS)*

It is generally installed on private vehicles and has inbuilt speed adaptation control system that reduces variation in acceleration and adapts speed. ADAS[23] also controls the positioning of vehicle in terms of lanes, other vehicles and obstacles. The overall benefit derived is reduction in fuel usage and fatality.

4.12 *Sensor Support with an Extended Viewing system*

Sensors can be used to detect an accident at a location and inform the nearby hospital services which reduces the waiting time for assistance post accident. In addition sensors and radar are being in built into cars which alerts drivers on blind spot and obstacles during night.

4.13 *Advanced Traveler Information systems*

The technology is not to be built into the privately owned vehicles only but also to the public transport system. The personal information systems may be portable for passengers travelling using public transport and in vehicle for car drivers. It may include data related to facilities (shops, malls, services), new travel solutions in addition to best route options display. But it incurs high cost and may also be hampered by limited needs of travellers to plan their activities.

4.14 *Smart Parking facility*

Appended to the list of issues discussed above that exploit ICT to its maximum which are in initial stages or being already adopted, a Smart Parking facility too can be developed. This could work as follows. Through the application installed on the mobile, we search an authorized parking lot of our destination of journey. Then a prior booking is made. When we reach the destination, we park our car in the space shown through the app. This could be done as a free service or through payment using services like payback, etc.

We hope you would agree that if each of the above usage of ICT is implemented in each city then what a wonderful experience travelling of self, goods or services could be for each one of us.

5. Conclusion

The world is urbanizing rapidly and innovations are reaching new heights too. But a long way we still have to travel to reach the goal faster. This can only be achieved through well connected networks of towns and cities which are sensor-enabled and possessing a "Human Sense- Robotic Intelligence" which is absolutely necessary as on today. Enhanced and effective mass transport solutions which are technology enabled, reduced car owners, compact cities and neighbourhoods (which shortens the trip distances and makes public transport more economically viable), car pooling, driverless cars which are fully adopted , high end applications which make travelling easier are some of the solutions which need to addressed.

In addition, each city has its own constraints in terms of buildings, income strata of the city, population of the city, location of offices and schools which also have to be taken care of in building the transport system to extract maximum benefits. There is still limited research on identification of whether significant improvements in particular parts of the network go along with overall network performance. Highly populated countries like India demand a proactive and a radical response.ICT in transportation sector can drastically improve the efficiency of logistics and also reduce the cargo transport emissions and can even reduce transportation needs by use of telecommunications and video conferencing.

It is near a decade that driverless cars were conceptualized but would near around a decade more to be adopted fully. In addition the future requires towns to be well connected to the cities and development of applications which are less expensive for users, are user friendly, applications that don't suffer from technological uncertainty or any legal liability. A clear and simple layer of the transport system which involves stakeholders and reduces the complexity inherent in privacy, legal framework needs to be devised. Moreover, adoption of the technology and innovations has to be enhanced so as to improve the "driving experience" and lead to a more sustainable transport system the future than we see today.

6. References

[1] World Energy Council 2007. *Transport Technologies and Policy Scenarios to 2050.*

[2] Steg L and Gifford R 2005 *J. Trans. Geo.* 13 p 59–69.

[3] Thomson I and Bull A 2002 *CEPAL Review.vol 76* Urban traffic congestion: its economic and social causes and consequences *pp 105-116.*

[4] Zuppo M C 2012 *Int. J. of Managing Info. Tech.4* pp 13-22

[5] Black R, W and Geenhuizen V M 2006 *European J. Transp. and Infra. Research.* 6 ICT Innovation and Sustainability of the Transport Sector p. 39-60.

[6] Leduc G 2007 Road Traffic data: Collection methods and applications.

[7] Kenney M and Curry J 2001. *Handbook on Worlds of E-Commerce: Economic, Geographical and Social Dimensions* In T. Leinbach and S. Brunn (eds.) (New York: Wiley), p 45-65

[8] Palmgren E and Zapico L J 2008 Multimedia Applications for a Sustainable Urban Lifestyle URL: www.csc.kth.se.

[9] *Towards a Sustainable Waterborne Transportation Industry 2011 Publication: EnviCom Task Group 2*

[10] Technologies that Enable Congestion Pricing: A Primer 2008 US Department of Transportation World Energy Council (2007)
[11] Go with the flow: Transportation information management solutions from IBM 2010
[12] Wheeler O J and Warf B 2000 Cities in the Telecommunications Age: The Fracturing of Geographies Taylor & Francis.
[13] Ezzel 2010 *Intelligent Transport Systems*. Information Technology and Innovation Foundation
[14] Andreev P, Salomon I and Pliskin N 2010. *Transportation Research C: Emerging Technologies* 18 pp 3-20.
[15] Nilles M J 2002 *Transportation Research Part A*. Traffic reduction by telecommuting: A status review and selected bibliography. Elsevier
[16] Blankshtain C G and Mindali R O 2016 *Int. J. Sust. Trans.* 10 Key research themes on ICT and sustainable urban mobility pp 9-17.
[17] Rouse M 2011 Driverless car Retrieved from http://whatis.techtarget.com/definition/driverless-car., p 1.
[18] Zanona M 2016 Retrieved from http://thehill.com/policy/transportation/302550-how-driverless-cars-can-reduce-pollution
[19] Manzalini A 2015 Enabling the self-driving car Retrieved from http://www.networkcomputing.com/author/36436253.
[20] Grgurević S H and Šarić Ž 2013. *Proc. Int. Virtual Research Conf. in Technical Disciplines* Slovakia. EDIS-Publishing institution of the University of Zilina. vol 1, p 142-47.
[21] Husnjak S, Forenbacher I and, Bucak I 2015 *Promet- Traffic & Transportation* 27 p 335-344.
[22] Goyal A and Bhatia R 2016 *IOSR J. Of Comp. Engg.* 18 pp 34-40
[23] Maerivoet S 2012 Congestion under Investigation.